Green Energy and Technology

Climate change, environmental impact and the limited natural resources urge scientific research and novel technical solutions. The monograph series Green Energy and Technology serves as a publishing platform for scientific and technological approaches to "green"—i.e. environmentally friendly and sustainable—technologies. While a focus lies on energy and power supply, it also covers "green" solutions in industrial engineering and engineering design. Green Energy and Technology addresses researchers, advanced students, technical consultants as well as decision makers in industries and politics. Hence, the level of presentation spans from instructional to highly technical.

Indexed in Scopus.

Indexed in Ei Compendex.

Muaz Mohd Zaini Makhtar · Hafiza Shukor ·
Abu Zahrim Yaser
Editors

Microbial Fuel Cell (MFC) Applications for Sludge Valorization

Springer

Editors
Muaz Mohd Zaini Makhtar
School of Industrial Technology
Universiti Sains Malaysia
Penang, Malaysia

Hafiza Shukor
Chemical Engineering Technology
Universiti Malaysia Perlis
Arau, Perlis, Malaysia

Abu Zahrim Yaser
Faculty of Engineering
Universiti Malaysia Sabah
Kota Kinabalu, Sabah, Malaysia

ISSN 1865-3529 ISSN 1865-3537 (electronic)
Green Energy and Technology
ISBN 978-981-99-1082-3 ISBN 978-981-99-1083-0 (eBook)
https://doi.org/10.1007/978-981-99-1083-0

This Springer imprint is published by the registered company Springer Nature Singapore Pte Ltd.
The registered company address is: 152 Beach Road, #21-01/04 Gateway East, Singapore 189721, Singapore

We dedicate this book to our families and friends……

Foreword by the Exercising the Function of the Vice-Chancellor, Universiti Sains Malaysia (USM)

Worldwide there is increasing awareness of the global challenges that face us all—ranging from preventing conflict, to feeding an expanding population, to helping the poor, to fostering health, and protecting the environment in particular global warming and ecosystem degradation. These challenges are at the heart of sustainable development—meeting the needs of the present without compromising the ability of future generations to meet their own needs. With the full realization that environmental problems are closely linked to economic and socio-cultural problems, a great change is required in the stewardship of the earth and the life on it to move away from paradigms that rely exclusively on concepts of continuous economic growth, corporate profit, and consumer avarice. We must redirect our intellectual enterprise to develop the capacity to understand, anticipate, and act on the basis of global challenges.

Convinced that building capacity for making decisions that consider the long-term future of the economy, ecology, and equity is a key task of education Universiti Sains Malaysia (USM) has embraced the vision of becoming a sustainability-led university of world-class standing as part of its APEX initiative. While the APEX award is a fitting recognition for the university'swide-ranging and remarkable accomplishments of the past, it is also a call to excel in addressing the sustainability challenges of the future. To achieve the broad APEX vision, USM has embarked on a range of missions which through their specific objectives and activities are expected to contribute to the achievement of the overall sustainability vision

"***Microbial Fuel Cell (MFC) Applications for Sludge Valorization***" is a book that gathers all the experts in renewable energy research from diverse biomass which would benefit all the readers who are eager to know more about MFC technology capability. USM is aware that MFC technology has attracted so many potential collaborators as the technology has gained numerous awards at an international exhibition and was awarded a research grant from the Ministry of Higher Education (MOHE) and the Ministry of Science, Technology and Innovation (MOSTI). It is such a great accomplishment for the team to publish the book under the prestigious publisher of Springer Nature's Books. This book also presents the visibility of networking between USM represented by Dr. Muaz Mohd Zaini Makhtar as the main editor together with Dr. Hafiza Shukor from **Universiti Malaysia Perlis (UniMAP)** and Assoc. Prof. Dr. Abu Zahrim Yaser from **Universiti Malaysia Sabah (UMS)** through the editing of the book. Publishing a research work or chapter in the book is also a great tool for making your research content more visible. This will increase the chances of your research being noticed, used, and having an impact that can increase your reputation and chances of success in your academic work. Congratulations to the APEX Young Scholars research team for the effort and I look forward to more book publications from USM experts that should greatly benefit many, especially lecturers, researchers, policymakers, and matters relating to environmental sustainability in Malaysia.

Thank you and enjoy reading the book.
Best wishes,
Prof. Dato' Gs. Dr. Narimah Samat
Vice Chancellor
Universiti Sains Malaysia (USM)
Gelugor, Malaysia

Foreword by Director of Centre for Innovation and Consultation (CIC) Universiti Sains Malaysia

The management of sludge from wastewater treatment facilities is one of the most critical environmental concerns in Malaysia, due to the very fast increase in sludge production as a result of sewerage extension, new installations, and upgrading of existing facilities. In Malaysia, approximately 327,479 m^3 of sludge was produced annually and is expected to reach upto10 million cubic meters of sludge by the year 2035. Therefore, sludge dewatering is the practice of minimizing waste by volume to prepare for its effective disposal. The dewatered sludge contains a significant amount of energy that is hidden in the form of biodegradable organic materials and it can be retrieved with appropriate technology. They are the most favorable substrate for bioconversion since they are abundant and renewable. Besides, the waste industries can improve their financial and environmental efficiency by utilizing the energy found in sludge. Instead of harming the environment, sludge can be used as a substrate for value addition. Indeed, it is necessary to develop a new emerging green technology to address this issue. Thus, this book brings together experts from a variety of disciplines; including pure and applied sciences, engineering, health, and social science to elaborate on the new emerging green technology innovation, most importantly on the microbial fuel cell (MFC).

MFC works by bioremediating the sludge and recovering the energy within it. The information in the book offers the reader a variety of viewpoints on how MFC can be utilized as an advanced waste treatment technique while simultaneously producing energy. The book then discussed people's perceptions in green technology and how that can be incorporated into school curricula through the science, technology, engineering, and mathematics (STEM) syllabus. This book is also foreseen as one of our efforts toward achieving Sustainable Development Goals 4 (quality education), 6 (clean water and sanitation), and 7 (affordable and clean energy). It might also serve as a reference guidebook for undergraduates and postgraduates pursuing study in wastewater treatment and biofuel engineering technologies.

I would like to express my sincere gratitude to all the authors who have worked on this book. I would like to convey my heartfelt thanks to the APEX Young Scholars, the School of Industrial Technology, the Center for Innovation and Consultation (CIC), the Research Center Management Office (RCMO), USM, and the publisher for providing direct and indirect assistance towards the publication of this book. I would like to express my sincere gratitude to Dr. Muaz Mohd Zaini Makhtar, Dr. Hafiza Shukor, and Assoc. Prof. Abu Zahrim Yaser for their rigorous editing. Finally, it is hoped that this book will be useful to our readers. We would value comments and suggestions

Prof. Ir. Dr. Srimala Sreekantan
Director
Centre for Innovation and Consultation (CIC)
Universiti Sains Malaysia (USM)
Gelugor, Malaysia

Preface

This book focuses on the current activities and research in microbial fuel cell (MFC) related to the simultaneous bioremediation of sludge and energy recovery. Jiei Kobe et al. reported that the sludge has the potential to be converted into valuable products The sludge generation rate and other site-specific factors influence the sludge treatment technology to be used. The findings from the chapter reveal that a variety of technology nowadays can be used to exploit the sludge such as into fuels ("Overview of Sludge in Waste Treatment Plant"). Nur Azzalia Kamaruzzaman et al. have provided an overview of the various effects of untreated and treated sludge on human health, the environment, plants, and animals. Depending on a few factors, the effects of sludge can be negative and positive. Plants and animals were also affected in terms of phytotoxicity and mortality respectively. However, it is equally important to note that sludge also has profound beneficial properties, as it contains high content of nitrogen, phosphorus, organic matter, and plant nutrients, which was the reason for its extensive use in agriculture ("Effects of Treated and Untreated Sludge Applications on Human Health, the Environment and Other Ecological Factors"). The scientists could exploit the sludge and harvest many advantages such isolated certain suitable microbe to be used in polluted areas for the bioremediation method. The usage of microbiological approach is a greener method for solving environmental pollution. Muhammad Najib Ikmal Mohd Sabri et al. has extended the adoption of environmentally friendly technology not limited to the bioremediation process, the fundamental knowledge of microbe presence allows the scientist to purify useful chemical compound which could be obtained via the implementation of certain species isolated from the sludge ("An Insight of Component and Typical Mechanism of Sludge Degrader Microbes in Dewatered Sludge"). A potential technology of sustainable and renewable energy that can be exploited is the MFC technology. While MFC promises a clean, zero-emission energy technology for a sustainable source of energy, it is constrained by costs, durability, complexity as well as operational safety. Despite these limitations, MFC technology is gaining traction globally as one of the most sustainable ways to treat wastewater, an alternative source of renewable energy, and by extension, reduce carbon intensity. Nevertheless, there is a need to understand public attitudes towards global issues and multiple energy technologies. Indeed, public acceptance is necessary for the successful implementation of an energy transition towards renewable energy sources, and this is further discussed by Aziatul Waznah Ghazali's research team ("Microbial Fuel Cells (MFC) as an Alternative Energy Source: The Perceptions and Attitudes Towards Sustainable and Renewable Energy in Malaysia"). "Application of Microbial Fuel Cell for Bioremediation of Sewage Sludge" authored by Farhan et al., summarized that MFC technology high energy content of sewage sludge waiting to be tapped, application of sewage sludge represents an emerging source of renewable energy, of which MFCs are primed to take advantage of given their numerous advantageous over other alternative approaches. "Microbial Fuel Cell Technology as Advanced Sewage Sludge Treatment" written by Muhammad Najib Ikmal Mohd Sabri et al. addresses the advantages of MFC as an advantage technology that has the ability to further treatment of sludge helped by electrogenic bacteria (EB) in transferring the electron to complete the redox potential thus created a high voltage for energy recovery. The research team led by Amira Suriaty Yaakop described the detail of EB with their role in bioremediate the sludge. The team also elaborates on the importance to have heterogenous EB hence fastening the bioremediation and energy ("Utilization of Electrogenic Bacteria Consortium for Sewage Sludge Treatment via Organic Compound Degradation").

Mohamad Farhan Mohamad Sobri et al. explained the application of sludge in MFC has extended several purposes ranging from simple mixed culture inoculum towards studies on the degree of bioremediation and enhancement of MFC components. To each aim, EIS represents a value-added approach for elucidation of the numerous internal resistances at play and the resultant effect each experiment has on the resistance components within. Such an approach offers a promising outlook for further understanding and enhancement of sludge-based MFC in the years to come ("Assessment of Sludge-Based Microbial Fuel Cell Performance via Electrochemical Impedance Spectroscopy"). Mohamad Danial Shafiq et al. described the Colloidal behaviour of sludge particles as a pivot key in the selection and determination of the biomass purification and recovery processes. The underlying origin of sludge colloidal stabilization can be described by the classical DLVO theory, and is largely dependent on the physical properties of the water bodies, key chemical components, and origin of the sludge particles. A proper understanding of sludge colloidal behaviour is vital to ensure that abundant biomass can be fully utilized for a more sustainable and healthy future ("Sewage Sludge Particle Surface Interactions: Technology and Purification Approaches"). Ana Masara Ahmad Mokhtar et al. described the fundamentals of AMR and the possibility for wastewater to act as a carrier of AMR in the environment. Compared to other conventional wastewater treatment procedures, MFCs can help minimize AMR by boosting antibiotic removal rates. Nonetheless, it is vital to improve the current design of MFCs and comprehend their operation, and doing so will reduce AMR transmission in the environment ("Antimicrobial-Resistant Microorganisms and the Possibility of Using Microbial Fuel Cell Technology to Reduce Their Transmission in the Environment"). Soliha Sanusi et al. have provided the economic perspective where it has been noted that MFCs are accepted for their benefits to replace the conventional electrical supply.

Still, the capital and operational cost is haunting it to spread the application in many industries. However, the market of MFCs as large as billions of monies were spent on catering to the water sludge. MFCs are widely accepted in many countries that plan to replace their scarce natural energy supply with other substitutions ("Microbial Fuel Cell (MFC) Innovation in Wastewater Treatment Plant: From Economic Perspective").

Nor Asniza Ishak et al. detailed up from the educational perspective of green technology's curriculum syllabus. The chapter emphasizes the importance of making environmental education (EE) well accepted among young generations as informal learning and close to nature and including helping change attitudes and mindsets of human resources, particularly among young generations ("The Environmental Education Element of Secondary School Science Curriculum in Malaysia: Enhancing Students Awareness in Microbial Fuel Cell (MFC) as Green Technology in Treating Dewatered Sludge"). The book ended with "Potential Biodegradable Product from Dewatered Sludge" by highlighting the clear path to improve the sustainability of the bioplastics market through the utilization of compounds extracted from dewatered sludge. The chapter presents the potential of many different types of sludge that can offer a promising nutrient that would reduce pollution, decrease oil consumption, improve environmental impact, and in some cases even improve bioplastic performance. Overall, the book comprehensively elaborates on the importance of valorization of sludge via MFC, therefore accomplishing the goals listed in Malaysia's Sustainable Development Goals.

Penang, Malaysia — Muaz Mohd Zaini Makhtar
Arau, Malaysia — Hafiza Shukor
Kota Kinabalu, Malaysia — Abu Zahrim Yaser

Acknowledgements

First of all, the editors acknowledge the following individuals for their time and efforts in assisting the editors with the reviewing of manuscripts. This book would not have been possible without the commitment of the reviewers.

- Faiznur Fuad
- Fauziah Hanim Ibrahim
- Kiew Peck Loo
- Gobi Kanadasan
- G. Sivarajah Ganesan
- Nur Kamilah Md Isa
- Izni Mariah Ibrahim
- Mohd Usman Mohd Junaidi
- Mohd Dinie Muhaimin Samsudin
- Nor Hawani Salikin
- Noreen Suliani Mat Nanyan
- Peer Mohamed Abdul
- Usaid Rehman Farooqi
- Stephanie Chan Yen San
- Tariq A. Almohamad
- Vel Murugan Vadivelu

Special thanks to the Ministry of Higher Education (MOHE), Ministry of Science, Technology & Innovation (MOSTI), Universiti Sains Malaysia (USM), and all external grant providers for the funding granted to our researchers which supports all the activities involved in their research. In addition, the establishment of Japan-Malaysia engagement via the collaboration of Centre for Innovation and Consultation (CIC) USM with Leave A Nest Group to bring the innovation from the university to the market through industrial engagement with the Japanese companies. The exchange of expertise between USM and R&D department of Japan biotechnology companies has leveraged the networking and strengthened the bond between the two countries. The linkages are broadened by the involvement of several other universities such Universiti Teknologi Malaysia (UTM) and Universiti Putra Malaysia (UPM).

Contents

Editors and Contributors

About the Editors

Muaz Mohd Zaini Makhtar is a senior lecturer in the Bioprocess Technology Division at the School of Industrial Technology, Universiti Sains Malaysia (USM). Currently he is a Deputy Director of USM Center of Innovation and Consultation (CIC). He is also a Head of Sustainable Development Goals #15: Life on Land under the Centre for Global Sustainability Studies (CGSS) in USM. Dr. Muaz completed his Bachelor of Engineering in Bioprocess Engineering at Universiti Malaysia Perlis (UniMAP) and obtained his MSc and PhD from the School of Chemical Engineering, USM. His core research interests are in bioprocess technology and bioconversion of waste, focusing on biological treatment technologies, microbial fuel cells (MFC) for waste treatment and recovery of energy, electricity. Dr. Muaz enthusiasm and dedication towards his research works have been reflected in his achievements in winning numerous scientific invention awards in international exhibitions and he also is the recipient of multiple research grants in various disciplines sponsored by the government of Malaysia.

Hafiza Shukor is a senior lecturer specializing in the areas of biomass conversion and fermentation technology in the Faculty of Chemical Engineering Technology, Universiti Malaysia Perlis (UniMAP) and Research Fellow from the Centre of Excellence for Biomass Utilization (CoEBU). Dr. Hafiza received her Ph.D. from the Faculty of Engineering and Build Environment, Universiti Kebangsaan Malaysia (UKM) under the field of Chemical and Process Engineering and she obtained her Bachelor Degree and MSc in Chemical Engineering from the School of Chemical Engineering, Universiti Sains Malaysia (USM). She was awarded one of the Young Scientist Awards in UniMAP as her research contributions include publications in indexed journals, international conference proceedings, and books. Her excitement in research can be translated into receiving several research exhibitions awards and grants from the university and government.

Abu Zahrim Yaser is Associate Professor in Waste Processing Technology at the Faculty of Engineering, Universiti Malaysia Sabah, Malaysia. He obtained his Ph.D. from Swansea University. Dr. Zahrim has published 10 books and more than 50 technical papers (Scopus). He has reviewed more than 200 articles (Publons) and his creative publication in the *Festival of Idea* has been viewed more than 1000 times. His inventions on composting systems have been adapted by Tongod, Nabawan, and Kundasang communities in Sabah. Dr. Zahrim was a visiting scientist at the University of Hull.

Contributors

Amir Muhammad Noh Amin Abdul Rahman School of Materials and Mineral Resources Engineering, Universiti Sains Malaysia, Gelugor, Penang, Malaysia

Muhammad Naziff Ahamad Said School of Materials and Mineral Resources Engineering, Universiti Sains Malaysia, Gelugor, Penang, Malaysia

Ana Masara Ahmad Mokhtar Bioprocess Technology Division, School of Industrial Technology, Universiti Sains Malaysia, Penang, Malaysia;
Green Biopolymer Coating and Packaging Centre, School of Industrial Technology, Universiti Sains Malaysia, Penang, Malaysia

Nur Fazreen Alias School of Materials & Mineral Resources Engineering, Engineering Campus, Universiti Sains Malaysia, Nibong Tebal, Pulau Pinang, Malaysia

Siti Amirah Alias School of Materials & Mineral Resources Engineering, Engineering Campus, Universiti Sains Malaysia, Nibong Tebal, Pulau Pinang, Malaysia

Rosnani Alkarimiah Engineering Campus, School of Civil Engineering, Penang, Malaysia

Sharon Fatinathan Faculty of Pharmacy and Health Sciences, Universiti Kuala Lumpur-Royal College of Medicine Perak (UniKL-RCMP), Ipoh, Perak, Malaysia

Aziatul Waznah Ghazali Faculty of Economics and Management, Universiti Kebangsaan Malaysia, Bangi, Selangor, Malaysia;
Accounting Research Institute, Universiti Teknologi MARA, Shah Alam, Selangor, Malaysia

Ooi Kah Hong School of Biological Sciences, Universiti Sains Malaysia, Minden, Penang, Malaysia

Md Sohrab Hossain Department of Fundamental and Applied Sciences, HICoE-Centre for Biofuel and Biochemical Research, Institute of Self-Sustainable Building, Universiti Teknologi PETRONAS, Perak Darul Ridzuan, Malaysia

Muhammad Hazwan Hussin School of Chemical Sciences, Universiti Sains Malaysia, Gelugor, Penang, Malaysia

Nor Asniza Ishak School of Educational Studies, Universiti Sains Malaysia, Bayan Lepas, Penang, Malaysia;
National Higher Education Research Institute (IPPTN), Bayan Lepas, Penang, Malaysia

Mohd Hafiidz Jaafar School of Industrial Technology, Universiti Sains Malaysia, Gelugor, Penang, Malaysia

Zulaikha Amirah Johari Faculty Economics and Management, Universiti Kebangsaan Malaysia, Bangi, Selangor, Malaysia;
Accounting Research Institute, Universiti Teknologi MARA, Shah Alam, Selangor, Malaysia

Nur Azzalia Kamaruzaman National Poison Centre, Universiti Sains Malaysia, Gelugor, Penang, Malaysia

Jiei Kobe School of Industrial Technology, Environmental Technology Division, Universiti Sains Malaysia, Gelugor, Penang, Malaysia

Ku Ishak Ku Marsilla School of Materials & Mineral Resources Engineering, Engineering Campus, Universiti Sains Malaysia, Nibong Tebal, Pulau Pinang, Malaysia

Siti Rohana Man School of Educational Studies, Universiti Sains Malaysia, Bayan Lepas, Penang, Malaysia

Mohamad Farhan Mohamad Sobri School of Industrial Technology, Bioprocess Technology Division, Universiti Sains Malaysia, Gelugor, Penang, Malaysia

Mohammed Zharif Asyrani Mohammed Alias Bioprocess Technology Division, School of Industrial Technology, Universiti Sains Malaysia, Penang, Malaysia

Nur Atiqah Mohd Abdul Rasik School of Industrial Technology, Bioprocess Technology Division, Universiti Sains Malaysia, Penang, Malaysia

Fatin Nur Izzati Mohd Fadzil Bioprocess Technology Division, School of Industrial Technology, Universiti Sains Malaysia, Penang, Malaysia

Muhammad Najib Ikmal Mohd Sabri School of Industrial Technology, Bioprocess Technology Division, Universiti Sains Malaysia, Gelugor, Penang, Malaysia

Muaz Mohd Zaini Makhtar School of Industrial Technology, Bioprocess Technology Division, Universiti Sains Malaysia, Gelugor, Penang, Malaysia;
Centre for Innovation and Consultation (CIC), Universiti Sains Malaysia, Gelugor, Penang, Malaysia;
Fellow of Centre of Global Sustainability Studies (CGSS), Universiti Sains Malaysia, Penang, Malaysia

Mazlin Mohideen Faculty of Pharmacy and Health Sciences, Universiti Kuala Lumpur-Royal College of Medicine Perak (UniKL-RCMP), Ipoh, Perak, Malaysia

Chen Sep Ngee Bioprocess Technology Division, School of Industrial Technology, Universiti Sains Malaysia, Penang, Malaysia

Kavita Pusphanathan School of Industrial Technology, Bioprocess Technology Division, Universiti Sains Malaysia, Penang, Malaysia

Nor Hawani Salikin Bioprocess Technology Division, School of Industrial Technology, Universiti Sains Malaysia, Penang, Malaysia

S. M. Salman School of Biological Sciences, Universiti Sains Malaysia, Minden, Penang, Malaysia

Soliha Sanusi Faculty of Economics and Management, Universiti Kebangsaan Malaysia, Bangi, Malaysia;
Accounting Research Institute, Universiti Teknologi MARA, Shah Alam, Malaysia

Mohamad Danial Shafiq Engineering Campus, School of Materials and Mineral Resources Engineering, Universiti Sains Malaysia, Gelugor, Penang, Malaysia

Nurul Atiqah Shamsuddin School of Industrial Technology, Bioprocess Technology Division, Universiti Sains Malaysia, Gelugor, Penang, Malaysia

Hafiza Shukor Centre of Excellence for Biomass Utilization, Faculty of Chemical Engineering Technology, Universiti Malaysia Perlis, Perlis, Malaysia

Husnul Azan Tajarudin School of Industrial Technology, Bioprocess Technology Division, Universiti Sains Malaysia, Gelugor, Penang, Malaysia

Amira Suriaty Yaakop School of Biological Sciences, Universiti Sains Malaysia, Minden, Penang, Malaysia

Abu Zahrim Yaser Faculty of Engineering, Universiti Malaysia Sabah, Kota Kinabalu, Malaysia

Overview of Sludge in Waste Treatment Plant

Jiei Kobe, Mohamad Danial Shafiq, Rosnani Alkarimiah, Abu Zahrim Yaser, Hafiza Shukor, and Muaz Mohd Zaini Makhtar

Abstract The treatment of wastewater in sewage treatment plants results in the generation of sewage sludge, which is a muddy residue that can exist in solid, semi-solid, or liquid form. Sewage sludge comprises a complex mixture of proteins, carbohydrates, detergents, phenols, and lipids, as well as harmful and dangerous organic and inorganic contaminants. The production of sewage is a result of the combination of domestic and industrial wastes, which typically contains over 99% water. The production of sewage is a result of contributions from various sources, including residential, institutional, commercial, and industrial facilities. This book chapter provides an overview of sludge in waste treatment plants, focusing on the terminology of wastewater treatment plants (WWTPs), the typical processes used in wastewater treatment, the constituents of sludge, rules and regulations governing WWTPs, and current methodologies employed for handling sewage sludge. The

J. Kobe
School of Industrial Technology, Environmental Technology Division, Universiti Sains Malaysia, 11800 Gelugor, Penang, Malaysia

M. D. Shafiq
Engineering Campus, School of Materials and Mineral Resources Engineering, 14300, Nibong Tebal Penang, Malaysia

R. Alkarimiah
Engineering Campus, School of Civil Engineering, 14300, Nibong Tebal Penang, Malaysia

A. Z. Yaser
Faculty of Engineering, Universiti Malaysia Sabah, 88400 Kota Kinabalu, Malaysia

H. Shukor
Centre of Excellence for Biomass Utilization, Faculty of Chemical Engineering Technology, Universiti Malaysia Perlis, 02600 Arau Perlis, Malaysia

M. Mohd Zaini Makhtar (✉)
School of Industrial Technology, Bioprocess Technology Division, Universiti Sains Malaysia, 11800 Gelugor, Penang, Malaysia
e-mail: muazzaini@usm.my

Centre for Innovation and Consultation (CIC), Universiti Sains Malaysia, 11800 Gelugor, Penang, Malaysia

M. Mohd Zaini Makhtar et al. (eds.), *Microbial Fuel Cell (MFC) Applications for Sludge Valorization*, Green Energy and Technology,
https://doi.org/10.1007/978-981-99-1083-0_1

chapter provides insight into the complexities of managing sewage sludge, which is a crucial aspect of wastewater treatment, as it poses environmental and health hazards. The chapter highlights the importance of effective management of sewage sludge, emphasizing the need for sustainable approaches that ensure safe disposal and minimize environmental impact.

Keywords Sludge biosolid · Biomass conversion · Bioprocessing · Affordable and clean energy · Renewable energy

1 Introduction

1.1 Terminology

In a wastewater treatment plant (WWTP), sludge is a common solid by-product composed of heavy metals, and organic and inorganic substances originating from municipal and industrial wastes [1]. WWTP combines various multi-processes of chemical, physical, and biological methods for organic waste degradation, phosphorus, and nitrogen removal, and mitigation of pathogen risk before being released to the environment [2, 3]. In general, reclaimed wastewater is clean for agricultural irrigation and contains a slightly higher concentration of dissolved solids than the source water [4]. The waste residue generated from the wastewater treatment process is known as sludge [2].

In a general sense, sludge is a watery solid yield from wastewater treatment that contains suspended particles of mainly organic matter. The exact composition of sludge is primarily dependent on the origin of the sludge, where carbon and other organic matters dominate the overall composition. Physical observations and characterizations revealed that sludge has a highly viscous and mud-like texture, whereas the rheological properties of sludge obey the non-newtonian behavior [5]. The density of activated sludge solids varies (approximately 1.02 to 1.06 g/cm^3) [6] and is typically close to the density of water (1.0 g/cm^3). Although, many works reported that the characteristic of sludge is complex and majorly influenced by the sludge composition, affecting the physico- and biochemical properties of the sludge [7]. Figure 1 exhibits the air-dried sludge appearance. Table 1 tabulates the main sludge origins, its major components, and final uses after treatment.

Apart from its origin, sludge can be also categorized by its processing stage during waste treatment. Primary sludge (PS) is generated via a mechanical process during the screening of insoluble substances such as grit, grease, and scum from wastewater prior to coagulation and sedimentation processes [18]. The screening process yields water and settleable solids of organic matter that are volatile, highly putrescible, and odorous [19]. Primary sludge is readily and easily to be digested anaerobically. Secondary sludge (SS) is treated via biological means and is sometimes referred to as activated sludge (AS). The organic matter in secondary sludge is fed

Fig. 1 Air-dried sludge. Reprinted with permission from [8] (Permission: CC-BY)

Table 1 Major types of sludge based on their origin

Main sludge origins	Major composition	End uses
Sewage sludge (SS) from municipalities, residential, industrials, and street runoffs WWTP and industrial wastewater sludge from manufacturing and production industries	Heavy metals [9, 10], pathogens, Nitrogen, Phosphate [11]	Agricultural fertilizer [12]
Drinking water treatment sludge (DWTS)	Fine sand, Alumina, Silica, Ferric Oxide, and Lime [13]	Landfill disposal [14]
Fecal sludge from septic tanks	Ash, Carbon, Nitrogen, Hydrogen [15]	Soil conditioner, and renewable energy production [16, 17]

to microorganisms and used as a nutrient substrate for microbial cultivation under aerobic conditions [20]. During this process, the fed microorganisms grow upon the oxidizable materials in wastewater and produce suspended flocs [21]. Treated and dried sludge is widely used as agricultural fertilizers or is disposed of in landfills [22].

Figure 2 shows the typical flowchart of wastewater and sludge treatment processes [23]. The production of primary sludge is owing to mechanical treatment and separation processes, where secondary sludge is produced via a biological treatment. Sludge treatment is an essential procedure to ensure the safe use of reclaimed sludge. The chosen method of sludge treatment is controlled by the technology availability, cost efficiency, and final uses of the treated sludge [24] (Table 2).

Factually, sludge is a mixture of concentrated solids and liquid, where the disposal of these concentrated solids is essential from the wastewater treatment plant. The biosolid-liquid separation process is a vital procedure in wastewater treatment plants to remove suspended solids of high amounts of COD, nitrogen, phosphorus, and organic and inorganic substituents in a high volume of water body [28]. The disposed

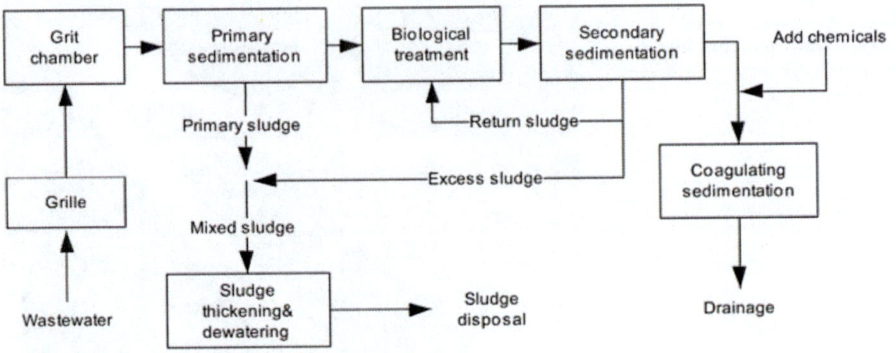

Fig. 2 Typical flowchart of wastewater and sludge treatment. Reprinted with permission from [23] (Permission: CC-BY)

Table 2 Differences between primary and secondary sludges

	Primary sludge	Secondary sludge
Content	High carbon-to-nitrogen ratio [25] Two predominant types of organic compounds are protein and carbohydrate [26]	Complex activated floc structure [18]
Drawbacks	Large particle sizes, thus biological degradation can be a challenge, and pretreatment is needed [27]	Aeration is required for the formation of activated sludge

of sludges are commonly dewatered or thickened to increase the solid content ultimately reducing hassles during transportation and disposal. During the sludge thickening procedure, the presence of bound water exists, but it is minimized, and the sludge is still behaving like liquid [29]. The thickening process employs the gravitational method, where the disparity between solid and liquid densities is conceptualized to achieve solid compaction [29, 30]. This sludge treatment is also commonly referred to as the gravity-thickening process [31]. The thickening process of sludge also can be achieved by the centrifugal method. This method is performed using a high-speed rotating cylindrical vessel that migrates the solid particles to the vessel's wall controlled by centrifugal force [32]. Another method that is commonly used to thicken activated sludges is air flotation. This method floats sludge particles by air under pressure or vacuum, thus separating the solid content from the bound liquid [33].

Meanwhile, sludge dewatering is a process of eliminating water from the sludge without evaporation and the yield of this process is often referred to as dewatered sludge (DS). The filtration process is aided by chemical conditioning to produce flocs before the dewatering process [34]. The common conditioners used are synthetic organic polymers and metal ions. The underlying concept of the conditioning process is based on the coagulation of colloidal particles in water [35]. Thermal conditioning such as direct heating and wet air oxidation can also be utilized before the dewatering process, while the freezing/thawing method can be efficient in cold climates [34].

2 Impact of Global Wastewater Regulations on Environment and Health

Regulations and policies on wastewater treatment are essential and ubiquitous across the globe, especially in urbanized areas. The regulations set by lawmakers are targeted to address and mitigate environmental and health concerns arising from the detrimental effects caused by untreated wastewater. Saimy and Yusof (2013) compiled the statistic outlined by the Food and Agricultural Organization (FAO) that revealed the dominancy of daily water usage is for agricultural uses (62%), followed by industrial and domestic uses at 21% and 17% respectively [3]. United Nations disclosed that 80% of wastewater goes back to ecosystems without treatment. Population growth and rapid global urbanization intensify the need for efficient wastewater treatment with minimal downsides for holistic water supply and distribution.

As the world is moving towards the circular economy, wastewater management and operations are also directing activities towards sustaining the environment and health. The wastewater discharges to surface waters and municipal sewage plants are mandated to obey the Effluent Guidelines as outlined by the United States Environmental Protection Agency (EPA). The water quality guidelines concerning health standards are also outlined by the World Health Organization (WHO). These standards are vital points for policies implemented to govern, reduce, and prevent inputs of hazardous and risky substances and pollutants penetrating aquatic ecosystems [36].

Table 3 tabulates general impacts and targeted impacts of wastewater regulations in some countries. The specific regulations are tailored based on the origin of the water and the end uses of the discharged, treated, or reclaimed water. Water quality criteria (WQC) is designed based on scientific research data predominantly for water and wastewater management for the benefit of health and the environmental [38]. However, most countries outlined proper WQC where several main factors are taken into consideration such as the number of chemicals and pollutants that can cause a toxicological effect on the water. Although, the development of WQC and regulations is essential to cater to current trends of emerging and new pollutants such as microplastic and microfiber particles. Figure 3 illustrates the economic and environmental assessment of four approaches in outlining guidelines for discharged wastewater quality. The approaches are [37]:

A. Allowed concentrations of pollutants in wastewater, and the efficiency reduction rates achieved at national, regional, and local demographics
B. Uniform quality standards for reclaimed wastewater throughout the countries
C. Receiver's water quality that takes account of environmental standards that must not be jeopardized due to the discharge of treated wastewater

Table 3 General targeted impacts and impacts of wastewater regulations in some countries

Wastewater regulations	Country/region (references)	Targeted impacts/impacts
Urban Wastewater Treatment Directive	European Union (EU)(European Commission)	• Improvement of water quality from the reduction of organic matter and various pollutants in treated wastewater • The directive needs to address new concerns of pharmaceutical and microplastic pollutants in the wastewater system
Effluent Water Quality Standards for Sewage System	Japan (Water Environment Partnership in Asia)	• Specific pH, biological oxygen demands, iodine consumption and, cyanide amount based on facilities scale
Environmental Protection Act	Australia (Department of Environment and Science)	• Minimization of contaminants strength • Minimization of water used
Environmental Quality Act-Industrial Effluent	Malaysia (Department of Environment)	• Industrial and mixed effluent at specific minimum concentration parameters and substances such as heavy metals, free chlorine, sulfide, oil, and grease
Municipal wastewater	Switzerland [37]	• Minimum amount of biological and chemical oxygen demands, total suspended solids, and three nutrients; ammonium, nitrates, and orthophosphates

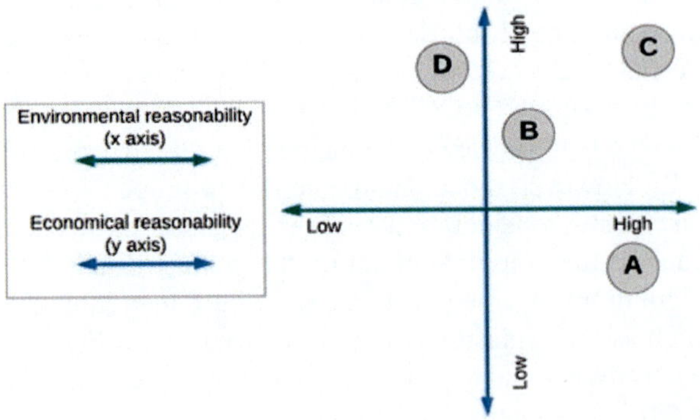

Fig. 3 Economic and environmental assessments of current methodological approaches to designing quality requirements of discharged wastewater Reprinted with permission from [37] (Permission: CC-BY)

D. Technological standards suggest the use of technological procedures without allocating the permissible values of pollutants in reclaimed wastewater

3 Health and Environmental Concern in Sludge Management

Despite the nutrient-enriched (N and P) sludge and their amenability for different applications, they are seen generally as deposits of organic contaminants and heavy metals. Public concerns of sludge application exist for these potentially toxic elements or organic compounds' long-term effects. The risk of sludge application to individuals and the environment is also small if appropriately handled under European Union directives' strict measures and rules. However, when unauthorized vegetable cultivation occurs on soils containing sewage sludge or irrigated with sewage sludge admixtures, excessive accumulation of heavy metals in agricultural soils can lead to increased heavy metal uptake by crops, affecting food quality and humans. Sewage sludge components indicated a high carcinogenic risk for humans according to RISK value.

World health organization issues specific guidelines that state direct or indirect contact to wastewater and sludge results in microbial and chemical contaminants which frequently cause negative health outcomes. Various pathogenic bacterial and viral strains can cause health problems such as diarrhea, respiratory infections, and skin infections. Human pathogenic bacteria discharged in the fecal matter are generally found in biosolids. *Salmonella spp.*, *Listeria spp.*, *Escherichia coli* (enterotoxigenic and enteropathogenic varieties), Campylobacter spp., Clostridium spp., and Yersinia spp. are among the intestinal pathogenic bacteria detected in biosolids. A high percentage of these bacteria are both human pathogenic and zoonotic, which implies they may infect both humans and animals. Furthermore, these creatures have a strong ability to adapt to changes in their environment throughout time.

In developing countries, reusing wastewater for agricultural use is a widely used practice. Although this technique may help farmers maintain their livelihoods, there are health problems linked with the treatment and reuse of wastewater and sludge. There is a risk of eating vegetables grown on these plots that the vegetables appear to be healthy and growing well despite heavy metals accumulations in these vegetables that far exceed the maximum amounts considered safe for human consumption.

In general, excessive nitrogen compounds applied to land can have an adverse effect on aquatic systems, causing eutrophication and the death of aquatic vegetation and wildlife. Algal blooms are a symptom of eutrophication, as are clogged water treatment process filters, a foul odour and taste to drinking water, animal and human health issues, and ecological and economic obstacles.

The risk of chemicals in sewage sludge being applied to land was investigated by a Working Group created by the World Health Organization's Regional Office for Europe. Due to bioaccumulation from the soil by food plants, cadmium appears to be the most significant pollutant. If sludge applications lie between the nitrogen needs of the crops, other metals in sludge are insufficient to cause health issues. Sewage sludge exhibited significant amounts of Zn, Cu, and Cr, according to the data [14].

According to the mean pollution index, Cu, Zn, As, and Cr in sewage sludge were all linked to probable ecological issues (PI). Ibuprofen, 17-ethinylestradiol, and 17-estradiol in sewage sludge pose a significant ecotoxicological risk to aquatic and terrestrial ecosystems, according to an environmental risk assessment. After the disposal of wastewater in the sludge disposal onto soils, a significant decrease of the ecotoxicological risk occurs. Current concern about the environmental impact of releasing sewage sludge on agricultural land has focused attention on the broad range of organic pollutants that can enter sewage treatment operations and end up in biosolids for disposal. Organic pollutants have the potential to harm the soil, crop, and animal systems in areas where biosolids are applied to the ground. However, because of the low amounts present and the numerous transformations (particularly biodegradation) that may occur in soils, the relative danger of organic compounds in biosolids is typically regarded to be minimal.

4 Sludge Constituents

Most of the sludges are the by-products of municipal wastewater treatments. The sludge content is approximately 50% organic and 50% inorganic material such as silica. In general, the physicochemical characteristics of the biosolids contain 20% content of fat, 50 carbohydrate content (sugar, starch, and fiber), 30 to 40% content of organic matter, 3% total nitrogen, 1.5% total phosphorus, 0.7% total potassium content, 10 to 20% C/N ratio, high levels of heavy metal ions: Cu, Zn. The heat value (Ho) of the dry sludge is about 12.000 kJ/kg. However, sludge generally varies in characteristics and contains organic and inorganic chemicals, toxic metals, and pathogens and depends on treatment technology and retention time in wastewater facilities.

5 Physical Characteristic

The important physical properties of sludges are floc size, density, fractal dimension, specific gravity (1.0 equal to water), solids fraction as the relative fraction of solids, sludge volume index (SVI), water in the slurry [19, 21]. After the process of mechanical dewatering, sludges are masses showing a lumpy structure and a bulk density in the range of 650–800 kg m^{-3}. The dry solid fraction of biosolids ranges between 2 and 12% by weight and is one of the crucial parameters when considering waste-to-energy management. The fractal dimension, flocs size, and filament index are the main parameters related to Zone settling velocity (ZSV) and the sludge volume index (SVI) is determined by the parameters (fractional dimension, flocs size, and filament index). The sludge characteristics of SVI and ZSV largely depend on the filament index of sludge flocs and the fractal dimension.

6 Chemical Characteristic

Chemical parameters are associated with the existence of nutrients and toxic compounds Table shows a typical chemical composition of untreated and digested sludge. The total solid content of sewage sludge contains dissolved and suspended solids. The organic chemicals in sludge volatile solids (VSs) are reduced during the sludge being heated to 550 °C (1,022°F) during oxidation. Pathogen concentration and its control are the main reasons to adjust pH in sewage sludge. Sludge with low pH (<pH 6.5) increases the leaching of heavy metals and that with high pH (>pH 11) kills many bacteria. Sludge application rates depend on nutrient concentrations. Excessed levels of nutrients from high sludge application rates potentially lead to environmental pollution of groundwater and surface water, which should be avoided. Amounts of metals in sludge may change due to the different sources of sludge and some metals may be toxic to humans, animals, and plants at high levels [23].

7 Inorganic Nonmetallic Constituents

Sludge's inorganic contents, as assessed by ash content, typically range from 30 to 60%. Local soil and sediment materials, stormwater drains, inorganic residues in human feces (e.g., relatively high levels of SiO_2 seen in foods originating from plant material; 1–4%), cosmetics, and other products washed down residential drains are the main inorganic components of sludges. The bulk of essential macronutrients, such as nitrogen and phosphorus, were found to be present in their organic and inorganic forms, respectively. Organic and inorganic C, organic N, and inorganic Ca, P, and Mg fractions were consistent during the whole study period. However, Inorganic N, organic P, K, and the other metals showed fluctuation in concentration. The biggest differences for trace elements were Cd, Zn, Cu, Ni, and Pb. The bulk of odorants in sludge were volatile sulfur compounds, and these odor emissions from biosolids hinder their beneficial reuse through land application. Table 4 shows a comparison of the physiochemical properties of biosolids in different countries.

8 Metallic Constituents

Heavy metal is a term commonly used for a group name of metals and semimetals and is usually defined as having an atomic number greater than 20 or 21 that is associated with contamination and potential toxicity to animals or plants. Common

Table 4 A comparison of the physicochemical properties of biosolids of different countries [1]

Properties	India	China	Australia	Spain
pH	6.16–7.5	6.86–8.73	4.4–8.3	7.1–8.1
EC(ms cm^{-1})	2.28–2.7	0.667–5.01	1.6–7.9	1.2–3.9
Org.C (%)	5.52–12.6	–	–	–
TotalN (%)	1.6–1.73	2.23–6.50	0.60–2.5	3–4.1
TotalP (%)	0.49–1.3	1.06–2.18	0.28–0.83	2–3.6
Av.P (mg kg^{-1})	132–716.7	–	–	13,900
TotalK (%)	0.8–1.26	0.46–0.62	0.18–0.45	0.24–0.47
Ex.K (mg kg^{-1})	208.9	593	–	–
Ex.Na (mg kg^{-1})	483	–	–	–
Ex.Ca (mg kg^{-1})	154.1	–	–	–
Total metals				
Fe (mg kg^{-1})	6059–14,390	0.46–2.40	13,824–18,026	31,200
Ni (mg kg^{-1})	47.17–60	52.5–202	166	<25–71
Mn (mg kg^{-1})	186.2–260	0.35–537	173	165–233
Zn (mg kg^{-1})	161–2050	0.21–1350	210–3060	560–1100
Pb (mg kg^{-1})	28.5–240	49.1–186	323	43–219
Cr (mg kg^{-1})	35.5–60	52.8–288	308	1–210
Cd (mgkg^{-1})	32.3–154.5	2.23–7.61	0.70–13.6	<0.2–3
Cu (mgkg^{-1})	186–330	0.27–975	92–1996	149–230

elements of heavy metals are Cu, Zn, Co, Ni, Pb, Hg, Cd, Cr, Se, and As. The main contaminants in biosolids are heavy metals. Industrial wastewater is often the major source of these contaminants. Wastewater from surface treatment processes (e.g., electroplating, galvanizing) is a potential source of metals like Cu, Zn, Ni, and Cr while industrial products may be disposed of as wastes at the end of their life. The main urban inputs are drainage waters, business effluents (e.g., car washes, dental uses, and other enterprises), and traffic-related emissions brake linings, tires, asphalt wear, vehicle exhausts, petrol/oil leakage, etc.) that are carried with stormwater into the sewage system. The analysis on sludge showed that metals concentrations in biosolids Mn>Zn>Ba>Cu>B>V>Cr>Ni>Pb>As>Co>Mo>Hg>Cd (Table 5).

9 Aggregate Organic Constituents

About 50% of biosolid's components are organic material and the other 50% are inorganic material. The organic percentage of sludge contains fats, proteins, carbohydrates, lignin, amino acids, sugar, celluloses, humic materials, and fatty acids. The organic matter content of secondary sludge (62–82%) is usually more than that of

Table 5 Metal concentration in common wastewaters (mg/L) [2]

Name	Symbol	Municipal treatment plant
Aluminum	Al	
Antimony	Sb	
Arsenic	As	0–0.0019
Barium	Ba	
Bismuth	Bi	
Cadmium	Cd	0–0.0033
Calcium	Ca	
Chromium	Cr	0.04–0.56
Cobalt	Co	
Copper	Cu	0.079–0.58
Gold	Au	
Iron	Fe	0.48–3.9
Lead	Pb	0–0.039
Magnesium	Mg	
Manganese	Mn	0.067–1.16
Mercury	Hg	0–0.0002
Molybdenum	Mo	
Nickel	Ni	0.0067–0.77
Potassium	K	
Silver	Ag	0–0.0014
Sodium	Na	
Strontium	Sr	
Tin	Sn	0–0.028
Vanadium	V	
Zinc	Zn	0.26–0.75

primary sludge (47–70%) because many insoluble inorganic matters are removed by primary sedimentation. The organic matter in sludge contains two spatially and chemically main components originated in plant material degraded partially and bacterial residues, respectively.

The exponential development of synthetic organic compounds for industrial use has resulted in an increase in the presence and concentration of organic pollutants in effluent sewage and sludge for the past 50 years. The occurrence and concentration of organic pollutants in biosolids are influenced by the character of wastewater, numerous local point sources, the physiochemical features of certain organic compounds, and the functioning of wastewater treatment plants. In general, industrial sewage has larger levels of organic pollutants than domestic sewage. There are

143,000 chemicals for industrial purposes registered in the European Union and potentially observed in sludge. Here are the ranked selected chemicals based on human toxicity and negative impacts on the environment and the result showed the ranking in decreasing order of priority as below (the highest score is 11).

- perfluorinated chemicals (10)
- polychlorinated alkanes (9), polychlorinated naphthalenes (9)
- polybrominated diphenyl ethers (7), organotins (7), triclosan (7), triclocarban (7)
- benzothiazoles (6)
- antibiotics and pharmaceuticals (5)
- synthetic musks (3)
- bisphenol A (2), quaternary ammonium compounds (2), steroids (2)
- phthalate acid esters (1), polydimethylsiloxanes (1)

The analysis of organic pollutants in sludge reveals several difficulties because extraction produces a complex mix of organic compounds, and the different organic fractions must be separated before analysis. The organic molecules are attached chemically and/or physically to the sludge in the solid phase and need to be removed using somewhat harsh reagents and methods before being analyzed. Furthermore, contamination concentrations are frequently low, and sometimes even below the limitations of current analytical procedures. As a result, before analysis, preconcentration is frequently required. Soxhlet extraction, rotary evaporation concentration, and column chromatography cleanup are all common treatments.

10 Sludge Reclamation and Reuse

Contaminants and pathogens abound in sludge, posing a major hazard to the environment. The sludge treatment comprises the addition of a microorganism inactivator. The biggest issue with this treatment is that inactivating all germs and microorganisms takes a long time. Sludge treatment describes the process used to manage and dispose sludge produced during sewage treatment. In this context, sludge is referring to residual, semi-solid material left from industrial/municipal wastewater. Meanwhile, sewage treatment refers to the process of removing contaminants from wastewater. The household wastewater treatment process uses physical–chemical and biological methods to remove particles, organic compounds, pathogens, and nutrients from sewage, resulting in two major products: treated water and sewage sludge. The treatment and disposal of sewage sludge account for up to 60% of total costs, with transportation costs dominating. In the past, the wastewater treatment plant (WWTP) (Fig. 4) was designed to remove floating, settleable, and suspended solids. As well as soluble organic materials (BOD, COD, and TOC) and pathogenic bacteria from the water.

Sludge treatment and disposal usually includes several activities such as collection of sludge, transportation of sludge, processing of sludge, and disposal of sludge as shown in Fig. 5. Most sludge treatment and disposal efforts are focused on removing

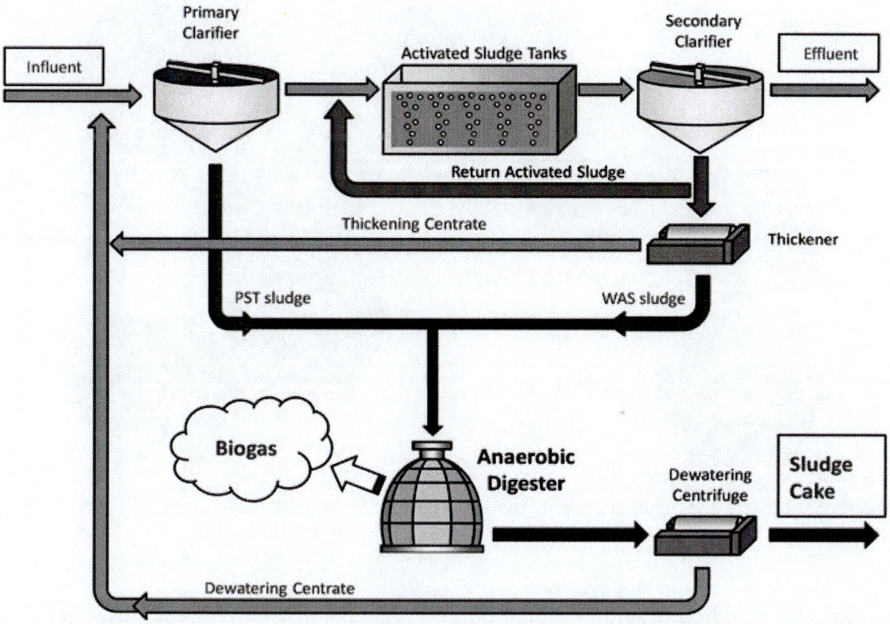

Fig. 4 Conventional WWTP activation process. Reprinted with permission from [38] (Permission: CC-BY)

liquids and decomposing organic and inorganic chemicals to limit the volume of solids that must be handled. The aims of sludge treatment, disposal, and reuse are as follows:

i. Volume reduction: Includes thickening and dewatering process.
ii. Elimination of pathogenic germs: If used in agriculture as fertilizer or compost
iii. Stabilization of organic substances: For gas production, reduction of dry content, and improvement of dewatering
iv. Recycling of substances: reduction of odor, production of gas, fertilizer.

11 Treatment Methods and Current Status

The treatment of sludge is quite expensive for wastewater treatment plants. Because of the presence of dangerous pollutants and environmental pollution, sludge disposal is also a difficult task. To address these issues, a variety of sludge treatment techniques have been developed, including composting, landfilling, anaerobic digestion (AD), pyrolysis, and incineration. AD is the most widely employed of these technologies due to its numerous advantages, which include low environmental effect, low cost, minimal solid residue, and high bioenergy production, among others [40]. Environmental protection procedures for sludge treatment and disposal are critical due to the

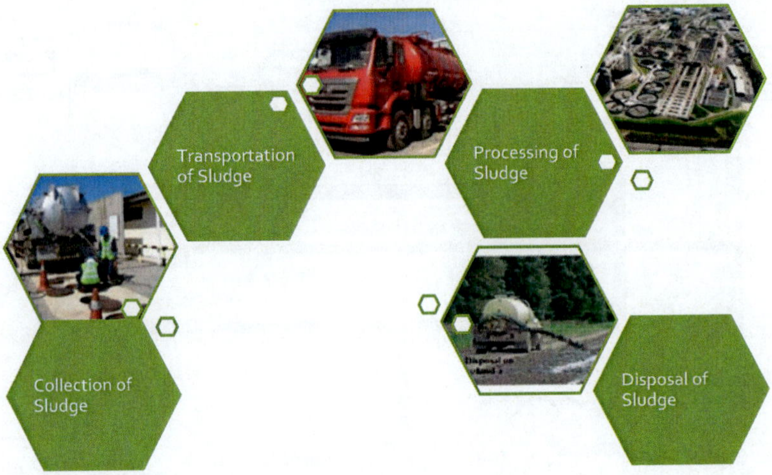

Fig. 5 Sludge treatment and disposal activities. Reprinted with permission from [39] (Permission: CC-BY)

presence of residual organic pollutants, toxic metals, and pathogenic microbes that can cause health problems and must be removed. At the same time, they consume a significant amount of energy (with accompanying environmental consequences), with the cost of sludge treatment accounting for nearly half of the total operating costs of WWTPs [41].

For treating sludge production or reducing its production entirely, numerous physical, chemical, and biological techniques have been invented. For the disposal of excessive sludge, the most often employed methods are incineration, landfilling, ocean dumping, reuse in agriculture (either directly or after composting), and reuse in the manufacture of cement, bricks, and asphalt [42]. The costs of gas scrubbing for air pollution control, the release of heavy metals into the environment, and the need for incineration should all be considered when choosing the best sludge management strategy. Incineration should be used when there are large WWTPs or when the quality of the sludge is not good enough for land application, or when there is a lot of sludge and it can't be used on the land, for example. However, there has been a lot more attention paid to the environment and strict laws, which has led to the thought of other ways to reduce and treat sludge. This has led to the thought of other ways to reduce and treat sludge. Figure 6 shows the Possibilities for reclaiming materials and energy from wastewater sludge based on selective treatment technologies.

To limit the amount of non-processed sludge entering the environment, most countries have adopted sludge-processing systems. The basic sludge treatment principles are (i) volume reduction, (ii) stabilization, (iii) energy and nutrient recovery, and (iv) beneficial use of end products. To achieve the goals of primary sludge treatment principles, the usual process for the essential treatment consists of primary operation, thickening, stabilization, conditioning, dewatering, drying, and combustion.

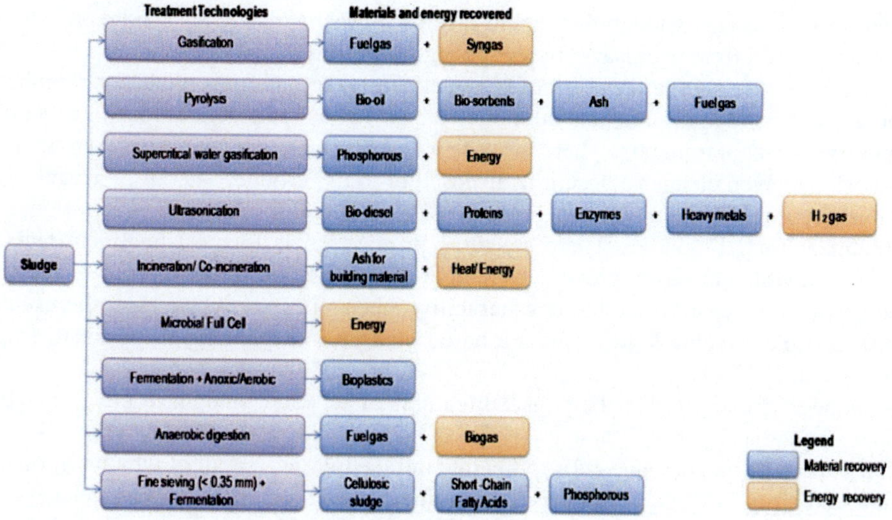

Fig. 6 Material and energy recovery for sludge treatment processes [42]. Reprinted from "A review on wastewater sludge valorization and its challenges in the context of circular economy," *Journal Clean Production*, vol. 228, pp. 244–263, 2019

Although there are numerous treatment options available, anaerobic digestion (AD) has always played and continues to play a critical role because it allows for pathogen removal, the conversion of volatile solids into biogas (and then into energy), as well as the production of stable sludge for mechanical dewatering. Stabilizing sludge, reducing organic compounds, and recovering bioenergy in the form of biogas are all benefits of an anaerobic digestion process. Another by-product of anaerobic digestion is anaerobically digested (AD) sludge, which is generated in addition to biogas. soil retention, organics, and nutrients are all improved by using AD sludge as a soil amendment for land application [43].

During anaerobic digestion (AD), biogas is produced that contains a high concentration of methane and can be utilized to heat the tank as well as to power motors or microturbines for other on-site processes. This method can be used to generate enough energy in large treatment plants to generate more electricity than is required by the equipment. In the anaerobic process, methane is produced, which is a significant advantage. The process can take up to 30 days, which is a significant drawback, as is the large initial investment necessary. Because of the drawbacks of the AD process, the researcher comes out with the development of pretreatment technologies to improve the AD process. The example of pretreatment methods before undergoing AD process are thermal hydrolysis, alkaline pretreatment, ultrasonic pretreatment, ozonation, alkaline hydrolysis, enzymatic lysis, freezing and thawing, mechanical disintegration, high-pressure homogenizers, ultrasound, microwave irradiation, and photocatalytic pre-treatment. These pretreatments basically, aim to break microbial cell walls, liberate extracellular, and intracellular organic molecules, and reduce the solid retention time necessary for sludge digestion [42].

Because AD sludge has a lot of water, it can make land unsuitable for farming because of its high organic and odor content, as well as the potential for toxic metal and organic contaminants. There are a few technologies that can be used to make AD sludge even cleaner before it is disposed of to the ground. Therefore, currently most of the researchers looking forward to the current and future prospects of post-treatment for AD sludge. There are three primary goals for post-treatment technologies, which are to improve dewaterability, stabilize sludge, and promote metal solubilization. To enhance the dewaterability mechanisms of AD sludge, the usual post-treatment technologies used are based on employing chemical, physical, and biological methods.

In addition, to altering the physical features of activated sludge, reactive radical species such as hydroxyl radicals (•OH) and sulfate radicals (SO4•) can break EPS fractions and even damage cells to liberate intracellular water, all of which promote dewaterability [43]. Both hydroxyl and sulfate radicals, on the other hand, necessitate the use of additional chemicals, which invariably raises the expense of sludge treatment. Meanwhile, to enhance sludge stabilization and reduction, the usual technologies that are being used are physiochemical treatment, biological treatment, and combination of physiochemical and biological. Lime and fly ash are two chemicals that are usually used to stabilize AD sludge. But the use of lime or fly ash may result in an increment in total sludge production [44].

On the other hand, anaerobic or aerobic conditions can be used for biological post-treatment for AD sludge. Aerobic digestion is another sludge stabilization process that depends on system temperature and retention period. The process is either mesophilic or thermophilic, with mesophilic conditions being the most often used. But thermophilic is by far the most satisfactory. The study done by [45] revealed that by adopting the thermophilic aerobic treatment (>50 °C) with sludge retention time (SRT) of 16 days may reduce the volatile solid (VS) up to 54% and efficiently reduce the human bacteria pathogen from 2.42% to 0.77%. Nevertheless, when the temperature rises above 35 °C, ammonium nitrogen builds up in the system because the nitrification and denitrification processes are slowed down. This will set back the bacteria's activity and stabilization of the post-sludge treatment [46]. However, Wang and team [47] found that the acidic condition (pH < 5) of aerobic post-treatment at the ambient temperature of AD also may improve AD sludge management in terms of reduction and stabilization [48].

To improve the biodegradability of sludge, a combination of physiochemical and biological treatment can be applied. When used in conjunction with thermal treatment, alkalis allow for significant solubilization and improvement in biodegradability performance [38]. Hence, the usual combination of physiochemical and biological treatment has been studied using different methods such as alkaline, heat, ozone, or ultrasound. However, this combination of techniques need to be further study because it may require technical and economic practicability. As example, when using chemical-based procedures, additional chemicals, such as ammonium, nitrite, and lime, must be added to the sludge before it can be treated biologically. The use of ultrasound, ozone, and heat requires a lot of energy. Additional equipment is required for each of these physiochemical techniques, which increases expense and obscures management.

12 New Trends and Concerns

As an alternative to traditional treatment technology that focuses on dewatering, reduction, and stabilizing operations, researchers are now focused on treatment to remove and reclaim harmful or precious metals from sludge that has accumulated over time. The presence of heavy metals such as Zn, Ni, Pb, Hg, Cr, Cu, and Cd in sewage sludge constrains its usage for land application due to the possibility of soil and groundwater pollution, which can further impair human and animal health [49]. For this reason, many procedures for recovering heavy metals have been investigated by some researchers, with the ultrasonication process being one among them. In the end, the data revealed that the recovery rate was greater than 90%. In order to extract the metals, other researchers also have turned to thermal treatments such as pyrolysis, gasification, and microwave.

The sludge dewatering through H_2O_2 lysis and ultrasonication and recycled for energy by torrefaction found that rich protein and hydrocarbon can be recovered as a nutrient source for the activated sludge unit [50]. Among other effects, they conclude that the proposed technique has environmental and economic benefits, including the ability to handle sludge disposal issues, create biochar at a reasonable cost for renewable energy production, reduce global climate change, and recover supernatant for use as fertilizer feedstock. Meanwhile, after studying the possible bioavailability of Cu, Cd, Pb, and Zn in sewage sludge, [50] discovered that pyrolysis increases the stability of these metals when the temperature is raised to particular levels. On the other hand, some metals like Cu, Zn, and Pb can be retrieved in the char produced by sewage sludge gasification, while Hg and Cd are depleted from the sewage sludge and end up in different downstream flows [42].

Sulfides, oxides, hydroxides, silicates, and other organic chelate complexes are the most common forms of metals found in activated sludge (AD sludge). Ph is the most important factor affecting the solubility of precipitates. Some of the techniques that are widely used to remove these contaminants from sludge are chemical leaching and bioleaching. The use of the chemical leaching method to recover metals from ad sludge has been implemented at several large-scale wastewater treatment facilities. Sulfuric acid is used in this process. However, this technique comes with several drawbacks, including high running costs and the development of secondary contamination because of the usage of chemicals in its production. Because it is a more affordable and environmentally friendly method, the usage of bioleaching techniques is becoming increasingly popular. The usual techniques for bioleaching use either organic acids-based bioleaching or sulfur-, iron- and ammonium-based bioleaching.

13 Future Trends in Technology

Recent and existing technologies for sludge treatment, reclamation, and reuse are mainly focused on the dewaterability properties of sludge. It was very seldom researched and proven that those procedures could achieve other parallel objectives, such as solids reduction, stability, and metal solubilization. Meanwhile, due to stringent environmental regulations and public pressure, the use of conventional sludge disposal methods such as landfilling, ocean disposal, land application, and incineration is being curtailed. As a result, selecting an effective and environmentally friendly method of sludge management is a difficult task for wastewater treatment authorities. The circular economy, in which sludge is no longer considered a waste but rather a useful resource for energy and material recovery, is a good example.

For example, the future AD sludge treatment methods are expected to be multi-faceted in nature, given the current state of the art. Examples of promising technical routes are proposed including iron-based accelerated oxidation and acidic aerobic digestion. Although both systems could meet the three objectives in a single procedure, they are both still hindered by a number of restrictions. [43] concluded that for iron-based advanced oxidation, the optimal conditions for achieving the three objectives (treatment, reclamation, and reuse) concurrently must be investigated. Additionally, economic feasibility must be determined by upscaling studies. Meanwhile, the dewaterability of acidic sludge digestion processes has never been examined previously. Additionally, improving the capacity and linking Ferrous, Sulfur, and Nitrogen oxidation are both basic and practical concerns.

The transformation of sludge into a useful resource opens up a new, more sustainable path for sludge management that goes further than treatment and proper disposal. Because of its multi-functional working mechanisms, which include electrooxidation, electrodialysis, and electromigration, the electrochemical technique is attracting scientific attention for sludge resource recovery. When it comes to the recovery of sludge materials, the electrodialysis procedure is gaining popularity. One of its main features is excellent recovery efficiency. The in-situ electro-dialytic extraction of ammonia from the digestate can further improve the process stability by reducing the amount of toxic waste produced. Even though numerous hurdles, including energy consumption, total cost, process efficacy, and system stability, enhanced operation with novel electrode materials design provides light on the electrochemical process's future applicability in sludge treatment and resource recovery [51].

The mobility and bioavailability of organic and inorganic pollutants in sewage sludge can be enhanced by some of the sludge treatment methods; as a result, when the final products are utilized for land application, they can pose a threat to humans and ecosystems. Hence, [52] suggested that the use of modern spectroscopic analytical techniques to investigate the geochemical transformation of pollutants during the processing of sewage sludge may need to be conducted [54]. Furthermore, to prevent potential toxicity to humans and animals, the long-term effects of continued application of biosolids on the accumulation of pollutants such as potential toxic elements (PTEs) and poly- and perfluoroalkyl substances (PFAS), as well as their subsequent uptake by plants, must be studied. Specifically, this area of research is significant because treated sewage sludge (biosolids) is becoming an increasingly important source of carbon and nutrients for agricultural lands as well as for land that is being replanted after mining operations.

The fact that these developing technologies were implemented, and more importantly, that they provided good recovery yields, constitutes a crucial first step in the framework of the circular economy. Many wastewater treatment plants (WWTPs) still have difficulties in recovering materials and energy from urban biorefineries, owing to the complexity of the processes and the high costs associated with them. Future experiments should take a long-term strategy, integrating technical, environmental protection, health, legal, economic, and social components into a single cohesive system.

14 Conclusion

In conclusion, the sludge generated during the sewage treatment process must be handled with care and disposed of properly due to its potential to be converted into valuable products. The rate of sludge generation and site-specific factors determine the appropriate sludge treatment technology to be employed. This book chapter has covered the terminology of wastewater treatment plants (WWTPs), the typical processes used in wastewater treatment, and the constituents of sludge. Additionally, it has explored various technologies that can be used to exploit sludge for the production of fuels, syngas, fertilizers, and fine chemicals. Among the available technologies, Microbial Fuel Cell (MFC) is a promising option as it can be operated at ambient temperature and pressure, while providing dual functions of bioremediation and energy recovery. MFC has shown great potential as a sustainable approach for sludge treatment, offering the possibility of generating renewable energy while reducing the environmental impact of sludge disposal. The implementation of MFC technology in sludge treatment can contribute to achieving the goals of sustainable development and circular economy. Overall, this book chapter highlights the importance of effective sludge management and the need for innovative and sustainable approaches to address the challenges associated with sludge treatment.

Acknowledgements The authors would like to thank the Universiti Sains Malaysia for the financial support of this study via APEX Era Grant (1001/PTEKIND/881004). The authors have declared no conflict of interest for the manuscript.

References

1. Safie NN, Zahrim AY (2020) Recovery of nutrients from sewage using zeolite-chitosan-biochar adsorbent: current practices and perspectives. J Water Process Eng 40(December):101845, 2021. https://doi.org/10.1016/j.jwpe.2020.101845
2. Yaser AZ et al (2022) Composting and anaerobic digestion of food waste and sewage sludge for campus sustainability: a review. Int J Chem Eng 2022. https://doi.org/10.1155/2022/6455889
3. Shukor H, Yaser AZ, Shoparwe NF, Mohd Zaini Makhtar M, Mokhtar N (2022) Biosorption Study of Methylene Blue (MB) and Brilliant Red Remazol (BRR) by coconut dregs. Int J Chem Eng 2022. https://doi.org/10.1155/2022/8153617
4. Molognoni D, Devecseri M, Cecconet D, Capodaglio AG (2017) Cathodic groundwater denitrification with a bioelectrochemical system. J Water Process Eng 19. https://doi.org/10.1016/j.jwpe.2017.07.013
5. Meng J, Feng X, Dai Z, Liu X, Wu J, Xu J(2014) Adsorption characteristics of Cu(II) from aqueous solution onto biochar derived from swine manure. Environ Sci Pollut Res 21(11). https://doi.org/10.1007/s11356-014-2627-z
6. Lawrence GB, Burns DA, Riva-Murray K (2016) A new look at liming as an approach to accelerate recovery from acidic deposition effects. Sci Total Environ 562. https://doi.org/10.1016/j.scitotenv.2016.03.176
7. Muchuweti M, Birkett JW, Chinyanga E, Zvauya R, Scrimshaw MD, Lester JN (2006) Heavy metal content of vegetables irrigated with mixtures of wastewater and sewage sludge in Zimbabwe: implications for human health. Agric Ecosyst Environ 112:41–48
8. Martín J, Camacho-Muñoz D, Santos JL, Aparicio I, Alonso E (2012) Occurrence of pharmaceutical compounds in wastewater and sludge from wastewater treatment plants: removal and ecotoxicological impact of wastewater discharges and sludge disposal. J Hazard Mater 239–240. https://doi.org/10.1016/j.jhazmat.2012.04.068
9. World Health Organization (2006) Guidelines for the safe use of wastewater, excreta, and greywater. World Health Organization.
10. Arthurson V (2008) Proper sanitization of sewage sludge: a critical issue for a sustainable society. Appl Environ Microbiol 74(17). https://doi.org/10.1128/AEM.00438-08
11. Fuhrimann S et al Health risk assessment along the wastewater and faecal sludge management and reuse chain of Kampala, Uganda: a visualization [Online]. https://www.youtube.com/watch?v=CQEC3d4iE_A&spfreload=10
12. Lananan F et al (2014) Symbiotic bioremediation of aquaculture wastewater in reducing ammonia and phosphorus utilizing Effective Microorganism (EM-1) and microalgae (Chlorella sp.). Int Biodeterior Biodegradation 95(PA). https://doi.org/10.1016/j.ibiod.2014.06.013
13. Karunanithi R et al (2015) Phosphorus recovery and reuse from waste streams. Adv Agron 131:173–250. https://doi.org/10.1016/bs.agron.2014.12.005
14. Safie NN, Zahrim AY (2021) Recovery of nutrients from sewage using zeolite-chitosan-biochar adsorbent: Current practices and perspectives. J Water Process Eng 40:101845. https://doi.org/10.1016/J.JWPE.2020.101845
15. Rogers HR (1996) Sources, behaviour and fate of organic contaminants during sewage treatment and in sewage sludges. Sci Total Environ 185(1–3). https://doi.org/10.1016/0048-9697(96)05039-5
16. Bhandari A, Xia K (2005) Hazardous organic chemicals in biosolids recycled as soil amendments. Water Pollut. https://doi.org/10.1007/b11440
17. Elliott HA (2003) Land application of sewage sludge and biosolids. J Environ Qual 32(5). https://doi.org/10.2134/jeq2003.1927
18. Haynes RJ, Murtaza G, Naidu R (2009) Chapter 4 inorganic and organic constituents and contaminants of biosolids. Implications for land application. Adv Agron 104:165–267. https://doi.org/10.1016/S0065-2113(09)04004-8
19. Xu G, Brattebø H (2014) Analysis of sewage sludge recovery system in eu-in perspectives of nutrients and energy recovery efficiency, and environmental impacts

20. Kumar V, Chopra AK, Kumar A (2017) A review on sewage sludge (biosolids) a resource for sustainable agriculture. Arch Agric Environ Sci 2(4). https://doi.org/10.26832/24566632.2017.020417
21. Fytili D, Zabaniotou A (2008) Utilization of sewage sludge in EU application of old and new methods-A review. Renew Sustain Energ Rev 12(1). https://doi.org/10.1016/j.rser.2006.05.014
22. Dean RB, Suess MJ (1985) The risk to health of chemicals in sewage sludge applied to land*. Waste Manage Res 3(3): 251–278
23. Sommers LE, Nelson DW, Yost KJ (1976) Variable nature of chemical composition of sewage sludges J Environ Qual 5(3). https://doi.org/10.2134/jeq1976.00472425000500030017x
24. Jaynes WF, Zartman RE (2005) Origin of talc, iron phosphates, and other minerals in biosolids. Soil Sci Soc Am J 69(4). https://doi.org/10.2136/sssaj2004.0207
25. Terry RE, Nelson DW, Sommers LE (1979) Carbon cycling during sewage sludge decomposition in soils. Soil Sci Soc Am J 43(3). https://doi.org/10.2136/sssaj1979.03615995004300030013x
26. Fisher RM, Alvarez-Gaitan JP, Stuetz RM (2019) Review of the effects of wastewater biosolids stabilization processes on odor emissions. Crit Rev Environ Sci Technol 49(17). https://doi.org/10.1080/10643389.2019.1579620
27. Brisolara KF, Bourgeois J (2019) Biosolids and sludge management. Water Environ Res 91(10). https://doi.org/10.1002/wer.1212
28. Magee HY, et al (2018) U.S. nationwide reconnaissance of ten infrequently monitored antibiotics in municipal biosolids. Sci Total Environ 643. https://doi.org/10.1016/j.scitotenv.2018.06.206
29. Sharma B, Sarkar A, Singh P, Singh RP (2017) Agricultural utilization of biosolids: a review on potential effects on soil and plant grown. Waste Manage 64. https://doi.org/10.1016/j.wasman.2017.03.002
30. Bergbäck B, Johansson K, Mohlander U (2001) Urban metal flows—a case study of Stockholm. Review and conclusions. Water, Air, Soil Pollut Focus 1:3–24
31. Comber SDW, Gunn AM (1996) Heavy metals entering sewage-treatment works from domestic sources. Water Environ J 10(2). https://doi.org/10.1111/j.1747-6593.1996.tb00023.x
32. Sörme L, Lagerkvist R (2002) Sources of heavy metals in urban wastewater in Stockholm. Sci Total Environ 298(1–3). https://doi.org/10.1016/S0048-9697(02)00197-3
33. Alvarez EA, Mochón MC, Sanchez JCJ, Rodríguez MT (2002) Heavy metal extractable forms in sludge from wastewater treatment plants. Chemosphere 47(7). https://doi.org/10.1016/S0045-6535(02)00021-8
34. Solís GJ, Alonso E, Riesco P, Distribution of metal extractable fractions during anaerobic sludge treatment in Southern Spain WWTPS. Water Air Soil Pollut 140(1–4). https://doi.org/10.1023/A:1020128605211
35. Smernik RJ, Oliver IW, Merrington G, Advanced solid-state carbon-13 nuclear magnetic resonance spectroscopic studies of sewage sludge organic matter. J Environ Qual 32(4). https://doi.org/10.2134/jeq2003.1523
36. Bodzek D, Janoszka B (1999). Comparison of polycyclic aromatic compounds and heavy metals contents in sewage sludges from industrialized and non-industrialized region. Water Air Soil Pollut 111(1–4). https://doi.org/10.1023/a:1005066327712
37. Clarke BO, Smith SR (2011) Review of 'emerging' organic contaminants in biosolids and assessment of international research priorities for the agricultural use of biosolids. Environ Int 37(1). https://doi.org/10.1016/j.envint.2010.06.004
38. Trzcinski AP (2019) Advanced biological, physical and chemical treatment of waste activated sludge. Taylor & Francis Group, Boca Raton
39. Spinosa L et al (2011) Sustainable and innovative solutions for sewage sludge management. Water (Basel) 3(2):702–717. https://doi.org/10.3390/w3020702
40. Nizami AS et al (2017) Developing waste biorefinery in Makkah: a way forward to convert urban waste into renewable energy. Appl Energy 186:189–196. https://doi.org/10.1016/j.apenergy.2016.04.116

41. Anjum M, Al-Makishah NH, Barakat MA (2016) Wastewater sludge stabilization using pretreatment methods. Process Saf Environ Prot 102:615–632. https://doi.org/10.1016/j.psep.2016.05.022
42. Gherghel A, Teodosiu C, De Gisi S (2019) A review on wastewater sludge valorisation and its challenges in the context of circular economy. J Clean Prod 228:244–263. https://doi.org/10.1016/j.jclepro.2019.04.240
43. Wang Z, et al (2021) Post-treatment options for anaerobically digested sludge: current status and future prospect. Water Res 205(June). https://doi.org/10.1016/j.watres.2021.117665
44. Cho SR, Jeong ST, Kim GY, Lee JG, Kim PJ, Kim GW (2018) Evaluation of the carbon dioxide (CO2) emission factor from lime applied in temperate upland soil. Geoderma 337(March):742–748, 2019. https://doi.org/10.1016/j.geoderma.2018.10.007
45. Min Jang H, Choi S, Shin J, Kan E, Mo Kim Y (2018) Additional reduction of antibiotic resistance genes and human bacterial pathogens via thermophilic aerobic digestion of anaerobically digested sludge. Bioresour Technol 273(September):259–268, 2019. https://doi.org/10.1016/j.biortech.2018.11.027
46. Yuan H, Xu C, Zhu N (2014) Disinhibition of the ammonium nitrogen in autothermal thermophilic aerobic digestion for sewage sludge by chemical precipitation. Bioresour Technol 169:686–691. https://doi.org/10.1016/j.biortech.2014.07.016
47. Lin YL, Zheng NY, Wang HC (2022) Sludge dewatering through H2O2 lysis and ultrasonication and recycle for energy by torrefaction to achieve zero waste: an environmental and economical friendly technology. Renew Sustain Energy Rev 155(1):111857. https://doi.org/10.1016/j.rser.2021.111857
48. Wang Z, et al (2021) Acidic aerobic digestion of anaerobically-digested sludge enabled by a novel ammonia-oxidizing bacterium. Water Res 194, 2021. https://doi.org/10.1016/j.watres.2021.116962
49. Tyagi VK, Lo SL (2013) Sludge: a waste or renewable source for energy and resources recovery? Renew Sustain Energy Rev 25(71):708–728. https://doi.org/10.1016/j.rser.2013.05.029
50. He ZM, Zhai YD, Li YB, Yang CT, Chen F, Fan L, Peng XP, Fu WF (2010) The fate of Cu, Zn, Pb and Cd during the pyrolysis of sewage sludge at different temperatures. Environ Technol 31(5):567–574
51. Zeng Q, Huang H, Tan Y, Chen G, Hao T (2021) Emerging electrochemistry-based process for sludge treatment and resources recovery: a review. Water Res 209(December):117939. https://doi.org/10.1016/j.watres.2021.117939
52. Hoang SA et al (2021) Treatment processes to eliminate potential environmental hazards and restore agronomic value of sewage sludge: a review. Environ Pollut 293(November):2022. https://doi.org/10.1016/j.envpol.2021.118564
53. Hernandez T, Moreno JI, Costa F (1988) Characterization of sewage sludge humic substances. Biol Wastes 26(3). https://doi.org/10.1016/0269-7483(88)90163-2

Effects of Treated and Untreated Sludge Applications on Human Health, the Environment and Other Ecological Factors

Nur Azzalia Kamaruzaman, Mohd Hafiidz Jaafar, Mazlin Mohideen, and Sharon Fatinathan

Abstract The use of treated and untreated sewage sludge in agriculture has been a widespread practice due to its potential benefits. However, the presence of different contaminants in the composition of sludge such as microplastics, pharmaceutical active substances, heavy metals, organic pollutants and pathogens are eliciting various hazards. In relation to short-term and long-term impact on human health, there are reported adverse events which included toxicity, genotoxicity, mutagenicity and carcinogenicity. In addition, the environment is implicated by events of marine and freshwater eutrophication, potentially harmful nutrient leaching and emission of greenhouse gases which leads to climate change. Animals and marine life are also affected negatively by this practice such as abnormal growth, reproductive anomaly and accumulation of heavy metals and toxins, all of which ultimately affect mortality. Contradictory to its negative effects, sewage application is practised to improve soil productivity and fertility due to its high concentrations of organic matter and plant nutrients. This will increase plant yield, especially in fruit and vegetable production as well as dairy pasture and forestry. It is important to understand that despite the negative implications of sludge treatment in agriculture, it is still considered an excellent fertilizer and a soil conditioner to sustain an optimized growth. Therefore, appropriate sludge treatment is imperative to ensure its safe application to the land.

Keywords Agriculture · Fertilizer · Toxicity · Heavy metals · Microplastics · Sludge treatment

N. A. Kamaruzaman (✉)
National Poison Centre, Universiti Sains Malaysia, 11800 USM Gelugor, Penang, Malaysia
e-mail: azzalia@usm.my

M. H. Jaafar
School of Industrial Technology, Universiti Sains Malaysia, 11800 USM Gelugor, Penang, Malaysia

M. Mohideen · S. Fatinathan
Faculty of Pharmacy and Health Sciences, Universiti Kuala Lumpur-Royal College of Medicine Perak (UniKL-RCMP), 30450 Ipoh, Perak, Malaysia

M. Mohd Zaini Makhtar et al. (eds.), *Microbial Fuel Cell (MFC) Applications for Sludge Valorization*, Green Energy and Technology,
https://doi.org/10.1007/978-981-99-1083-0_2

1 Introduction

Sewage sludge is the inevitable by-product of wastewater treatment and a sludge management system treats the sludge to reduce its volume and for stabilization. Sludge (untreated and treated) is widely used in agricultural land, landfills and incineration [1, 2]. Its use in agricultural land can improve the physical property of poorly structured soil. Furthermore, in the European Union, 40% of the sludge produced is used to increase soil organic matter [3, 4]. In addition, sludge with high content of nitrogen and phosphorus is used as fertilizer. Apart from agriculture application, sludge with low organic matter and high organic substance is used in sanitary landfills. The sludge used in sanitary landfills must exhibit stability either with very slight or very slow degradation without emitting any putrefying odour [5]. In addition, sludge that is rich in carbon, hydrogen, and nitrogen can also be used as a fuel source via biogas degeneration or carbonization [6], whereby biogas produced can be used to generate electricity and heat [7].

Although many countries are using untreated and treated sewage sludge in several applications, there are concerns about the presence of various contaminants. According to Collivignarelli et al. [3], some of the adverse effects of sludge when used in various applications are the release of odour compounds, greenhouse gas emission and an increase in the level of persistent toxins [3]. Due to these concerns, there are regulations in many countries that require sufficient treatment to reduce the contaminants and to enhance the biological and physical stability of sludge [1].

The presence of different contaminants in sludge is largely dependent on the wastewater source, dewatering conditions, and treatment procedures employed to produce the treated sludge. Different sources of wastewater consist of different types of pollutants and need to be treated accordingly. Moreover, sludge generated from wastewater treatment plants should consist of a very minimal amount of pollutants. This is because many of the pollutants have the potential to cause mutagenic, teratogenic and carcinogenic effects in humans [8].

The usage of chemicals during the chemical conditioning process of sludge imposes environmental and human health risks. As an example, the usage of polyacrylamide (PAM) as a flocculent introduces acrylamide, a toxic monomer into the sludge. This increases the risk for agricultural application. Furthermore, the usage of chemical coagulants such as polymeric ferric chloride (PFC) and polymeric aluminium chloride (PACl) poses the risk of increasing dioxin concentration in the flue gas during the incineration of dewatered sludge [9]. Additionally, during chemical conditioning, toxic gases released during the chemical reactions at the wastewater treatment plant can pollute the air and disrupt human activities in the vicinity of the wastewater treatment plant [9].

The different technologies used to produce treated sludge can alter the sludge's structure, increase the nutrient release, and influence the speciation of heavy metals contained in the sludge [9]. Heavy metals were discovered to migrate and convert as a result of the sludge condition process. The leaching of heavy metals from dewatered sludge into the soil can be dangerous to humans when the amount of

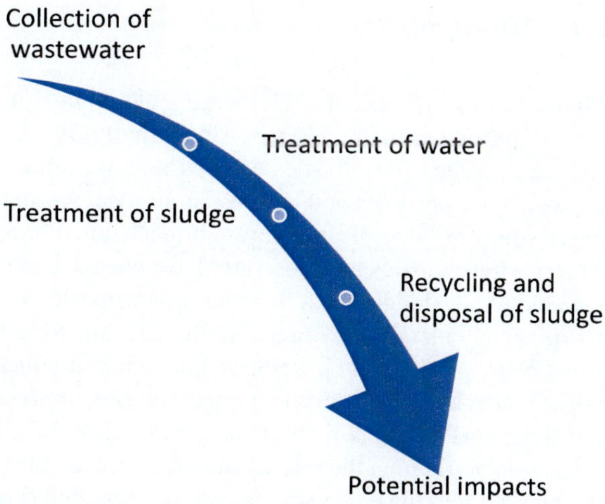

Fig. 1 Chronology of sludge, from origin to treatment, uses and potential impacts to humans and the ecology. Adapted from [11]

heavy metals entering the food chain is higher than the limit that the human body can metabolize [6].

Even though sludge brings benefits through its various applications, the composition of sludge has to be strictly monitored. Therefore it is important to understand the processes as depicted in Fig. 1, as various factors such as the origin of sludge, treatment of sludge and applications of sludge will ultimately determine the impacts imposed by the sludge [10]. This chapter outlines the possible toxic components present in the sludge, and the numerous implications of untreated and treated sludge on human health and ecology.

2 Potentially Toxic Composition of Sludge

The composition of sludge is ever-changing as the composition is highly dependent on the source of the wastewater and the treatment technologies used to produce the treated sludge. Many researchers have carried out analyses to identify the composition of sewage sludge and treated sludge. Regulations are in place in many countries to ensure that the composition of sludge used in different applications is safe for the environment and poses no harm to human health. The most commonly found toxic elements in sludge are microplastics, pharmaceutical active substances, heavy metals, organic pollutants and pathogens [12].

2.1 Microplastics

Microplastics are one of the emerging pollutants and the presence of microplastics in the environment is increasing rapidly due to the overuse of plastic materials in everyday life. Plastics are divided into primary plastics which are found in personal care products and clothes, and secondary plastics which are formed by the degradation of large plastics through mechanical erosion or photo-oxidative degradation. Microplastics in wastewater have been found to diffuse into the environment via sludge, as sludge tends to retain microplastics [13, 14]. Microplastics can also adsorb heavy metals and organic pollutants, allowing them to be transported into the aquatic environment [15]. Some of the commonly identified microplastics in sludge are polyproylene (PP), acrylic, polyamide (PA), polyester (PE), polycarbonate (PC), polystyrene (PS) and polyethylene terephthalate (PET) [15]. Although treated sludge is capable of trapping the microplastics, it is plausible for the trapped microplastics to be released into the environment over an extended time. This is confirmed by Keller et al. [14] who reported that microplastics were retained in the sludge, but nanosized plastics were still released from the sludge into the environment [14]. Therefore, further treatment of the sludge is required to ensure that the number of microplastics present in the sludge is kept at a very minimal level.

2.2 Pharmaceutical Active Substances

In recent years, the presence of pharmaceutical active substances in the environment has triggered concern because these substances possess ecotoxicological risks once present in the environment. Pharmaceutical active substances can leach from sludge into the soil and finally enter the surface water [4]. Plants can take up these pharmaceutical active substances and introduce these contaminants to humans via involuntary intake [16, 17]. Based on mass balance studies, it was found that antibiotics, antihypertensives, lipid-regulating agents and cardiovascular drugs removed from wastewater were absorbed by the sludge [18]. Ivanová et al. [4] reported that wastewaters also consisted of illicit drugs such as amphetamine, cocaine, benzoylecgonine [4]. In addition, Mejías et al. [18] reported that only selected pharmaceutical active substances and metabolites present in the sludge were reduced upon treatment [18]. Thus, the usage of untreated and treated sludge consisting of pharmaceutical active substances in agricultural land should be done with caution as studies have shown that some pharmaceutical active substances exhibited potential toxic effects on the soil [18, 19] (Fig. 2).

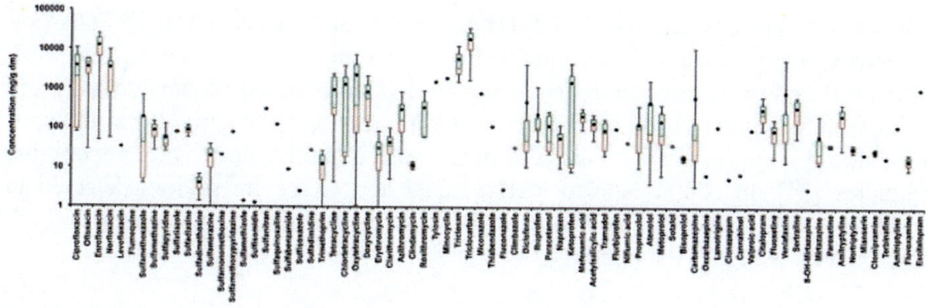

Fig. 2 Box-and-whisker plot on concentrations of pharmaceutical active substances in sludge [18] (CC by 4.0)

2.3 *Heavy Metals*

Researchers reported that sludge consists of high levels of heavy metals such as arsenic, cadmium, chromium, copper, lead and zinc. Currently, regulations in many countries require heavy metal testing to be carried out on sludge before it is used for various applications. Chen et al. [6] reported that dewatered sludge collected from 32 wastewater treatment plants in Japan had high concentrations of heavy metals especially copper, zinc and nickel [6]. Guo and Wen [20] stated that the usage of inorganic coagulants such as aluminium and iron salts to remove phosphorus and to increase dehydration during the dewatering process might inadvertently increase the presence of these elements in the dewatered sludge [20]. You et al. [21] emphasized that the toxicity of heavy metals was linked to the morphology of the heavy metal [21]. Heavy metals in the oxidizable and residual states are considered more stable and are less harmful because these heavy metals are not easily absorbed by organisms. However, heavy metals in the exchangeable, acid-soluble and reducible states are found to be unstable and are easily absorbed by organisms. Therefore, it is important to assess the presence of heavy metals in the sludge before utilization to prevent human and ecological risks.

2.4 *Organic Pollutants*

Organic pollutants such as polycyclic aromatic hydrocarbons (PAHs), polychlorinated biphenyl (PCBs), and organochlorine pesticides (OCPs) are classified as persistent organic pollutants which are commonly found in sludge [5]. These pollutants which are similar to heavy metals can accumulate in the soil and pollute the crops. Different cities and countries have different distributions of persistent pollutants in sewage sludge. PAHs are most commonly found in wastewater, are reactive, and can transform into oxygenated PAHs (OPAHs) and nitrated PAHs (NPAHs). These substitutes are given prominence due to their high toxicity in comparison to the parent

PAHs [8]. Long-term exposure to PAHs can cause lung cancer risk [22]. PCBs, on other hand, have been banned worldwide for over twenty years. However, some products are found to produce PCBs such as paint, silicone-based rubber products and sealants in furniture [23]. Meanwhile, OCPs which are used for agricultural and non-agricultural purposes such as pest control has been demonstrated to be an endocrine disruptor [24]. In comparison to PAHs, PCBs and OCPs are seldom detected in landfill sludge. However, the impact of both PCBs and OCPs on humans and the environment is highly damaging.

2.5 *Pathogens*

Wastewater treatment plants use biological processes to produce a large amount of anaerobically digested sludge, which will be dewatered before being used in applications. The anaerobically digested sludge has the potential to emit odour and contain pathogens [25]. Pathogenic microorganisms such as protozoa, viruses and parasites can cause diseases if humans are exposed to them directly or indirectly. In addition to pathogens, antibiotic-resistant bacteria are also enriched after the anaerobic digestion of pre-treated sludge [26]. Therefore, this increases the risk of sludge spreading pathogens in the environment via landfill or agriculture application. Wang et al. [25] reported that coliforms and *E. coli* exhibited the highest regrowth rate during incubation after various treatment processes [25]. This is because *E. coli* can tolerate low water content than other bacteria, thereby *E. coli* was able to regrow in treated sludge. Therefore, effective treatment is required to reduce pathogens prior to sludge application.

3 Impact on Human Health

Due to various potentially toxic pollutants present in untreated and treated sludge, it is important to investigate the risks that sludge application may impose to humans. Health hazards in terms of short-term and long-term toxicities should be identified to ensure that the practice of utilizing sludge for various applications is safe, and adequate treatment of sludge is necessary for a sustainable agricultural practice and disposal method.

3.1 *Toxicity Exposure*

The risk of toxicity from sludge towards human health is a critical issue that needs to be addressed. Due to the content of the sludge, there are probabilities that it can be a toxic agent that may be harmful to public health through a few pathways, either directly or indirectly. Therefore, sludge management options and technologies are being applied to focus on managing human toxicity (non-carcinogenic and carcinogenic) as well as ecotoxicity [27]. There are many sludge treatment approaches being applied in the industry, such as dewatering of mixed sludge, lime stabilization of dewatered sludge, anaerobic digestion of mixed sludge, dewatering of anaerobically digested sludge, and incineration of dewatered anaerobically digested sludge. Dewatered sludge has the potential to be utilized in products that are beneficial to industries and communities. One of the examples is the utilization of dewatered sludge into fertilizers [28–30].

However, through the application of fertilizers, the content of heavy metals and organic compounds in dewatered sludge may expose humans to higher risks of hygiene and toxicity [30]. Due to the potential harmful risk to health and the environment, its marketability is lower than normal fertilizer [29]. Human toxicity is one of the significant indicators being itemized when conducting a life cycle assessment on sludge management [29]. The process of sludge toxicity impacting human health is discussed in this part of the chapter by looking from the perspectives of source, pathway and receptor.

In this process, sludge is the source of toxicity. Sludge produced by wastewater treatment contains heavy metals, putrescible content and pathogenic hazards which may be harmful to humans and the ecosystem [9, 31, 32]. The fate of pollutants depends on multiple factors including [33]:

i. plant operational parameters
ii. physicochemical parameters of pollutants
iii. biochemical sludge parameters.

The concept of a pathway leading to toxicity exposure to humans, in general, involves the chemical fate from an emission before being exposed to humans [34]. Human exposure can be divided into the following routes; inhalation, ingestion, dermal absorption or injection. In the case of sludge, the pathway to human toxicity involves toxic agents that have been emitted to the environment before being exposed to humans. The potential of human exposure could occur as the pollutants or toxic agents are released into the environment at the disposal stage and utilization of resources [29].

The final component is the receptor. For the receptor, which is humans, it is important to look at the vulnerability and the sensitivity factors when discussing this aspect. A receptor can be categorized as a group that requires protection, in this case, from toxicity. Different routes of exposure (dermal absorption, inhalation or ingestion) might affect different parts or organs of the body. Toxicity exposure may lead to carcinogenic and non-carcinogenic issues in humans. Moreover, the severity of toxicity exposure depends on toxicological potency (dose response). Detailed toxicological effects should be taken into consideration, as the following [35]:

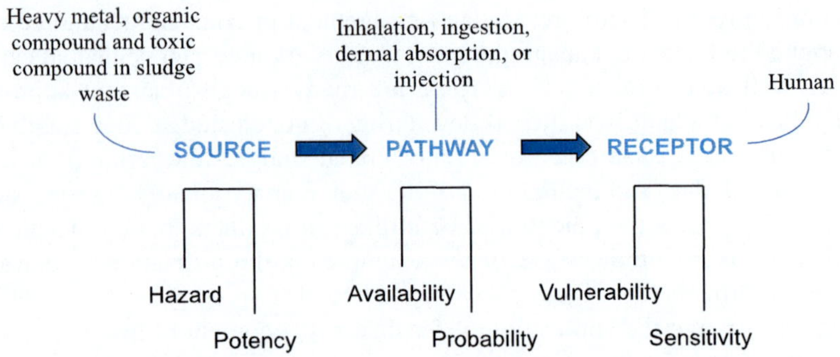

Fig. 3 Source-pathway-receptor model for sludge toxicity exposure in humans

i. Uptake, distribution, metabolism and excretion of the substance in the human
ii. The effects of the substance
iii. Dose-response relationships
iv. The biological mechanisms by which the substance exerts its effect.

A simple diagram model depicting the process of sludge toxicity in humans is shown in Fig. 3.

3.2 Genotoxicity and Mutagenicity Effects

Increasing world population and widespread urbanization lead to vast amounts of human waste. Thus, efficient wastewater treatment is required to produce sewage sludge, which requires further treatment. While treated sludge is applied on land, concerns arise about its effects on human health and the ecosystem [2]. Due to the possible toxic compositions of various types of untreated and treated sludge, many studies have looked into its potential to pose genotoxicity and mutagenicity effects on humans.

According to the World Health Organization (WHO), genotoxicity, also known as genetic toxicity, is a broader term that includes DNA damage, which may be reversible by DNA repair cellular processes, results in cell death, or may not result in permanent alterations in the content of surviving cell. Mutagenicity, a component of genotoxicity, is referred to as a process of inducing a permanent change in the genetic material (e.g. DNA, RNA) of an organism (e.g. human) which can lead to heritable changes to functions [36].

A study was looking into conventional activated sludge from different municipal wastewater treatment plants (WWTPs) in the north of Italy. For WWTP of domestic wastewater with agro-food industrial discharge, the samples showed no mutagenic activity in *Salmonella typhimurium* strains (with and without exogenous S9 metabolic activation), as shown from the Ames test. However, high toxicity was reported, which could potentially mask the effect of mutagenic activity due to a lack of bacterial growth. In addition, no genotoxicity was reported using the *Allium cepa* root test. On the other hand, the samples showed DNA damage of the exposed human leukocytes, as evaluated by the Comet test. Similarly, toxicity was observed, albeit at higher doses. For WWTP which treated winery effluents, slight mutagenic effect was reported for *Salmonella typhimurium* TA98 strain in the absence of S9 metabolic activation, thus indicating potentially detoxifying action of S9. While significant DNA damages were reported from the comet test, *Allium cepa* test showed no genotoxicity effect [37].

Another study which was conducted in Sao Paulo, Brazil was looking into the genotoxicity effects of decontaminated sewage sludge by monitored natural attenuation [38]. The natural processes in detoxifying sewage sludge included mainly degradation by on-site microorganisms as well as volatilization, transformation, dilution and dispersion [39]. It was found from the *Salmonella*/microsome assay that sludge samples after 12 months of treatment showed no mutagenic activity in *Salmonella typhimurium* strains TA98 and TA100, both in the absence and presence of S9 metabolic activation. However, mutagenic activity was observed for the sludge samples which were treated at 0, 2 and 6 months. Micronucleus test which was conducted on HepG2 cells (human-derived hepatoma) showed no significant genotoxic effects for sludge samples treated at 6 and 12 months, though a significant genotoxic effect was reported for sludge samples treated at 0 and 6 months. This indicated the importance of completing the detoxifying process of sludge samples and that natural attenuation at 6 months or longer was most optimized to ensure the safe release of sludge to the environment [38] (Fig. 4).

In addition to *in vitro* findings, an *in vivo* study was conducted whereby Wistar rats were utilized and the genotoxicity effects of treated sludge samples from a sewage treatment plant were observed through micronucleus test and comet assay. Results indicated that independent of rat gender and concentration of the treated sludge samples, no significant increase of micronucleated polychromatic erythrocytes (MNPCEs) in the femoral bone marrow and DNA damage in peripheral blood

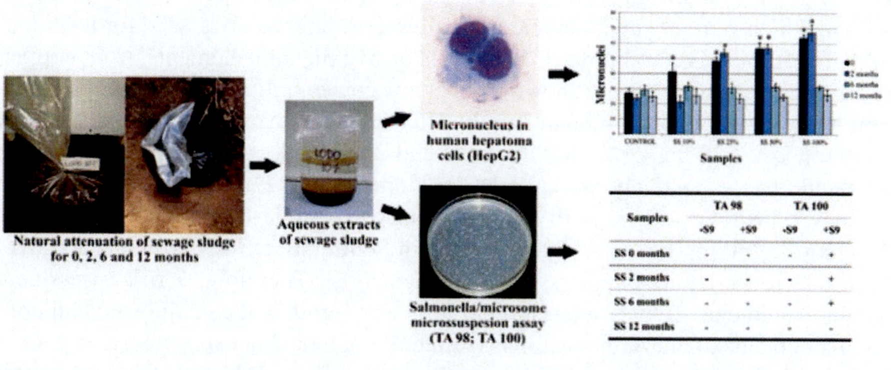

Fig. 4 Graphical representation of natural attenuation of sewage sludge for various time periods, which showed genotoxic and mutagenic effects in human hepatoma cells (HepG2) and *Salmonella typhimurium* respectively in different settings. Reprinted from [38], with permission from Elsevier

leucocytes was observed. Moreover, no influence on the proliferation of normal bone marrow erythroid was reported as the ratio of polychromatic erythrocytes (PCEs) and normochromatic erythrocytes (NCEs) for the treated sludge was not significantly different from the negative control. Further analysis of the metal levels showed below the standard levels [40].

A few other studies have highlighted the importance of treating sludge for the purpose of decontamination prior to disposal in the environment. The mutagenic effect was significantly reduced in treated sludge [41, 42], and was 2–3 times less toxic than its untreated counterpart [43]. Mutagenic substances may persist in soils for a long time, though different types of sludge have different genotoxic or mutagenic potentiality, which affects different degradation periods for which the toxic substances may decline to basal levels [38].

This section described various degrees of genotoxicity and mutagenicity findings for different types of untreated and treated sludge samples using varying test systems. It is therefore important to understand that the difference in results may depend on the type of influent of the sludge samples and the type of treatment that was conducted [10, 44]. In addition, with more than 150 assay systems developed to measure various endpoints of genotoxicity on a full spectrum of organism, from bacteria, to human cells and experimental animals, the complexity of these tests may affect the findings [45]. Therefore, studies so far have investigated these effects on a case-by-case basis [40].

3.3 Carcinogenic Potential

Carcinogenesis is the induction of cancer which originates from the accumulation of genomic alterations, whereby such alterations may be genotoxic or non-genotoxic [46]. It is another important element of long-term toxicity that needs to be investigated. Activated sludge from MWTPs from the Eastern Cape Province of South Africa, which were of suburban and agricultural origins were assessed for its potentially hazardous risks to humans. The levels of five metallic elements, iron, copper, cadmium, lead and zinc were shown to be at low concentrations and below the detection levels for hazards to human health. This followed the recommended limits of 1500 mg kg^{-1}, 39 mg kg^{-1}, 300 mg kg^{-1} and 2800 mg kg^{-1} for copper, cadmium, lead and zinc respectively as stated in the United States Environmental Protection Agency (USEPA) [47], as well as maximum permissible levels of 450 mg kg^{-1}, 5 mg kg^{-1}, 150 mg kg^{-1} and 700 mg kg^{-1} in South Africa [48]. Iron has no target limit due to its nature as an essential element [48]. Therefore, a risk assessment had indicated that all five metals did not pose any significant carcinogenic and non-carcinogenic health hazards to humans, either by oral or skin exposures. However, it is interesting to note that the metallic levels, though low, are higher than wastewater and river water samples. This is unsurprising as 80–90% of heavy metals are known to accumulate in sludge samples. In addition, as heavy metals bioaccumulate, efforts must be made into reducing their levels upon treatment at MWTPs prior to release into the environment to ensure long-term safety for all consumers [49].

Based on the study by Bertanza et al. [37], tumour potential of conventional activated sludge from WWTPs was tested on IAR203 hepatic cells. The sludge samples showed carcinogenic potential by significantly inhibiting gap junction-mediated intercellular communication. However, the carcinogenicity was not as marked as the positive control used, which was TPA, a reference tumour promoter. In *in vitro* cell transformation assay, a significant number of malignant foci or transformed cells were reported in comparison to negative control, though the mean values were significantly lower than positive control 3-MCA. This provided evidence of potentially carcinogenic characteristic of activated sludge [37].

4 Ecological Implications from Agricultural Use

Contradicting the presence of various hazardous agents in sludge as discussed earlier in this chapter, sludge also contains a myriad of beneficial substances that prompt its use in enhancing agriculture practice. The following sections described the effects of sludge application in agriculture and disposal from the perspective of ecology.

4.1 Environmental Factor

The environmental issue of sludge activities arises from wastewater treatment, solid waste and sludge management. The treated sewage water from sludge management is usually discharged to the nearest water ecosystem. Most of the treated sewage water regularly undergoes monitoring and assessment of water quality and is observed to be relatively clean. However, some studies still showed the long-term adverse effects of sewage water treatment [50–52] as not all of the applied water treatments managed to reduce the organic load that increased the nutrient compounds in the water ecosystem [27, 52]. This scenario will potentially cause marine and freshwater eutrophication.

Eutrophication is a condition whereby the environment is enriched with nutrients like phosphorus, nitrogen and other plants nutrients [53]. Freshwater eutrophication is usually associated with increasing nitrogen, while marine eutrophication is associated with increasing phosphorus content [50]. The gradual increase of plant nutrients will cause algae bloom that triggers the structural change of the water ecosystem [54]. Eutrophication is a serious environmental problem as it results in the deterioration of water quality. When there is a significant increase of algae surface of the water, sunlight is blocked from reaching the bottom of the water, thus significantly reducing the rate of photosynthesis [54]. This occurrence will reduce the oxygen concentration in the water, thereby causing environmental imbalance due to its negative impact on the water ecosystem. A simple diagram illustrating the eutrophication process caused by sludge (applied as fertilizer) is shown in Fig. 5.

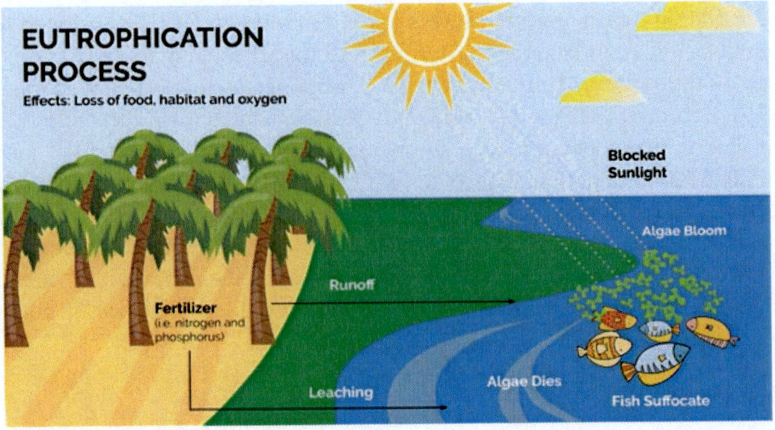

Fig. 5 Eutrophication process by fertilizer (from sludge). Algae bloom prevents sunlight from reaching the bottom of the water, thus reducing photosynthesis and subsequently decreasing oxygen levels. This ultimately kills plants and fish living in the water. Adapted from [55]

Apart from aquatic eutrophication, the rich organic matter in sludge has a risk of nutrient leaching, which can impact soil biodiversity [56]. The waste sludge is usually recycled as fertilizer in agriculture as the sludge is rich in nitrogen, phosphorus and total organic compound, which are suitable for plants [51]. The macronutrient in sewage sludge can increase the rate of growth of plant, however, in some cases, the sewage sludge might harm the soil. The two frequent toxic compounds that have been observed in the sludge is polychlorinated dibenzo-p-dioxins (PCDDs) and polychlorinated dibenzofurans (PCDFs) [51]. These two compounds have mutagenic, carcinogenic and immunotoxin effects on living organisms and increase soil phytotoxicity. In addition, land application of sewage sludge may lead to the accumulation of pathogenic and parasitic organisms in the soil [57]. However, the number of these organisms can be significantly reduced with proper treatment of sludge prior to application to the land [58].

Furthermore, the disposal of solid waste in sludge contributes to the emission of greenhouse gases during the treatment and recycling processes [59]. The increase in greenhouse gas emissions contributes to global warming and climate change. Most studies reported that sewage sludge potentially emitted greenhouse gases, methane and nitrous oxide [59, 60]. Although methane and nitrous oxide emissions are relatively smaller than the emission of carbon dioxide, these two greenhouse gasses have multiplied potency for global warming and climate change [61, 62]. Methane is the second most abundant greenhouse gas with a shorter life span than carbon dioxide, and this gas has been found to be 100 times more extensive in warming the earth than carbon dioxide [62]. Additionally, nitrous oxide is 300 times more extensive than carbon dioxide in warming the atmosphere [61].

Climate change is a serious environmental problem as it affects the global ecosystem. Increasing global temperature causes a rise in sea levels, frequencies of droughts and forest fires and triggers extreme weather conditions [63, 64]. Subsequently, climate change disrupts food production, water supply, human health, human activity, natural resources and economics [65–67].

Contradictory to the negative implications of sludge application, it is also important to note that sludge is extensively used in agriculture as it may improve soil productivity and fertility due to its high concentrations of organic matter and plant nutrients. Studies have shown that sludge application also significantly increased microbial biomass in the soil as well as important soil enzymes such as arylsulfatase, acid phosphatase and alkaline phosphatase, albeit utilizing low metal sludge and at an appropriate rate. This is because the highest rate of sludge application (200 $t{\cdot}ha^{-1}$) caused a significant reduction of the soil's functional community diversity. Furthermore, the organic matter of sewage sludge is generally high at more than 50% dry matter. Therefore, this caused a profound positive impact on soil physical properties and soil conditioning effect by improving aeration and water infiltration as well as increasing soil aggregate formation stability [57, 68].

4.2 *Plant Yield*

Treated sludge from MWTP is usually disposed of, and one sustainable and economical method for the disposal is reusing the sludge in agriculture as an organic fertilizer. This method is widely practised by developing and developed countries worldwide [57]. Sewage sludge has been reported to contain nitrogen, phosphorus and organic matter, which are beneficial for plants. Dewatered sludge has been reported to release nitrogen slowly, thus benefiting crops over a relatively long period of time. Phosphorus content has also been shown to be available at 50% concentration in the year of application. In addition, liquid-anaerobically digested sludge was reported to possess a high content of ammonia nitrogen, which is readily available to plants and is particularly useful in grassland. Furthermore, organic matter from dewatered sludge was found to improve soil structure and its water-retaining capacity, thus improving the growth of plants [69].

In fruits and vegetable production, organic manure alone is insufficient to sustain optimized growth. A study in Jordan found that lettuce growth was increased when applied with 40 $tons{\cdot}ha^{-1}$ sewage sludge, which was a concentration equivalent to the recommended fertilizer rate. However, higher plant metal content was also reported [70]. Another study reported increased lettuce growth in terms of yield, head circle, plant height and leaf number as well as high nitrogen, phosphorus, potassium and magnesium levels when the soil was applied with 20, 40 and 80 $t{\cdot}ha^{-1}$ of sewage sludge [71]. In the Philippines, upland rice yield was increased when the land was applied with sewage sludge, whether alone or in combination with nitrogen fertilizer [72]. In Hawaii, maize biomass was increased when the land was treated with aerobically digested sludge, though the biomass was reduced when two undigested sludges were applied instead. This is because a higher rate of sludge caused phytotoxicity on the roots containing reducible manganese nodules [73]. In the UK, 26% of the crops treated with liquid sludge showed a significant increase in crop yield, through soil structure improvement. However, bed-dried sludge treated on clay soils caused 6–10% yield reduction of wheat grain, probably from excessive nitrogen content which caused crop lodging [58]. In Spain, sewage sludge use was a suitable replacement for chemical fertilizer, due to its potential benefits in fertilizing 3–4-year-old citrus trees. However, it was also reported that its use must be cautioned against horticultural crops such as soft fruits or vegetable crop [74]. It was revealed that cauliflower, cabbage and potato showed growth deficit and phytotoxicity in some parts of the plants when the soil was applied with sewage sludge for long period of time, most likely from high content of heavy metals (e.g. chromium accumulation in potato) [75].

Similarly, sewage sludge application has been found to be beneficial for the growth of dairy pasture and forestry. Sludge has valuable plant nutrients such as nitrogen, phosphorus, iron, calcium, magnesium and various other macro and micronutrients, which are important for cow's growth and milk production, though farmers should adhere to holding periods in which animals should be prevented from grazing sewage sludge-treated pasture to protect the animals from pathogens and risks of chemical contamination [57, 76, 77]. In forestry, sewage sludge application improved the soil's physical, chemical and biological properties as well as fertility, thus creating a favourable condition for improved vegetation in the existing forests [10]. Though sludge is valuable to enhance crop growth due to its high organic matter content and rich macro and micronutrients, caution must be practised to prevent the presence of heavy metals and organic compounds in the food chain, and these pollutant load can also be phytotoxic [57].

4.3 Effect to Animals and Marine Life

All animal species which are in contact with sewage sludge are somewhat affected negatively. The discharge of sewage sludge into tropical coastal seas releases hundreds of compounds, the most common ones included freshwater, inorganic nutrients (ammonium, nitrite, nitrate, and phosphate), pathogens, endocrine disrupters, suspended solids, sediments, heavy metals and toxins [78, 79]. Coral mortality is affected by the toxicity effects caused by coral diseases and coral bleaching. In addition, nutrient enrichment is enhancing algal growth and phytoplankton shading, which increase coral reefs' competitive ability for space and affect calcification rates, thereby destroying corals and removing the foundation species [80].

The presence of heavy metals in the water system leads to the accumulation of metals in plankton, algae and smaller prey, which are consumed by fish, leading to the build-up of heavy metals in fish tissues, in a process called biomagnification. Accumulation of toxic amounts of heavy metals can poison the fish. In addition, other animals, including humans may eat these fish with heavy metals, and the chain of heavy metal is transferred to the consumers. Furthermore, microplastics in sewage sludge are affecting marine life such as fish, mammals and crustaceans as these animals may eat them, living in or on them, or getting tangled in this litter [81].

Another study found that sewage sludge application to pasture caused significant spermatogenic abnormalities in adult male sheep when they were exposed to the sludge in utero and post-natal (until weaning and post-weaning). This was observed from their testes, as germ cell number reduced significantly, thus affecting sperm count and fertility among these sludge-exposed animals [82]. Reports have also revealed that DDT, one of the toxins contained in sewage sludge was affecting peregrine falcons in the US, causing thin eggshells that rendered incubation extremely difficult. In addition, amphibians were malformed due to the exposure of hormone-disrupting chemicals in sewage sludge. Toxins were also found to accumulate in worms and insects after prolonged exposure to sludge, whereby subsequently mammals and birds ingesting these animals would accumulate the toxins as well in their bodies [83].

5 Conclusion

This chapter described the various effects of untreated and treated sludge on human health, environment, plants and animals. Depending on a few factors, the effects of sludge can be negative and positive. The toxic composition of sludge consists of microplastics, heavy metals, organic pollutants, pharmaceutical active substances and pathogens, thus there were reports of toxicity, genotoxic, mutagenic and carcinogenic effects on human health. In addition, sludge application affected the environment in terms of aquatic eutrophication, soil phytotoxicity and climate change. Plants and animals were also affected in terms of phytotoxicity and mortality respectively. However, it is equally important to note that sludge also has profound beneficial properties, as it contains high content of nitrogen, phosphorus, organic matter and plant nutrients, which was the reason for its extensive use in agriculture. Plant growth was shown to be enhanced as sludge is an excellent fertilizer and a soil conditioner to sustain optimized growth and in some cases, no toxicity was reported among humans. Due to its potential, sludge application on land is part of a strategic and sustainable method of agricultural productivity. Therefore it is important to follow appropriate guidelines in treating sewage sludge prior to land application, either for agricultural use, disposal, or both, in order to maximize benefits and minimize health hazards to humans and the ecology. Lastly, understanding the effects of sludge application is imperative to ensure that sludge valorization can be optimized for a cost-effective and environmentally friendly sludge reduction and resource recovery method.

Acknowledgements This work is financially supported by APEX ERA Research Grant of Universiti Sains Malaysia (1001.PTEKIND.881006).

References

1. Canziani R, Spinosa L (2019) Sludge from wastewater treatment plants. In: Industrial and municipal sludge: emerging concerns and scope for resource recovery. Elsevier Inc, 3–30 p
2. Havukainen J, Saud A, Astrup TF, Peltola P, Horttanainen M (2022) Environmental performance of dewatered sewage sludge digestate utilization based on life cycle assessment. Waste Manag [Internet] 137(Nov 2021):210–221. Available from: https://doi.org/10.1016/j.wasman.2021.11.005
3. Collivignarelli MC, Abbà A, Frattarola A, Miino MC, Padovani S, Katsoyiannis I et al (2019) Legislation for the reuse of biosolids on agricultural land in Europe: overview. Sustainability 11(21):1–22
4. Ivanová L, Mackuľak T, Grabic R, Golovko O, Koba O, Staňová AV et al (2018) Pharmaceuticals and illicit drugs – a new threat to the application of sewage sludge in agriculture. Sci Total Environ 634:606–615
5. Guangyin Z, Youcai Z (2017) Pollution control and recycling of sludge in sanitary landfill. In: Pollution control and resource recovery. Elsevier, 275–352 p
6. Chen M, Oshita K, Mahzoun Y, Takaoka M, Fukutani S, Shiota K (2021) Survey of elemental composition in dewatered sludge in Japan. Sci Total Environ 752:141857
7. Đurđević D, Blecich P, Jurić Ž (2019) Energy recovery from sewage sludge: the case study of Croatia, vol 12. Energies 12(10):1927
8. Saber AN, Zhang H, Islam A, Yang M (2021) Occurrence, fates, and carcinogenic risks of substituted polycyclic aromatic hydrocarbons in two coking wastewater treatment systems. Sci Total Environ 789:147808
9. Wu B, Dai X, Chai X (2020) Critical review on dewatering of sewage sludge: influential mechanism, conditioning technologies and implications to sludge re-utilizations. Water Res 180:115912
10. European Commission (2001) Disposal and recycling routes for sewage sludge. Part 1 - sludge use acceptance report [Internet]. Office for Official Publications of the European Communities, Luxembourg. Available from: https://ec.europa.eu/environment/archives/waste/sludge/pdf/sludge_disposal1.pdf
11. European Commission (2001) Disposal and recycling routes for sewage sludge - Part 3: scientific and technical sub-component report [Internet]. European Commission DG Environment–B/2. Available from: https://ec.europa.eu/environment/archives/waste/sludge/pdf/sludge_disposal3.pdf
12. Silva S, Rodrigues JA, Coelho MR, Martins A, Cardoso E, Cardoso VV et al (2021) Occurrence of pharmaceutical active compounds in sewage sludge from two urban wastewater treatment plants and their potential behaviour in agricultural soils. Environ Sci Water Res Technol 7(5):969–982
13. Liu X, Yuan W, Di M, Li Z, Wang J (2019) Transfer and fate of microplastics during the conventional activated sludge process in one wastewater treatment plant of China. Chem Eng J 362(Jan):176–182
14. Keller AS, Jimenez-Martinez J, Mitrano DM (2020) Transport of nano- and microplastic through unsaturated porous media from sewage sludge application. Environ Sci Technol 54(2):911–920
15. Alavian Petroody SS, Hashemi SH, van Gestel CAM (2021) Transport and accumulation of microplastics through wastewater treatment sludge processes. Chemosphere 278:130471
16. Dalahmeh SS, Thorsén G, Jönsson H (2022) Open-air storage with and without composting as post-treatment methods to degrade pharmaceutical residues in anaerobically digested and dewatered sewage sludge. Sci Total Environ 806:151271
17. Silva PRP, Barbisan LF, Dagli MLZ, Saldiva PHN (2012) Sewage sludge does not induce genotoxicity and carcinogenesis. Genet Mol Biol 35(3):657–663
18. Mejías C, Martín J, Santos JL, Aparicio I, Alonso E (2021) Occurrence of pharmaceuticals and their metabolites in sewage sludge and soil: a review on their distribution and environmental risk assessment. Trends Environ Anal Chem 30:e00125

19. Camotti Bastos M, Soubrand M, Le Guet T, Le Floch É, Joussein E, Baudu M et al (2020) Occurrence, fate and environmental risk assessment of pharmaceutical compounds in soils amended with organic wastes. Geoderma 375(Mar):114498
20. Guo J, Wen X (2020) Performances and mechanisms of sludge dewatering by a biopolymer from piggery wastewater and application of the dewatered sludge in remediation of Cr(VI)-contaminated soil. J Environ Manage 259(May):109678
21. You M, Hu Y, Yan Y, Yao J (2021) Speciation characteristics and ecological risk assessment of heavy metals in municipal sludge of huainan, china. Molecules 26(21):1–12
22. Ali N (2019) Polycyclic aromatic hydrocarbons (PAHs) in indoor air and dust samples of different Saudi microenvironments; health and carcinogenic risk assessment for the general population. Sci Total Environ 696:133995
23. Hombrecher K, Quass U, Leisner J, Wichert M (2021) Significant release of unintentionally produced non-Aroclor polychlorinated biphenyl (PCB) congeners PCB 47, PCB 51 and PCB 68 from a silicone rubber production site in North Rhine-Westphalia, Germany. Chemosphere 285:131449
24. Shen B, Wu J, Zhan S, Jin M (2021) Residues of organochlorine pesticides (OCPs) and polycyclic aromatic hydrocarbons (PAHs) in waters of the Ili-Balkhash Basin, arid Central Asia: concentrations and risk assessment. Chemosphere 273:129705
25. Wang M, Chen H, Liu S, Xiao L (2021) Removal of pathogen and antibiotic resistance genes from waste activated sludge by different pre-treatment approaches. Sci Total Environ 763:143014
26. Min Jang H, Choi S, Shin J, Kan E, Mo KY (2018) Additional reduction of antibiotic resistance genes and human bacterial pathogens via thermophilic aerobic digestion of anaerobically digested sludge. Bioresour Technol 2019(273)(Sept):259–268
27. Yoshida H, ten Hoeve M, Christensen TH, Bruun S, Jensen LS, Scheutz C (2018) Life cycle assessment of sewage sludge management options including long-term impacts after land application. J Clean Prod 174:538–547
28. Cao B, Zhang T, Zhang W, Wang D (2021) Enhanced technology based for sewage sludge deep dewatering: a critical review. Water Res 189:116650
29. Lishan X, Tao L, Yin W, Zhilong Y, Jiangfu L (2018) Comparative life cycle assessment of sludge management: a case study of Xiamen, China. J Clean Prod 192:354–363
30. Rastetter N, Gerhardt A (2017) Toxic potential of different types of sewage sludge as fertiliser in agriculture: ecotoxicological effects on aquatic, sediment and soil indicator species. J Soils Sediments 17(1):106–121
31. Adamcová D, Vaverková MD, Břoušková E (2016) The toxicity of two types of sewage sludge from wastewater treatment plant for plants. J Ecol Eng 17(2)
32. Lee LH, Wu TY, Shak KPY, Lim SL, Ng KY, Nguyen MN et al (2018) Sustainable approach to biotransform industrial sludge into organic fertilizer via vermicomposting: a mini-review. J Chem Technol Biotechnol 93(4):925–935
33. Yoshida H, Christensen TH, Guildal T, Scheutz C (2015) A comprehensive substance flow analysis of a municipal wastewater and sludge treatment plant. Chemosphere 138:874–882
34. Rosenbaum RK, Huijbregts MAJ, Henderson AD, Margni M, McKone TE, Van De Meent D et al (2011) USEtox human exposure and toxicity factors for comparative assessment of toxic emissions in life cycle analysis: sensitivity to key chemical properties. Int J Life Cycle Assess 16(8):710–727
35. Krewitt W, Pennington DW, Olsen SI, Crettaz P, Jolliet O (2002) Indicators for human toxicity in life cycle impact assessment. Toward Best Available Practice for Life Cycle Impact Assessment 123
36. World Health Organization (2020) Hazard identification and characterization: toxicological and human studies (Section 4.5 Genotoxicity, 2nd edn) [Internet]. Available from: https://cdn.who.int/media/docs/default-source/food-safety/publications/section4-5-genotoxicity.pdf?sfvrsn=8ec3434_2
37. Bertanza G, Boniotti J, Ceretti E, Feretti D, Mazzoleni G, Menghini M et al (2021) Environmental footprint of wastewater treatment: a step forward in the use of toxicological tools. Int J Environ Res Public Health 18(13):6827

38. Mazzeo DEC, Fernandes TCC, Marin-Morales MA (2016) Attesting the efficiency of monitored natural attenuation in the detoxification of sewage sludge by means of genotoxic and mutagenic bioassays. Chemosphere 163:508–515
39. Megharaj M, Ramakrishnan B, Venkateswarlu K, Sethunathan N, Naidu R (2011) Bioremediation approaches for organic pollutants: a critical perspective. Environ Int [Internet] 37(8):1362–1375. Available from: https://doi.org/10.1016/j.envint.2011.06.003
40. de Solano MLM, de Lima PLA, Luvizutto JFL, Silva PRP, de Umbuzeiro GA, de Camargo JLV (2009) In vivo genotoxicity evaluation of a treated urban sewage sludge sample. Mutat Res - Genet Toxicol Environ Mutagen 676(1):69–73
41. Gajski G, Oreščanin V, Garaj-Vrhovac V (2011) Cytogenotoxicity of sewage sludge leachate before and after calcium oxide-based solidification in human lymphocytes. Ecotoxicol Environ Saf 74(5):1408–1415
42. Ramírez WA, Domene X, Ortiz O, Alcañiz JM (2008) Toxic effects of digested, composted and thermally-dried sewage sludge on three plants. Bioresour Technol [Internet]. 99(15):7168–7175. Available from: https://www.sciencedirect.com/science/article/pii/S0960852408000084
43. Roig N, Sierra J, Nadal M, Martí E, Navalón-Madrigal P, Schuhmacher M et al (2012) Relationship between pollutant content and ecotoxicity of sewage sludges from Spanish wastewater treatment plants. Sci Total Environ [Internet] 425:99–109. Available from: https://www.sciencedirect.com/science/article/pii/S0048969712003506
44. Schowanek D, Carr R, David H, Douben PET, Hall JE, Kirchmann H et al (2004) A risk-based methodology for deriving quality standards for organic contaminants in sewage sludge for use in agriculture—conceptual Framework. Regul Toxicol Pharmacol 40(3):227–251
45. Brusick DJ, Gopalan HNB, Heseltine E, Huismans JW, Lohman PHM (1992) Hazard identification. In: United Nations Environment Programme: assessing the risk of genetic damage. Hodder & Stoughton, Botswana, pp 21–24
46. Jacobs MN, Colacci A, Corvi R, Vaccari M, Aguila MC, Corvaro M et al (2020) Chemical carcinogen safety testing: OECD expert group international consensus on the development of an integrated approach for the testing and assessment of chemical non-genotoxic carcinogens. Arch Toxicol [Internet] 94(8):2899–2923. Available from: https://doi.org/10.1007/s00204-020-02784-5
47. US EPA (2002) Supplemental guidance for developing soil screening levels for superfund sites. Office of Solid Waste and Emergency Response (OSWER) [Internet]. United States Environmental Protection Agency. Available from: https://nepis.epa.gov/Exe/ZyPDF.cgi/91003IJK.PDF?Dockey=91003IJK.PDF
48. Department of Water Affairs Republic of South Africa (2010) Department of Water Affairs Strategic Plan 2010/11–2012/13
49. Agoro MA, Adeniji AO, Adefisoye MA, Okoh OO (2020) Heavy metals in wastewater and sewage sludge from selected municipal treatment plants in Eastern Cape Province, South Africa. Water (Switzerland) 12(10):2746
50. Pradel M, Reverdy AL, Richard M, Chabat L, Pradel M, Reverdy AL et al (2014) Environmental impacts of sewage sludge treatment and disposal routes: a life cycle assessment perspective. 4th Eur Cofenrence Sludge Manag 9
51. Tarpani RRZ, Alfonsín C, Hospido A, Azapagic A (2019) Life cycle environmental impacts of sewage sludge treatment methods for resource recovery considering ecotoxicity of heavy metals and pharmaceutical and personal care products. J Environ Manage 2020(260)(Aug):109643
52. Liu W, Iordan CM, Cherubini F, Hu X, Fu D (2021) Environmental impacts assessment of wastewater treatment and sludge disposal systems under two sewage discharge standards: a case study in Kunshan, China. J Clean Prod 287:125046
53. Dodds WK, Smith VH (2016) Nitrogen, phosphorus, and eutrophication in streams. Inl Waters 6(2):155–164
54. Zohdi E, Abbaspour M (2019) Harmful algal blooms (red tide): a review of causes, impacts and approaches to monitoring and prediction. Int J Environ Sci Technol 16(3):1789–1806
55. EarthHow (2022) How does eutrophication work? Causes, process and examples [Internet]. Available from: https://earthhow.com/eutrophication-causes-process-examples/

56. Urbaniak M, Wyrwicka A, Tołoczko W, Serwecińska L, Zieliński M (2017) The effect of sewage sludge application on soil properties and willow (Salix sp.) cultivation. Sci Total Environ 586:66–75
57. Usman K, Khan S, Ghulam S, Khan MU, Khan N, Khan MA et al (2012) Sewage sludge: an important biological resource for sustainable agriculture and its environmental implications. Am J Plant Sci 3(12):1708–1721
58. Pescod MB (1992) Wastewater treatment and use in agriculture - FAO irrigation and drainage paper 47, vol. 47. Food and Agriculture Organization of the United Nations, Rome
59. Willén A, Rodhe L, Pell M, Jönsson H (2016) Nitrous oxide and methane emissions during storage of dewatered digested sewage sludge. J Environ Manage 184:560–568
60. Willén A, Jönsson H, Pell M, Rodhe L (2016) Emissions of nitrous oxide, methane and ammonia after field application of digested and dewatered sewage sludge with or without addition of urea. Waste Biomass Valor 7(2):281–292
61. Griffis TJ, Chen Z, Baker JM, Wood JD, Millet DB, Lee X et al (2017) Nitrous oxide emissions are enhanced in a warmer and wetter world. Proc Natl Acad Sci USA 114(45):201704552
62. Howarth R (2015) Methane emissions and climatic warming risk from hydraulic fracturing and shale gas development: implications for policy. Energy Emiss Control Technol 3:45–54
63. Kundzewicz ZW (2016) Extreme weather events and their cnsequences. Pap Glob Chang IGBP, Sciendo 23(1):59–69
64. Littell JS, Peterson DL, Riley KL, Liu Y, Luce CH (2016) A review of the relationships between drought and forest fire in the United States. Glob Change Biol 22(7):2353–2369
65. Islam MS, Kabir MH, Ali MS, Sultana MS, Mahasin M (2019) Farmers' knowledge on climate change effects in agriculture. Agric Sci 10(3):386–394
66. Mele M, Gurrieri AR, Morelli G, Magazzino C (2021) Nature and climate change effects on economic growth: an LSTM experiment on renewable energy resources. Environ Sci Pollut Res 28(30):41127–41134
67. De Blois J, Kjellstrom T, Agewall S, Ezekowitz JA, Armstrong PW, Atar D (2015) The effects of climate change on cardiac health. Cardiology 131(4):209–217
68. Wortman CS, Iqbal J (2021) Sewage sludge utilization for crop production [Internet]. Nebraska Extension, pp 1–4. Available from: https://extensionpublications.unl.edu/assets/pdf/g1939.pdf
69. Chenini F (2010) Overview of new practices in the reclaimed water reuses in the Mediterranean countries. In: Barceló D, Petrovic M (eds) Waste water treatment and reuse in the Mediterranean Region. Springer, Berlin Heidelberg, pp 43–62
70. Mohammad MJ, Athamneh BM (2004) Changes in soil fertility and plant uptake of nutrients and heavy metals in response to sewage sludge application to calcareous soils. J Agron 3(3):229–236
71. Sönmez F, Bozkurt MA (2006) Lettuce grown on calcareous soils benefit from sewage sludge. Acta Agric Scand Sect B — Soil Plant Sci [Internet] 56(1)(Feb 1):17–24. Available from: https://doi.org/10.1080/09064710510005813
72. Manguiat IJ (1997) Sewage sludge: turning an environmental pollutant into an agricultural resource. In: 2nd Professional Lecture Delivered in Department of Soil Science. University of Philippines, Los Banos
73. Hue N V, Ranjith SA (1994) Sewage sludges in Hawaii: chemical composition and reactions with soils and plants. Water Air Soil Pollut [Internet] 72(1):265–283. Available from: https://doi.org/10.1007/BF01257129
74. García-Agustín P, Flors V, Cerezo M, Romero E, Lapeña L (2004) Aquifer contamination by nitrogen after sewage sludge fertilization. Bull Environ Contam Toxicol 72(2):344–351
75. Chatterjee C, Dube BK (2005) Impact of pollutant elements on vegetables growing in sewage-sludge-treated soils. J Plant Nutr [Internet] 28(10)(Oct 1):1811–1820. Available from: https://doi.org/10.1080/01904160500251175
76. Ross AD, NSW (1991) Agriculture. Guidelines for the use of sewage sludge on agricultural land, 3rd edn. NSW Agriculture, Sydney.
77. Holmes CW, Wilson GF (1984) Milk production from pasture [Internet]. Butterworths, 319 p (Butterworths agricultural books). Available from: https://books.google.com.my/books?id=OoxuQgAACAAJ

78. Shahidul Islam M, Tanaka M (2004) Impacts of pollution on coastal and marine ecosystems including coastal and marine fisheries and approach for management: a review and synthesis. Mar Pollut Bull [Internet] 48(7):624–649. Available from: https://www.sciencedirect.com/science/article/pii/S0025326X03005459
79. Pantsar-Kallio M, Mujunen S-P, Hatzimihalis G, Koutoufides P, Minkkinen P, Wilkie PJ et al (1999) Multivariate data analysis of key pollutants in sewage samples: a case study. Anal Chim Acta [Internet] 393(1):181–191. Available from: https://www.sciencedirect.com/science/article/pii/S0003267099002871
80. Wear SL, Thurber RV (2015) Sewage pollution: mitigation is key for coral reef stewardship. Ann N Y Acad Sci 1355(1):15–30
81. Blaettler KG (2018) The effects of sewage on aquatic ecosystems. Sciencing.com [Internet]. Available from: https://sciencing.com/effects-sewage-aquatic-ecosystems-21773.html
82. Bellingham M, Mckinnell C, Fowler PA, Amezaga MR, Zhang Z, Rhind SM et al (2012) Foetal and post-natal exposure of sheep to sewage sludge chemicals disrupts sperm production in adulthood in a subset of animals. Int J Androl 35(3):317–329
83. Kowalski D (2018) Pathogens and toxins in sewage sludge and their effects on wildlife [Internet]. Sewage Sludge Victims. Available from: http://www.sludgevictims.com/effects-on-wildlife.html

An Insight of Component and Typical Mechanism of Sludge Degrader Microbes in Dewatered Sludge

Muhammad Najib Ikmal Mohd Sabri, Nurul Atiqah Shamsuddin, Husnul Azan Tajarudin, Md Sohrab Hossain, and Muaz Mohd Zaini Makhtar

Abstract This book chapter provides insights into the potential of wastewater sludge and the characteristics of different types of sludge. Each type of sludge has unique characteristics, microorganism consortium, and reactions that occur within the sludge. The behavior of sludge, typical microorganism degraders, and reactions involved in the natural process transform complex substrates into simpler ones. The presence of microbial degraders is crucial for the exploitation of sludge valorization for future sustainability. The chapter explores the components and typical mechanisms of sludge degrader microbes in dewatered sludge. The understanding of the microbial degraders present in sludge is essential for the development of sustainable approaches to sludge management. The exploitation of sludge valorization has the potential to provide renewable energy sources, contribute to the circular economy, and reduce the environmental impact of sludge disposal. This book chapter highlights the importance of microbial degraders in the transformation of complex substrates into simpler ones and the need for sustainable approaches to exploit the potential of sludge valorization.

Keywords Sludge biosolid · Biomass conversion · Bioprocessing · Affordable and clean energy · Renewable energy

M. N. I. Mohd Sabri · N. A. Shamsuddin · H. A. Tajarudin
School of Industrial Technology, Bioprocess Technology Division, Universiti Sains Malaysia, 11800, Gelugor, Penang, Malaysia

M. S. Hossain
Department of Fundamental and Applied Sciences, HICoE-Centre for Biofuel and Biochemical Research, Institute of Self-Sustainable Building, Universiti Teknologi PETRONAS, 32610 Seri Iskandar, Perak Darul Ridzuan, Malaysia

M. Mohd Zaini Makhtar (✉)
School of Industrial Technology, Bioprocess Technology Division, Universiti Sains Malaysia, 11800 Gelugor, Penang, Malaysia
e-mail: muazzaini@usm.my

Centre for Innovation and Consultation, Universiti Sains Malaysia, 11800 Gelugor, Penang, Malaysia

M. Mohd Zaini Makhtar et al. (eds.), *Microbial Fuel Cell (MFC) Applications for Sludge Valorization*, Green Energy and Technology,
https://doi.org/10.1007/978-981-99-1083-0_3

1 Introduction

Wastewater that has been treated and refined from the wastewater treatment plant (typically consisting of preliminary, primary and secondary treatment) has the potential to be valorized into valuable bioproducts such as bioenergy and biomaterials. The end waste at the treatment site is in the form of dewatered sludge (biosolid) which is composed of a lot of nutrient composition and tonnes of microbes. Table 1 tabulates the differences between sludge, sewage sludge, activated sludge and leachate.

Table 1 Comparison of sludge, activated sludge, sewage sludge, and leachate

	Sludge	Activated sludge	Sewage sludge	Leachate
Definition	• Formed during both primary sewage treatment and secondary treatment	• Sludge form and growth in the biological treatment process that is composed of microorganisms • Agitated and aerated	• Sludge that is produced in form of residual and semi-solid for both municipal and industrial wastewater • Also known as biosolids	Liquid squeezed out from the waste as well as the water (solvent) which infiltrates into the waste and percolates through it carrying substances dissolved from the waste (solutes)
Position/ Location	• Preliminary treatment; biological, chemical and physical • Secondary treatment; further biological process (anaerobic digestion)	• After the secondary treatment; the product • Closed biological reactors are known as anaerobic sludge digesters	• Located at the effluent • Municipal and industrial wastewater	• Dump area
Benefit	• Can be converted to biogas through anaerobic digestion process	• Microorganisms are used to consume organic matter in WW • Aeration is required in this treatment	• Good source of plant nutrients (macronutrients); soil conditioner or fertilizer • Electron transfer (microbial fuel cell) • Carbonization as energy generation	Fermented leachate can be used to recover and adsorb acetic and butyric acid
References	[4]	[5, 6]	[7]	[8]

2 Overall Reaction and Type of Microbe Degraders

2.1 Sugar Degrader

The source of sugar came from lignocellulosic material which can be found in plant-derived residue and waste such as paper mill sludge [1]. Research conducted by Ducan and team [2], found that the conversion of mill sludge to sugar later can be used as either isoprene or ethanol. Based on Yildiz et al. [3, 6], microorganisms are used due to their ability to remediate the sugar industry effluent. The application of microorganisms is eco-friendly because they do not require any chemicals during the sludge treatment. Basically, lactic acid bacteria (LAB) that are used as sugar degrader reacted can be monitored by the reduction of pH. There are several types of LAB strains that are used as sugar degraders, for example, *Lactobacillus plantarum*, *Lactobacillus casei*, and *Streptoccuslactis*. The production of lactic acid in the early fermentation stage suppressing the growth of putrefying bacteria while enhancing the availability of inorganic compounds which are being used by these lactic acid bacteria for growth and reproduction.

Besides, *Saccharomyces cerevisiae* is the most useful microorganism for ethanol production through alcoholic fermentation by metabolizing sugar in the absence of oxygen which leads to the production of ethanol and carbon dioxide [9].

The metabolic reaction of sugar degradation is further described below:

$$C_6H_{12}O_6 \rightarrow 2CH_3CH_2OH + 2CO_2 \quad (1)$$

where $C_6H_{12}O_6$ is glucose, CH_3CH_2OH is ethanol and CO_2 is carbon dioxide.

2.2 Amino Acid Degrader

Amino acid is a soluble monomer from a breakdown of complex organic matter dependent upon syntropic interaction of a consortium of microorganism in anaerobic digestion [10]. Amino acids vary significantly in size and structure and are fermented via different pathways to a range of products where these products are built up by amphoteric substances that contain amino and carboxyl groups. These amino acids are comprised of a four-step process of hydrolysis; amino acid fermentation, acid production and methanation of the anaerobic degradation process of proteins.

The degradation of amino acids produces organic compounds such as ammonia, carbon dioxide and small amounts of hydrogen and sulphur compounds. Amino acids are degraded in two ways that include deamination through a Stickland reaction; injection of two types of amino acids. One side of the amino acid (containing the majority of the carbon atoms) acts as an electron accepter, while the other (containing one or only a few carbon atoms) acts as an electron donor.

The reaction that takes place is the deamination by bacteria within the *Clostridium* species (obligatory species). The second type of amino acid decomposition occurs through the general fermentation process of single amino acids that requires the presence of hydrogen-utilizing bacteria. The fermentation of amino acids by the Stickland reaction; a chemical reaction that involves the coupled oxidation and reduction of amino acids to organic acids, is known to be the dominant reaction among these two types [11, 12].

Based on Table 2, there are five classifications of the bacteria based on their involvement in Stickland reactions and the amino acids that they typically utilize [11]. Group I bacteria are organisms that carry out the Stickland reactions. Fermentation process intermediately utilize proline and produce δ-aminovalerate, α-aminobutyrate or γ-aminobutyrate by these enzymes were accumulated with *Clostridial* species. While Groups II, III, IV and V do not carry Stickland reactions but ferment amino acids. These classifications mainly form obligate spore-formers (*Clostridial* species) and some non-sporing obligate anaerobes, for example, *Peptostreptococcus (Micrococcus) spp.*

Table 3 summarizes the amino acid metabolic degradation. All of the reactions are described either as Stickland or non-Stickland where there are five amino acids involved in Stickland reaction. These reactions can act either as an electron donor or electron acceptor.

2.3 Long-Chain Fatty Acid (LCFA) Degrader

Long-chain fatty acid (LCFA) is generated from the hydrolysis of lipids in sludge [13]. Fatty acids are organic molecules composed of a hydrophilic head, a carboxyl group and a hydrophobic aliphatic tail. The absence or presence of double bonds in the fatty acid aliphatic chain makes them saturated or unsaturated. Saturated and unsaturated LCFA are palmitate and oleate, respectively, thus they become the most abundant constituents [14]. The Prime way to identify the differences between saturated and unsaturated LCFAs are the presence of double bond in the fatty acid aliphatic chain, respectively. Table 4 showed the common unsaturated and saturated LCFA found in wastewater.

Hydrogen transfer between microorganisms plays a central role in LCFA degradation in methanogenic environments. This degradation through obligation syntrophic communities of proton-reducing acetogenic bacteria, converting LCFA to acetate and hydrogen/formate, *acetoclastic methanogenic archaea*, and hydrogen/formate-consuming methanogenic archaea as shown in Table 5.

The degradation of saturated LCFA follows the classic β-oxidation pathway while the unsaturated LCFA may require a preliminary step of hydrogenation or an alternative degradation pathway. The coculture of *Syntrophomonas and Methanospirillumhungatei* can degrade palmitate in LCFA [15, 16]. There are 14 fatty-acid-degrading syntrophic bacteria that have been obtained in pure culture and coculture with hydrogen-consuming microorganisms, all belong to *Syntrophomonadaceae* and

Table 2 Classification of anaerobic bacteria which degrade amino acids [11]

Group species		Enzyme production	Amino acids utilized	Characteristics
I	*C. bifermentans*	proteo/saccharolytic	proline, serine, arginine, glycine	organisms that carry out Stickland reaction
	C. sordellii	proteo/saccharolytic	leucine, isoleucine, valine	reaction
	C. botulinum types A, B, F	proteo/saccharolytic	ornithine, lysine, alanine,	prolineutilised by all species
	C. caloritolerans	–	cysteine, methionine, aspartate	δ–aminovalerate
	C. sporogenes	proteo/saccharolytic	threonine, phenylalanine	α-aminobutyrate and γ-aminobutyrate are produced
	C. cochlearium– one strain	specialist	tyrosine, tryptophan and glutamate	
	C. difficile	saccharolytic		
	C. putrificum	proteo/saccharolytic		
	C. sticklandii	specialist		
	C. ghoni	proteolytic		
	C. mangenotii	proteolytic		
	C. scatologenes	saccharolytic		
	C. lituseburense	proteo/saccharolytic		
	C. aerofoetidum	–		
	C. butyricum	saccharolytic		
	C. caproicum	–		
	C. carnofoetidum	–		
	C. indolicum	–		
	C. mitelmanii	–		
	C. saprotoxicum	–		
	C. valerianicum	–		
II	*C. botulinum types C*	proteo/saccharolytic	glycine, arginie, histidine and lysine	glycine is used by all species; δ-aminovalerate not produced
	C. histolyticum	proteolytic		

(continued)

Table 2 (continued)

Group species		Enzyme production	Amino acids utilized	Characteristics
	C. cochlearium– one strain	specialist		
	C. subterminale	proteolytic		
	C. botulinum types G	–		
	P. anaerobius	–		
	P. variabilis	–		
	P. micros	–		
III	*C. cochlearium– one strain*	Specialist	glutamate, serine, histidine,	δ–aminovalerate not produced;
III	*C. tetani*	Proteolytic	arginine, aspartate, threonine	histidine, serine and glutamate
	C. tetanomorphum	Saccharolytic	tyrosine, tryptophan and	used by all species
	C. lentoputrescens	–	cysteine	
	C. limosum	proteolytic		
	C. malenomenatum	specialist		
	C. microsporum	–		
	C. perfringens	proteo/saccharolytic		
	C. butyricum	saccharolytic		
	P. asaccharolyticus	–		
	P. prevotii	–		
	P. activus	–		
IV	*C. putrefaciens*	proteolytic	serine and threonine	δ-aminovalerate not produced
V	*C. propionicum*	specialist	alanine, serine, threonine, cysteine and methionine	δ-aminovalerate not produced

Syntrophaceae within the phyla *Firmicutes* and *Deltaproteobacteria*, respectively. During fatty acid degradation, these syntrophic bacteria are working together with hydrogenotrophicarchaea or hydrogen-consuming sulphate-reducing bacteria [14].

Table 3 Stoichiometry for amino acid fermentation (catholic reactions only) [11]

No	Reaction	Type
1	$C_6H_{13}O_2N$ (Leu) + 2 $H_2O \rightarrow C_5H_{10}O_2$ (3-methylbutyrate) + NH_3 + CO_2 + $2H_2$ + ATP	Stickland
2	$C_6H_{13}O_2N$ (Leu) + $H_2 \rightarrow C_6H_{12}O_2$ (4-methylvalerate) + NH_3	Stickland
3	$C_6H_{13}O_2N$ (Ile) + $2H_2O \rightarrow C_5H_{10}O_2$ (2-methylbutyrate) + NH_3 + CO_2 + $2H_2$ + ATP	Stickland
4	$C_5H_{11}O_2N$ (Val) + $2H_2O \rightarrow C_4H_8O_2$ (2-methylpropionate) + NH_3 + CO_2 + $2H_2$ + ATP	Stickland
5	$C_9H_{11}O_2N$ (Phe) + $2H_2O \rightarrow C_8H_8O_2$ (phenylacetate) + NH_3 + CO_2 + $2H_2$ + ATP	Stickland
6	$C_9H_{11}O_2N$ (Phe) + $H_2 \rightarrow C_9H_{10}O_2$ (phenylpropionate) + NH_3	Stickland
7	$C_9H_{11}O_2N$ (Phe) + $2H_2O \rightarrow$ C6H6 (phenol) + $C_2H_4O_2$ (acetate) + NH_3 + CO_2 + H_2 + ATP	Non-Stickland

2.4 *Valerate and Butyrade Degrader*

Butyrate and valerate are two compositions which can be found in a typical volatile fatty acid of an acidic anaerobic digestion reactor of sludge [17]. The degradation kinetics of normal and branched chain butyrate and valerate are important in protein-fed anaerobic systems, as a number of amino acids degrade to these organic acids.

Based on Table 6, the degradation for both *n*-buytrate and *n*-valerate is via β-oxidation to acetate and acetate + propionate, respectively. The organisms that are capable to degrade butyrate are *Syntrophaceae* sp, *Tepidanaerobacter* sp. and *Clostridium* spp. Typically, if one of these substrates can be degraded by these organisms then it may potentially degrade the others. *I*-butyrate is also oxidized by the same organisms, and reciprocal isomerism between the two forms of butyrate has been well established [18, 19]. Both *neo*-valerate and *i*-valerate are more complex and difficult to access in environmental situations, as they are lumped in gas chromatography measurements.

Clostridium bryantiisp. can oxidize *neo*-valerate to acetate and propionate via β-oxidation while *i*-valerate degrades to acetate as the only organic acid product [18].

2.5 *Propionate Degrader*

Abundance of *Smithella* spp. among Syntrophotbacterales indicates syntrophic degradation of propionate and butyrate. The syntrophy of bacteria (illustrated in Fig. 1) is responsible for carrying out degradation of amino acids, aromatic

Table 4 Saturated and unsaturated LCFA commonly found in wastewaters (shown as % of total LCFA) [14]

	LCFA common name (structure*)						
	Saturated LCFA				Unsaturated LCFA		
Wastewaters	Laureate (C12:0)	Myristate (C14:0)	Palmitate (C16:0)	Stearate (C18:0)	Palmitoleate (C16:1)	Oleate (C18:1)	Linoleate (C18:2)
Domestic sewage		2.2	16.4	8.1	0.9	30.5	29.2
Dairy wastewater			27.0	7.0		37.0	13.0

Table 5 Gibbs free energy changes for some of the acetogenic and methanogenic reactions (presumably) involved in syntrophic conversion of different fatty acids [14]

Reactant	Equation
Fatty acids oxidation reactions	
Linoleate (C18: 2)	Linoleate + $16H_2O \rightarrow$ 9acetate + $14H_2 + 8H^+$
	$CH_3(CHCH)COOH + 2H_2O \rightarrow 2CH_3COOH + 2H_2$
Oleate (C 18:1)	Oleate + $16H_2O \rightarrow$ 9acetate + $15H_2 + 8H^+$
Stearate (C 18: 0)	Stearate + $16H_2O \rightarrow$ 9acetate + $16H_2 + 8H^+$
Palmitate (C 16: 0)	Palmitate + $14H_2O \rightarrow$ 8acetate + $14H_2 + 7H^+$
Butyrate (C 4: 0)	Butyrate + $2H_2O \rightarrow$ 2acetate + $2H_2 + H^+$
Methanogenic reactions	
Hydrogen	$H_2 + 1/4HCO_3^- + 1/4H^+ \rightarrow 1/4CH_4 + 3/4H_2O$
Acetate	Acetate + $H_2O \rightarrow HCO_3^- + CH_4$

Table 6 Butyrate and valerate degradation reactions [18]

Reaction	Substrate	Reaction
1	*n*-buytrate	$CH_3CH_2CH_2OOH + 2H_2O \rightarrow 2CH_3COOH + 2H_2$
2	*i*-butyrate	$CH_3(CHCH)COOH + 2H_2O \rightarrow 2CH_3COOH + 2H_2$
3	*n*-valerate	$CH_3CH_2CH_2CH_2COOH + 2H_2O \rightarrow CH_3COOH + CH_3CH_2COOH + 2H_2$
4	*neo*-valerate	$CH_3CH_2(CHCH_3)COOH + 2H_2O \rightarrow CH_3COOH + CH_3CH_2COOH + 2H_2$
5	*i*-valerate	$CH_3(CHCH_3)CH_2COOH + CO_2 + 2H_2O \rightarrow 3CH_3COOH + H_2$
6	*i*-valerate	$CH_3(CHCH_3)CH_2COOH + 4H_2O \rightarrow 2CH_3COOH + CO_2 + 5H_2$

compounds and propionate and butyrate which ultimately leads to the formation of CH_4 [20].

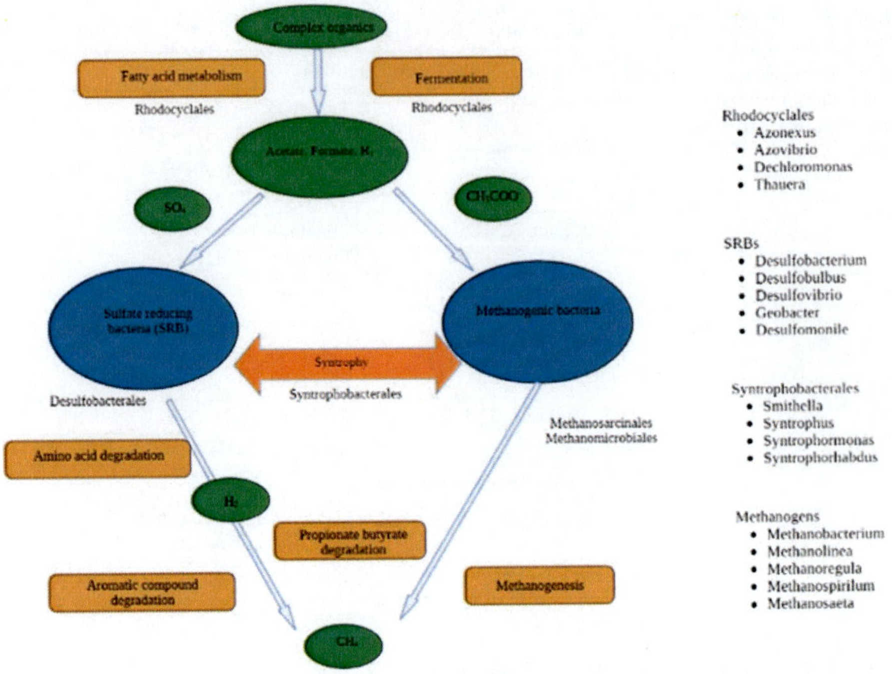

Fig. 1 Schematic representation of anaerobic carbon mineralization in sewage sludge with the microbial communities. Adapted from [20] (Created with Biorender)

2.6 Acetate Degrader

The source of acetate in sludge is originated from the conversion of volatile fatty acid in dark fermentation: acetogenesis [17]. Acetotrophic is a condition in which methyl groups are reduced by *Methanosarcinales* genus which uses simple compounds (acetate) for their growth. Acetotrophic methanogens are obligatory anerobes that transform acetate to methane and carbon dioxide. It was found that, during the anaerobic processing of sewage sludge and manure, the number of *Methanosaeta* genus increased with decreasing acetate in environment, simultaneously intensive growth of bacteria which are acetotrophic methanogens [21, 22]. Research conducted by Detman et al. [23] highlighted that *Methanosaeta* genus can be evaluated based on MAGs phylogenetic tree which shows *Methanothrix soehngenii* had the most abundant (12.1%) [23].

Stoichiometry reaction degradation of acetate:

$$\text{Acetate} + H_2O \rightarrow HCO_3^- + CH_4 \qquad (2)$$

where H_2O is water, HCO_3^- is bicarbonate and CH_4 is methane.

$$CH_3COO^- + SO_4^{2-} \rightarrow 2HCO_3^- + HS^- \quad (3)$$

where CH_3COO^- is acetate and SO_4^{2-} is sulphate.

2.7 Hydrogen Degrader

The anaerobic microorganisms produce hydrogenase enzyme which is capable to evolve and taking up hydrogen (H_2 [24]. Hydrogen production by fermentative microorganisms is an expectable method compared with photosynthetic bacteria due to its high utilization of organic compounds or wastes as substrate to produce hydrogen day and night. The production of molecular hydrogen (fermentation process) is generally associated with intracellular iron–sulphur protein, ferredoxin, which is an electronegative electron carrier [24]. The electrons transfer from ferredoxin to H^+ is catalyzed by hydrogenase enzyme. Two classes of fermentative bacteria are capable of producing hydrogen at a high rate and yield, including strictly anaerobic and facultative anaerobic bacteria. First *Clostridium butyricum* largely utilized in the biotechnological hydrogen production and secondly *Klebsiella pneumonia* typically a facultative anaerobic bacteria as nitrogen fixing [24].

Reaction 4summarized stoichiometry for both *Sporomusasphaeroides* and *Wolin-iela* for reduction of CO_2 to acetate.

Metabolism degradation of hydrogen:

$$H_2 + \text{fumarate} \leftrightarrow \text{Succinate} \quad (4)$$

Clostridium butyricum strict anaerobic bacterium, is known as a classical acid producer and usually ferments carbohydrates to butyrate, acetate, carbon dioxide, and molecular hydrogen [25, 26]. Based on Fig. 2, there are two pathways to produce hydrogen, one is via the cleavage of pyruvate to acetyl-CoA and the other to NAD^+ to generate $NADH_2$.

The production of 2,3-butaediol, ethanol and lactate from pyruvate by $NADH_2$ as a reductant, but not for H_2 [27].While *Klebsiella pneumonia;* a facultative anaerobic and nitrogen-fixing bacteria also has the ability to produce hydrogen in high quantities. Nitrogen is mainly associated for hydrogen production by *K. pneumonia.*

2.8 Sulphate Degrader

Sulphate ion ($S0_4^{2-}$) is one of the most universal anions occurring in rainfall, especially in air masses that have encountered metropolitan areas (During anaerobic conditions, sulphate is reduced to sulphide by sulphate-reducing bacteria [SRB]).

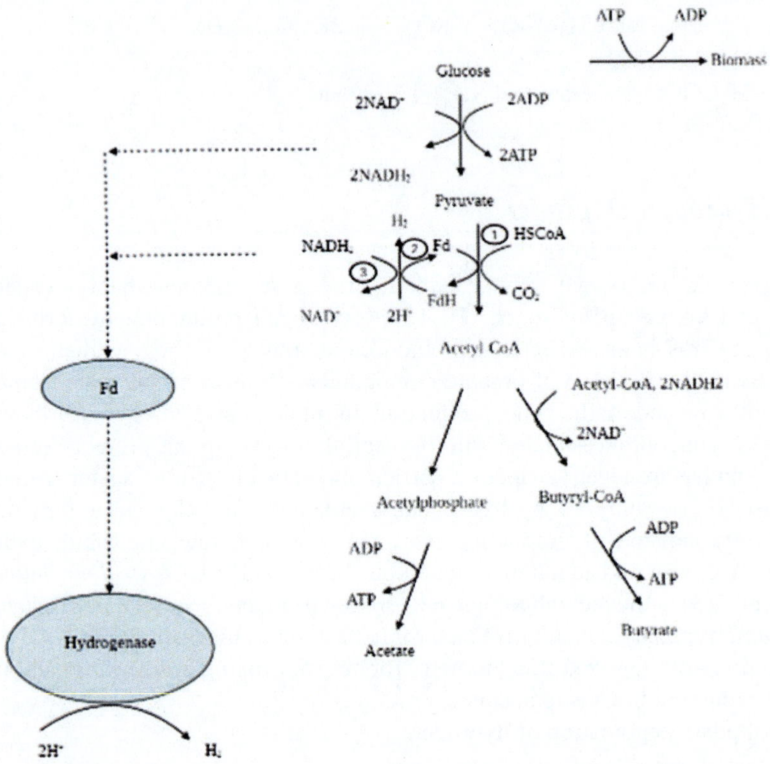

Fig. 2 Metabolic pathway of glucose by *Clostridiumbutyricum* under anaerobic conditions. (1) Pyruvate: ferredoxinoxidoreductase (PFOR); (2) Hydrogenase; (3) NADH: ferredoxinoxidoreductase. Adapted from 24 (Created with Biorender)

This SRB play a fundamental role as sulphate bioremediator through the conversion of sulphate to sulphide in the stabilization process [5]. Additionally they can compete with other anaerobic bacteria for a wide range of carbon sources and electron donors such as glucose, lactate, propionate, acetate, butyrate and ethanol. SRB found famously to grow at pH range 6–8 or called as neutrophilic condition [28]. Sulphate reducers that degrade carbon can be divided into two groups: (i) bacterial group that can completely degrade the carbon to carbon dioxide and (ii) bacterial group that catalyze partial carbon degradation to acetate which can be clearly figure in Table 7. The SRB can generate twice as much energy during the incomplete oxidation of lactate compared with its complete oxidation [29].

Table 7 Reduction of sulphate only partially oxidized [30]

Reaction
2**Lactate**$^-$ + $SO_4^2 \leftrightarrow$ 2**acetate**$^-$ $+ 2H_2O + 2CO_2 + S^{2-}$
2**Ethanol** + $3SO_4^2 \leftrightarrow 6H_2O + 4CO_2 + 3S^{2-}$
2**Malate**$^{2-}$ $+ 3SO_4^2 + 4H^+ \leftrightarrow 6H_2O + 8CO_2 + 3S^{2-}$

3 Conclusions

This book chapter provides an insight into the fundamental components of sewage sludge, including the natural microbe degraders present in the sludge. The knowledge of the microbial community in the sludge allows for the exploitation of the sludge and the isolation of suitable microbes for bioremediation purposes. The microbiological approach is a greener method for solving environmental pollution and has the potential to provide a sustainable solution.In addition to bioremediation, the chapter highlights the potential for the purification of useful chemical compounds from sewage sludge, such as expensive fatty acids that can be obtained through the isolation of certain species found in the sludge. This demonstrates the potential for the valorization of sludge in new emerging green technologies.One such technology is the microbial fuel cell (MFC), which requires a comprehensive and effective microbial degrader to accelerate the degradation process and increase the oxidation process, resulting in higher current density for energy recovery. The understanding of the microbial degraders in sewage sludge is essential for the development of effective and sustainable approaches to sludge management.Overall, this book chapter provides an insight into the components and typical mechanisms of sludge degrader microbes in dewatered sludge and highlights the potential for the exploitation of sludge valorization in sustainable approaches to sludge management. The utilization of natural microbe degraders can provide solutions to environmental pollution, produce valuable chemical compounds, and contribute to the development of new emerging green technologies such as the microbial fuel cell.

Acknowledgements The authors would like to thank the Universiti Sains Malaysia for the financial support of this study via APEX Era grant (1001/PTEKIND/881004). The authors have declared no conflict of interest for the manuscript.

References

1. Ahmed T, et al (2019) Biomass and bioenergy hydrothermal carbonization of lignocellulosic biomass for carbon rich material preparation: a review. Biomass Bioenergy 130(August):105384. https://doi.org/10.1016/j.biombioe.2019.105384
2. Duncan SM, Alkasrawi M, Gurram R, Almomani F, Wiberley-Bradfordand AE, Singsaas E (2020) Paper mill sludge as a source of sugars for use in the production of bioethanol and isoprene. Energies 13(18). https://doi.org/10.3390/en13184662

3. Doas N, Ghandour MA, Abd-alla MH (2018) Sludge reduction in wastewater of beet sugar industry using the effective microorganisms. In Abu Qurqas sugar factory. Egypt Sugar J 10:63–82
4. Oladejo J, Shi K, Luo X, Yang G, Wu T (2019) A review of sludge-to-energy recovery methods. Energies 12(1):1–38. https://doi.org/10.3390/en12010060
5. Tambo N, Kobayashi M, Thebault P, Haubry A (1982) Sludge treatment and disposal, vol 1, no 2/3. IWA
6. Yildiz BS (2012) 18—Water and wastewater treatment: biological processes. In Zeman FBT-MS (ed) Woodhead Publishing series in energy. Woodhead Publishing, Sawston, pp 406–428
7. Kumar V, Chopra AK, Kumar A (2017) A review on sewage sludge (biosolids) a resource for sustainable agriculture. Arch Agric Environ Sci 2(4):340–347. https://doi.org/10.26832/24566632.2017.020417
8. Talebi A, Razali YS, Ismail N, Rafatullah M, Azan Tajarudin H (2020) Selective adsorption and recovery of volatile fatty acids from fermented landfill leachate by activated carbon process. Sci Total Environ 707:134533. https://doi.org/10.1016/j.scitotenv.2019.134533
9. Raharja R, Murdiyatmo U, Sutrisno A, Wardani AK (2019) Bioethanol production from sugarcane molasses by instant dry yeast. IOP Conf Ser Earth Environ Sci, 230(1). https://doi.org/10.1088/1755-1315/230/1/012076
10. Anukam A, Mohammadi A, Naqvi M, Granström K, A review of the chemistry of anaerobic digestion: methods of accelerating and optimizing process efficiency, pp 1–19.
11. Ramsay IR, Pullammanappallil PC (2001) Protein degradation during anaerobic wastewater treatment: derivation of stoichiometry. Biodegradation 12(4):247–256. https://doi.org/10.1023/A:1013116728817
12. Park J, Park S, Kim M (2015) Anaerobic degradation of amino acids generated from the hydrolysis of sewage sludge, April. https://doi.org/10.1080/09593330.2013.863951
13. Singh S, et al (2020) Enhanced methanization of long-chain fatty acid wastewater at 20°c in the novel dynamic sludge chamber–fixed film bioreactor. Front Energy Res, 8:166 [Online]. https://www.frontiersin.org/article/10.3389/fenrg.2020.00166
14. Sousa DZ, Smidt H, Alves MM, Stams AJM (2009) Ecophysiology of syntrophic communities that degrade saturated and unsaturated long-chain fatty acids. https://doi.org/10.1111/j.1574-6941.2009.00680.x
15. Roy F, Samain E, Dubourguier HC, Albagnac G (1986) Synthrophomonas sapovorans sp. nov., a new obligately proton reducing anaerobe oxidizing saturated and unsaturated long chain fatty acids. Arch Microbiol 145(2):142–147. https://doi.org/10.1007/BF00446771
16. Hatamoto M, Imachi H, Ohashi A, Harada H (2007) Identification and cultivation of anaerobic, syntrophic long-chain fatty acid-degrading microbes from mesophilic and thermophilic methanogenic sludges. Appl Environ Microbiol 73(4):1332–1340. https://doi.org/10.1128/AEM.02053-06
17. Moestedt J, Westerholm M, Isaksson S, Schnürer A (2020) Inoculum source determines acetate and lactate production during anaerobic digestion of sewage sludge and food waste. Bioengineering, 7(1). https://doi.org/10.3390/bioengineering7010003
18. Batstone DJ, Pind PF, Angelidaki I (2003) Kinetics of thermophilic, anaerobic oxidation of straight and branched chain butyrate and valerate. Biotechnol Bioeng 84(2):195–204. https://doi.org/10.1002/bit.10753
19. Hatamoto M, Imachi H, Yashiro Y, Ohashi A, Harada H (2008) Detection of active butyrate-degrading microorganisms in methanogenic sludges by RNA-based stable isotope probing. Appl Environ Microbiol 74(11):3610–3614. https://doi.org/10.1128/AEM.00045-08
20. Sidhu C, Vikram S, Pinnaka AK (2017) Unraveling the microbial interactions and metabolic potentials in pre- and post-treated sludge from a wastewater treatment plant using metagenomic studies, 8(July): 1–10. https://doi.org/10.3389/fmicb.2017.01382
21. The Scientific World Journal (2017) Retracted: microbial ecology of anaerobic digesters: the key players of anaerobiosis. Sci World J 2017:3852369. https://doi.org/10.1155/2017/3852369
22. Griffin ME, McMahon KD, Mackie RI, Raskin L (1998) Methanogenic population dynamics during start-up of anaerobic digesters treating municipal solid waste and biosolids.

Biotechnol Bioeng 57(3):342–355. https://doi.org/10.1002/(SICI)1097-0290(19980205)57:3%3c342::AID-BIT11%3e3.0.CO;2-I

23. Detman A et al (2021) Evaluation of acidogenesis products' effect on biogas production performed with metagenomics and isotopic approaches. Biotechnol Biofuels 14(1):125. https://doi.org/10.1186/s13068-021-01968-0
24. Chen X, Sun Y, Xiu Z, Li X, Zhang D (2006) Stoichiometric analysis of biological hydrogen production by fermentative bacteria. Int J Hydrogen Energy 31(4):539–549. https://doi.org/10.1016/j.ijhydene.2005.03.013
25. Zigová J, Šturdík E, Vandák D, Schlosser S (1999) Butyric acid production by Clostridium butyricum with integrated extraction and pertraction. Process Biochem 34:835–843. https://doi.org/10.1016/S0032-9592(99)00007-2
26. Vignais P, Billoud B, Meyer J (2001) Vignais PM, Billoud B, Meyer J. Classification and phylogeny of hydrogenases. FEMS Microbiol Rev 25:455–501. https://doi.org/10.1111/j.1574-6976.2001.tb00587.x
27. Kurokawa T, Tanisho S (2005) Effects of formate on fermentative hydrogen production by Enterobacter aerogenes. Mar Biotechnol 7(2):112–118. https://doi.org/10.1007/s10126-004-3088-z
28. Tran TTT, Kannoorpatti K, Padovan A, Thennadil S (2021) Sulphate-reducing bacteria's response to extreme ph environments and the effect of their activities on microbial corrosion. Appl Sci 11(5):1–19. https://doi.org/10.3390/app11052201
29. Rubio-rincón F, et al (2017) Effects of electron acceptors on sulphate reduction activity in activated sludge processes, pp 6229–6240. https://doi.org/10.1007/s00253-017-8340-3
30. Wake LV, Christopher RK, Rickard PAD, Andersen JE, Ralph BJ (1977) A thermodynamic assessment of possible substrates for sulphate-reducing bacteria. Aust J Biol Sci 30(2):155–172. https://doi.org/10.1071/BI9770155

Microbial Fuel Cells (MFC) as an Alternative Energy Source: The Perceptions and Attitudes Towards Sustainable and Renewable Energy in Malaysia

Aziatul Waznah Ghazali, Soliha Sanusi, Zulaikha Amirah Johari, and Muaz Mohd Zaini Makhtar

Abstract Due to the exponential rate of waste production worldwide, the development of newly emerging technologies is rising and advancing proportionately. Entailed instalment of sophisticated sustainable and renewable energy development is constantly expanding to meet the socio-economic demands of communities and industries. Innumerable advanced technologies have been introduced to acquire for alternative energy sources and to minimise waste pollution while reducing emissions of greenhouse gases hence, combatting the impact of global warming. Among the energy conversion technologies includes gasification, pyrolysis, incineration, landfill, and bioelectrochemical technologies mainly microbial fuel cell (MFC), microbial electrolysis cells (MECs), and microbial electrosynthesis (MES). This chapter starts off with an overview of the current trend of sustainable and renewable energy in Malaysia. Next, the application of MFC is being discussed in the context of wastewater treatment and its potential application as an alternative source of renewable energy. Finally, the public acceptance and perceptions towards sustainable and renewable energy are also considered and discussed.

Keywords Microbial fuel cell · Alternative energy resources · Public acceptance · Renewable energy

A. W. Ghazali · S. Sanusi (✉) · Z. A. Johari
Faculty Economics and Management, Universiti Kebangsaan Malaysia, Bangi, Selangor, Malaysia
e-mail: solihasanusi@ukm.edu.my

Accounting Research Institute, Universiti Teknologi MARA, Shah Alam, Selangor, Malaysia

M. Mohd Zaini Makhtar
School of Industrial Technology, Bioprocess Technology Division, Universiti Sains Malaysia, 11800 Gelugor, Penang, Malaysia
e-mail: muazzaini@usm.my

Centre for Innovation and Consultation, Universiti Sains Malaysia, 11800 Gelugor, Penang, Malaysia

M. Mohd Zaini Makhtar et al. (eds.), *Microbial Fuel Cell (MFC) Applications for Sludge Valorization*, Green Energy and Technology,
https://doi.org/10.1007/978-981-99-1083-0_4

1 Introduction

As the world population grows, the demand for energy especially for electricity is also on the rise. Coal is the most widely used source of electricity today, accounting for 41% of total electricity generation. However, due to the high degree of pollution (water and air pollution during mining and air pollution when burning) and deplorable working conditions, certain energy source is not considered as sustainable source.

Generally, there are four types of renewable energy (RE) sources which are solar, wind, hydro, and biomass. Although these sources are renewable, their sustainability may be unfeasible. Sustainability is determined by three different parameters: environmental sustainability, social sustainability, and economic sustainability. According to the United Nations (UN), environmental sustainability is about acting in a way that ensures future generations have the natural resources available to live an equal, if not better, way of life as current generations. Being a net consumer of energy, if a renewable energy device costs more energy than it produces during its lifetime, it does not conform as being sustainable.

2 Revolution of Malaysia's Policies

On the other hand, social sustainability calls for safe working conditions and decent wages. As a whole, social sustainability revolves around the fact that we all live on the same planet and have access to enough food and energy to support ourselves. Therefore, it is essential to use the planet's resources effectively and efficiently. Economic sustainability is concerned with policies that promote long-term economic growth without jeopardising the community's social, environmental, and cultural characteristics. Greatly influenced by the aftermath of the 1973 and 1979 oil crises, Malaysian policymakers have shown a remarkable enthusiasm for developing timely and visionary energy plans and policies. Table 1 provides a summary of various energy policies in Malaysia.

Nevertheless, despite the government's effort to diversify the energy mix in the country, energy consumption in Malaysia is still excessively relying on the non-renewable energy sources compare to renewable energy sources. This could be owing to the current availability of natural resources and the high cost of renewable energy. Another reason could be a lack of strategic deployment of renewable energy resources. For example, the insufficient and unrealistic net-metering tariff for solar energy producers, the high priority placed on biomass energy production, the electrical system's incapacity to accept intermittency, and insufficient backup mechanisms to stabilise the electric grid. Figure 1 presents the total primary energy production while Fig. 2 presents the total primary energy supply to Malaysia by fuel type.

To date, the share of RE in Malaysia remained below the global and regional average [2]. As of the end of 2020, RE accounted for 23% of the national power

Table 1 A summary of various energy policies in Malaysia [1]

No	Year	Policies	Objectives
1	1974	Petroleum Development Act	Provide rights to PETRONAS for exploration, development, and production of petroleum
2	1975	National Petroleum Policy	To manage the downstream oil and gas industry through Petroleum Regulations 1974
3	1979	National Energy Policy	Three objectives, i.e., enhanced sufficiency, safety, and cost-effectiveness have been proposed along with the promotion of efficient energy utilisation to curb negative environmental issues
4	1980	National depletion policy	To prolong oil reserves to secure future energy requirements
5	1981	Four fuel diversification strategy	To enhance the lifetime of natural resources of Malaysia by introducing a balanced utilisation policy of oil, NG, hydro, and coal
6	2001	Five fuel diversification strategy	Renewable energy sources (most highlighted solar and biomass) have been included in the primary energy mix to improve the security of natural resources and the environment-friendly nature of RE technology
i	2005–2030	Hydrogen energy roadmap	Generation of hydrogen using RE resource, development of H2 network for H2 fuel cell vehicles
ii	2006	National biofuel policy	To promote palm oil demand and utilisation of biomass resources to produce electricity
iii	2010	National RE policy and action plan	Development of indigenous RE technologies and their enhanced utilisation

installed capacity compared to the global average of 37% and the Southeast Asia regional average of 30%. Hence, there is a crucial need to accelerate RE deployment in Malaysia to meet the committed RE and climate targets, by strengthening existing programmes and introducing new approaches, in parallel with the Government's practices in future-proofing existing electricity market regulations and power sector industry practices.

Recently, the Ministry of Energy and Natural Resources kicks off a new Malaysia Renewable Energy Roadmap is introduced to realise the Government's future sustainable and renewable energy aspiration with an anticipated cumulative investment of MYR 53 billion and 46,336 numbers of job opportunities. Malaysia Renewable

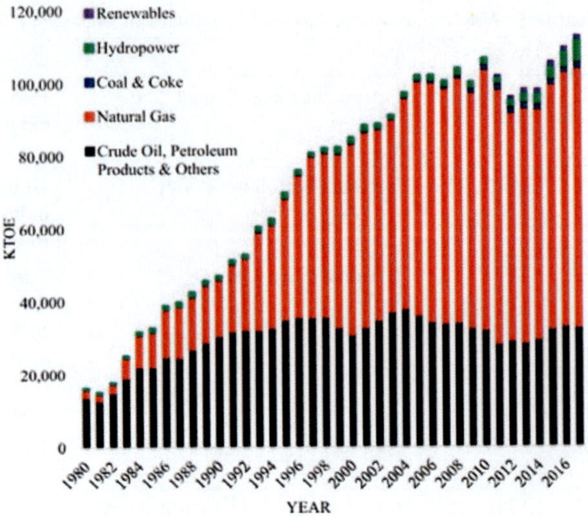

Fig. 1 Total primary energy production in Malaysia. Reprinted from Energy Strategy Reviews, Vol.35, Ahmad M.S., Ali M.S., Rahim N.A., Hydrogen energy vision 2060: Hydrogen as energy Carrier in Malaysian primary energy mix—Developing P2G case, 1–12, Copyright (2021), with permission from Elsevier

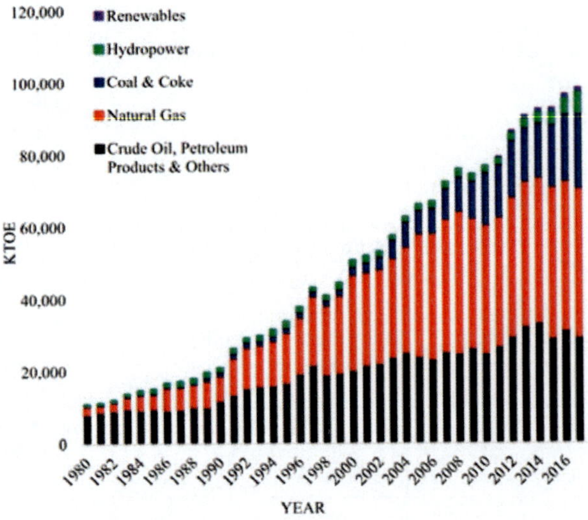

Fig. 2 Total primary energy supply to Malaysia by fuel type. Reprinted from Energy Strategy Reviews, Vol.35, Ahmad M.S., Ali M.S., Rahim N.A., Hydrogen energy vision 2060: Hydrogen as energy Carrier in Malaysian primary energy mix—Developing P2G case, 1–12, Copyright (2021), with permission from Elsevier

Energy Roadmap (MyRER) is a formulation of a strategic framework that aimed to achieve 31% RE share in the national capacity mix by 2025 and attain decarbonisation of the electricity sector by 2035.

The MyRER vision is upheld by 4 technology-specific pillars and 4 enabling initiatives. The strategic framework calls for concerted and coordinated actions from collaborations between various stakeholders in allowing Malaysia to tap into the huge potential made available through RE projects to promote improved economic, environmental, and social outcomes. With the said pillars, it is hoped to ensure the full capacity of the usage of renewable energy as well as the sustainability of these sources.

It is now widely accepted that the world's energy supply must become more sustainable. Thus, it is critical to ensure that the Malaysia Energy Vision 2060s milestones are backed up by state-of-the-art and faultless research schemes. At present, renewable energy accounts for less than 7.8% of the global primary energy supply including wind power, hydropower, solar energy, geothermal energy, and biomass. It is predicted that about half of the global energy demand could be met by renewable sources by the year 2050. Organics in wastewater used in MFCs could be an important renewable resource in the transition from fossil to sustainable fuels as discussed in the next section.

3 Microbial Fuel Cells (MFC) as an Alternative Energy Source

Sustainability has become a global challenge as natural resources such as fresh water and fossil fuels are depleting rapidly. It is imperative to have an effective and efficient water management to ensure its long-term sustainability and meet the needs of future generations [2]. Current wastewater treatment technology consumes between 1 and 3% of the total electrical energy output of a country [3], of which would require a 20% of public utilities' electrical energy consumption for its operation [4]. Moreover, energy consumption is expected to rise significantly soon. For example, the energy use by water utilities in Australia is estimated to grow between 130 and 200% above existing levels due to an expected population growth of 25% by 2030 [5]. Consequently, water treatment industries would have a significant carbon impact, posing a serious threat to the environmental sustainability.

Environmental sustainability encompasses all aspects of life. It is intimately related to humanity's future, defining how we should conserve and manage natural resources, air quality, water quality, and the ecosystems. It also aids in the prevention of environmental damage caused by technological advancement. Among the way to achieve environmental sustainability is to have effective wastewater treatment. Globally, around 80% of wastewater is being released back into the environment [6] without being treated or reused. This indicates that over 1.8 billion people in the world are consuming contaminated water, a potential health hazard. Hence, wastewater treatment is pivotal in fulfilling the growing demand for clean water in rapidly expanding cities, enhancing energy production and industrial development, and promoting sustainable agriculture.

Societal and environmental pressures over recent years have led to a growing movement for the industry to reduce its wastewater and treat it before discharge. Wastewater is now seen as a potential resource and its use, or recycling after suitable treatment, can provide economic and financial benefits. Societal and environmental pressures over recent years have led to a growing movement for the industry to reduce its wastewater and treat it before discharge. Wastewater is now seen as a potential resource and its use, or recycling after suitable treatment, can provide economic and financial benefits.

In Malaysia, there are 8147 sewage treatment plants and network pumping stations nationwide which are managed by the Indah Water Konsortium Sdn Bhd (IWK). Owned by the Minister of Finance Incorporated, IWK is Malaysia's national sewerage company which has been entrusted with the task of developing and maintaining a modern and efficient sewerage system for all Malaysians. The national operator of sewerage services operates and maintains more than 6800 sewage treatment plants across more than 19,400 km of sewer pipelines. Currently, IWK provides sewerage services to approximately 26 million population. IWK has been practicing the reuse of sewage by-products at 13 of their regional plants as an initial effort to conserve water, energy, and the environment. The reuse of the treated effluent is currently confined to internal housekeeping or non-potable use, such as sewage treatment plants compound cleaning, vehicle cleaning, and watering of plants for landscaping purposes.

Sewage treatment is the process of removing contaminants from wastewater, primarily from household sewage. It includes physical, chemical, and biological processes to remove these contaminants and produce environmentally safer treated wastewater. A by-product of sewage treatment is usually a semi-solid waste or slurry, called sewage sludge that has to undergo further treatment before being suitable for disposal or land application [3]. Overall, the processes of wastewater treatment consume a substantial portion of energy. Nevertheless, the wastewater influent contains a large amount of energy contents in the form of organics. Therefore, to maintain a sustainable society, it is clearly beneficial to efficiently extract energy from wastewater to compensate energy use of the treatment process which can be realised by anaerobic processes to produce methane, hydrogen, or bioelectricity.

Recently, a bioelectricity processing method known as microbial fuel cells (MFCs) has gained increased interest for their promising and potentially alternative renewable energy source. Initially, MFC emerged as an alternative energy-added wastewater treatment technology which can produce electricity during wastewater treatment. The recent advancement in the field of MFC provides promising technology not only to obtain energy but also to treat organic content wastewater at the same time. Moreover, the use of MFC as an alternative source for power generation is considered a reliable, clean, efficient process, which utilises renewable methods and does not generate any toxic by-product. MFC is also considered an environmentally friendly conversion technology due to its capability of converting the chemical energy stored in the organic matter to electricity assisted by microorganisms. Figure 3 shows the graphical application of MFC.

MFC is a process where microbes convert chemical energy generated by the oxidation of organic/inorganic molecules into adenosine triphosphate through sequential reactions. In this process, electrons are transported to a terminal electron acceptor to generate an electrical current. The MFC system consists of an anode and a cathode and an external load connecting the two electrodes. During MFC operation, the bacteria switch their electron-donating direction from the natural electron acceptor

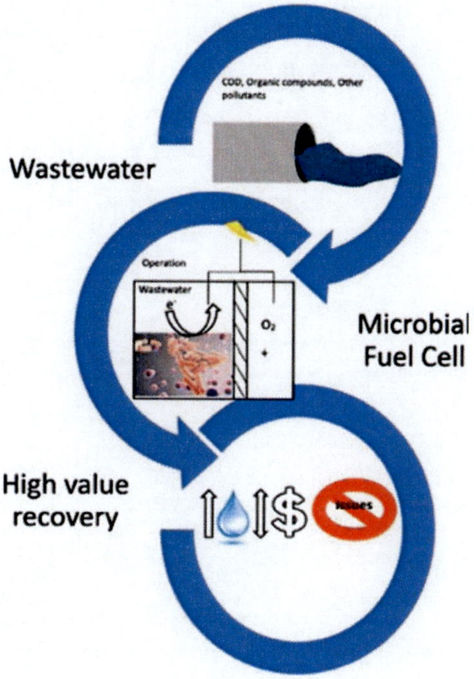

Fig. 3 Graphical application of MFC. Reprinted from Sci Total Environ, Munoz-Cupa C, Hu Y, Xu C, Bassi A., An overview of microbial fuel cell usage in wastewater treatment, resource recovery and energy production, Copyright (2021), with permission from Elsevier

in the bulk wastewater, mainly oxygen or nitrate, directly to the anode via a redox reaction of electron mediators on the anode surface. Figure 4 shows the concept of MFC towards the production of renewable energy.

MFC offers several advantages over other renewable energy. Mainly, its direct conversion of chemical energy within the substrate to electricity allows a higher energy conversion efficiency without any combustion process [7]. It uses low energy consumption while offering significant cost savings since the process of MFC does not require any energy input to drive the system for its operation. The system can be operated at ambient temperature, thus avoiding huge energy consumption to stabilise the temperature of the system. Additionally, the use of microorganisms in MFC for electricity generation eliminates the high cost of constructing the MFC compared to the cost of using a metal catalyst or extracted enzymes. Prior empirical studies have shown that energy from MFC is able to power up small devices like sensors, low-voltage capacitors, and small direct current (DC) motors.

Figure 5 depicts a comparison of the findings of the studies by Ge and He [8] and Li, Yu [9]. In the first part (section a), it shows that from Ge and He [8] findings, the input energy is 0.15 mW/m3 and the energy generated by MFC is 0.081 mW/m3. Meanwhile in the second part (section b), in the findings of Li, Yu [9], the energy required for the process is 0.076 mW/kg COD and a power of 0.026 mW/kg COD is produced. In comparison, there was an energy consumption of anaerobic sludge of 0.6 kW/kg COD indicating that MFC application for renewable energy is sustainable.

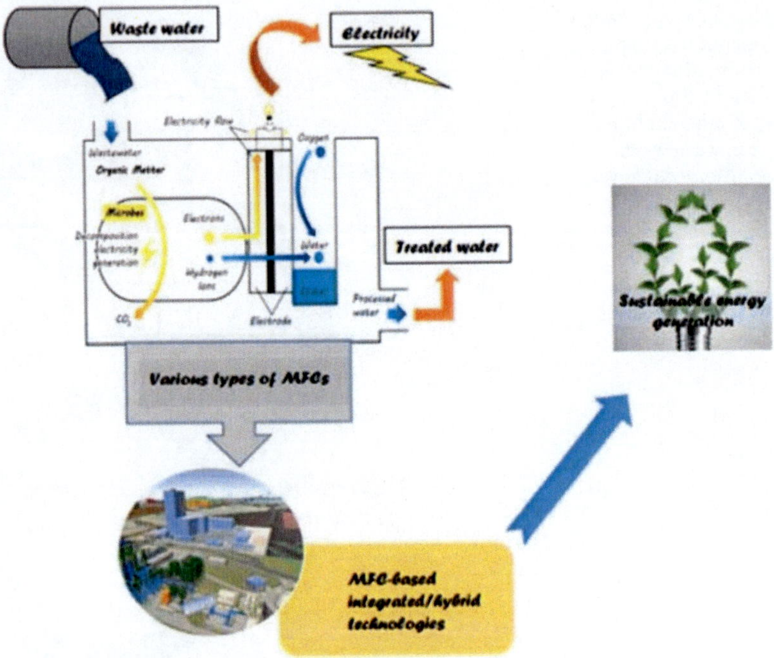

Fig. 4 The concept of MFC towards the production of renewable energy. Reprinted from Process Safety and Environmental Protection, A. Nawaz, Ikram ul Haq, K. Qaisar, B. Gunes, S.I. Raja, K. Mohyuddin, H. Amin., Microbial fuel cells: Insight into simultaneous wastewater treatment and bioelectricity generation., 357–373, Copyright (2022), with permission from Elsevier

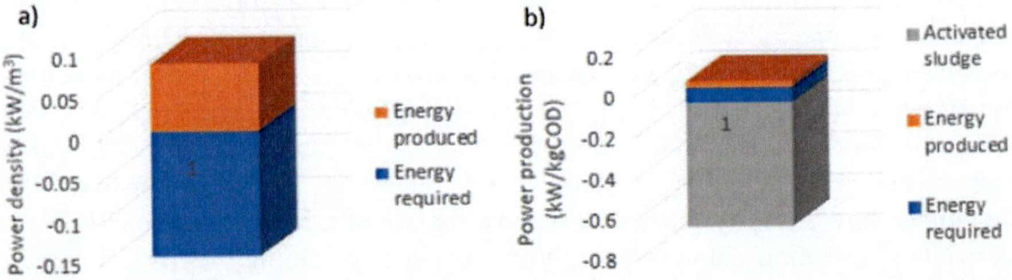

Fig. 5 Comparison of Energy Balance of MFC (a) Municipal wastewater with dual MFC (redrawn from Ge and He [8]; (b) municipal wastewater with single MFC and conventional activated sludge. Reprinted with permission from Li, Yu [9]

Furthermore, several studies have examined its application in wastewater treatment in terms of energy consumption and operational costs. The ability of the electroactive microbial communities to harvest and extract clean energy from waste organic sources rather than filtering the contaminants has allowed MFC to utilise many different types of wastewaters, ranging from domestic wastes, industrial wastes, agricultural waste, and animal waste. Furthermore, the fuels for MFCs are flexible in that they can be different kinds of biodegradable waste, reduced carbohydrates, glucose, sucrose, acetate, lactate, starch, and some other complex substrates in municipal or industrial wastewater, and in the form of liquid or sludge. In comparison with other wastewater treatment systems such as anaerobic digestion [10] or activated sludge [11], MFC offers a much lower cost to manage and dispose of sludge. Therefore, MFC is seen as an ideal and potential green technology that can generate clean and renewable electricity from the treatment of wastewater.

However, MFC technology is still in its infancy and has yet to be scaled up for mass operations due to several limitations and challenges. High internal resistance, high energy losses, high costs for construction and operation, and difficulty of scaling up are among the issues that need to be further addressed to facilitate an economically feasible operation of MFC. To overcome some of the present limitations, the combination of separation methods and new configurations for MFC needs to be investigated for further application in wastewater treatment for continued improvements in fuel cell design and efficiency as well scale-up with economically practical applications tailored to local needs. Future studies are needed to increase the power output to decrease the operational cost and have a better renewal value because of the high capital cost. Further investigation also should consider recovery or other high-value products such as nutrients or pharmaceutical products by MFC.

4 The Public Perceptions and Attitudes Towards Sustainable and Renewable Energy

Advances in new technologies and the desire to achieve a sustainable and safe energy supply, enable communities to transition from conventional to renewable resources. Although the transition to renewable energy is gaining popularity globally, its development is constrained by several challenges, most notably social resistance. Public perception of alternative energy sources is considered one of the most important factors influencing the development of renewable energy facilities and technologies. Other significant challenges include low acceptance in different social sectors, lack of knowledge, and negative public perceptions.

It is critical to understand the interdependence of technology, society, and government, as every change in one is likely to influence the others. As such, their interdependence affects the acceptance of the society in terms of change and technology, an aspect that plays a significant role in the evolution of renewable energy developments. According to Jasanoff [12], the relationship between science and technology is mutually beneficial, in that science has an effect on society and society has an effect on science. Therefore, entrepreneurs, scientists, and policymakers working on sustainable and renewable energy technology should continuously look for new ways to engage the public in innovation [13]. Research in sustainable and renewable energy should prioritise problem-solving, interdisciplinarity, and social inclusive studies [14, 15].

Renewable energy acceptance is shaped by three dimensions: market, socio-political, and community, thereby distinguishing a universal and homogenising sense of acceptance (often reflected in 'positive' national opinion polls) from its representation in policies and the varied reactions of local communities [16]. Public acceptance and perception can help to prevent, or at least anticipate, possible future controversies, to provide ideas for new and better products and services, and to increase the openness and transparency of scientific and technological developments, which is a prerequisite for a trust relationship between different actors [17]. Public involvement has also demonstrated that they can be proactive and come up with interesting suggestions, such as including energy and environmental issues in educational programmes, from elementary school to universities [18]. One tool that enables the connection between different actors involved in the development of a project is the communication of the environmental issues and energy technologies that shape public opinion, change policies, and affect our world [19].

Adequate communication techniques for renewable energy transition initiatives would transpire a positive or negative effect, depending on how those projects are being understood and accepted. Hence, it is critical to effectively communicate technical and scientific knowledge, keeping the pertinent information as simple to understand as feasible. Additionally, the timing of communication is also important, communication strategies should be implemented prior to initiating any project development. Volken, Xexakis [20] found that public acceptance can increase or decrease during the communication process for new technologies depending on the information load received. For example, in Switzerland, the acceptance of geothermal energy decreased with increasing knowledge, most likely as a result of learning about past accidental impacts, such as induced seismicity [21]. In another study, Pellizzone, Allansdottir [22] found that while the public's perception of renewable energies is positive and optimistic at the beginning, the public's favourable perception may decrease when the project is considered in the immediate environment, resulting in the phenomenon known as 'not in my backyard.'

Additionally, the media play a critical part in creating awareness and providing microbial fuel cell technology information to the public, as analysed by Stauffacher, Muggli [23], who highlighted the requirement of transparency and diligent monitoring of communication and public engagement. Numerous factors influence public opinion about microbial fuel cells as a sustainable and renewable energy, including the ability of the technology to solve the groundwater contamination, water resource sustainability, uncertainty regarding reversibility, seismic activity, cost–benefit ratio, level of knowledge about the technologies, dissemination channels, and the integration of various actors in project planning and public consultation exercises [24]. Renewable energy projects in the microbial fuel cell are generally discussed in different social sectors in terms of their environmental, economic, and political implications [17]. Therefore, creating awareness and providing adequate information of microbial fuel cell at the right time is crucial to ensure a continuous public support on the agenda of sustainable and renewable energy movement in Malaysia especially in this promising green technology.

As per sustainable development goal number seven on affordable and clean energy and 7(a) on enhancing international cooperation to facilitate access to clean energy research and technology by 2030, renewable energy development and improvement are critical for MFC to move forward. Investment from interested investors such as angle investors, government, society, NGO, and the public is most welcome to achieve this spirit. Cooperation and collaboration among those parties will swift the process tremendously.

Nevertheless, there is a need to understand public attitudes towards global issues and multiple energy technologies. Indeed, public acceptance is necessary for the successful implementation of an energy transition towards renewable energy sources. Although the global benefit of renewable energy is well known, some concerns still exist on their impact on local environment. Hence, the role of the government and media on rising the public awareness of sustainable and renewable energy is crucial. Supportive government policies and financial incentives such as subsidies, competitive pricing of energy technologies, and tax rebate would also improve public acceptance and perception of renewable energy adoption. Moving forward, it is reasoned that the policymaker should observe the development of a sustainable and renewable energy policies based on an effective coordination and collaboration with key stakeholders to ensure that the change towards sustainable and renewable energy is successful.

5 Conclusions

This chapter demonstrates that the transition from conventional ways of energy sources producers had slowly moved to renewable energy due to its scarcity. However, the transition from fossil fuels to renewable energy sources (microbial fuel cells) provides significant challenges and opportunities for various energy sectors. Although various efforts and strategies have been created to increase the adoption of renewables and other efficient technologies, including but not limited to information programmes and financial incentives, the concept of renewable energy in Malaysia is still in its infancy.

Continuous exploitation of non-renewable resources of energy would significantly impact the climate change and create vulnerabilities in future global energy security and sustainability, due to the fact that these resources will eventually decrease. Malaysia has a great potential to venture for sustainable and renewable energy as it is blessed with a vast supply of solar, wind, hydro, biogas, and biomass. Although these natural resources are readily available substantially, a dive towards the era of sustainable of renewable energy development is unavoidable.

A potential technology of sustainable and renewable energy that can be exploited is the MFC technology. While MFC promises a clean, zero-emission energy technology for a sustainable source of energy, it is constrained by costs, durability, complexity, as well as operational safety. Despite these limitations, MFC technology is gaining traction globally as one of the most sustainable ways to treat wastewater, alternative source of renewable energy, and by extension, reducing carbon intensity. Numerous large-scale research and development programmes are being funded globally to overcome these difficulties and to establish an industrial scale for MFC.

Acknowledgements The authors would like to thank the Universiti Sains Malaysia for the financial support of this study via APEX Era grant (1001/PGURU/881004).

References

1. Bujang AS, Bern C, Brumm T (2016) Summary of energy demand and renewable energy policies in Malaysia. Renew Sustain Energy Rev 53:1459–1467
2. Sustainable Energy Development Authority Malaysia (2021) Malaysia renewable energy roadmap: pathway towards low carbon energy system. Sustainable Energy Development Authority Malaysia (SEDA)
3. DOE (2014) The water-energy nexus: challenges and opportunities. U.S. Department of Energy
4. Kenway SJ et al (2019) Defining water-related energy for global comparison, clearer communication, and sharper policy. J Clean Prod 236:117502
5. Kenway SJ et al (2011) The connection between water and energy in cities: a review. Water Sci Technol 63(9):1983–1990
6. UN WWDR (2017) Wastewater: an untapped resource. UN Water
7. Rahimnejad M et al (2015) Microbial fuel cell as new technology for bioelectricity generation: a review. Alex Eng J 54(3):745–756
8. Ge Z, He Z (2016) Long-term performance of a 200 liter modularized microbial fuel cell system treating municipal wastewater: treatment, energy, and cost. Environ Sci Water Res Technol 2(2):274–281
9. Li W-W, Yu H-Q, He Z (2014) Towards sustainable wastewater treatment by using microbial fuel cells-centered technologies. Energy Environ Sci 7(3):911–924
10. He W et al (2016) The effect of flow modes and electrode combinations on the performance of a multiple module microbial fuel cell installed at wastewater treatment plant. Water Res 105:351–360
11. Ren L, Ahn Y, Logan BE (2014) A two-stage microbial fuel cell and anaerobic fluidized bed membrane bioreactor (MFC-AFMBR) system for effective domestic wastewater treatment. Environ Sci Technol 48(7):4199–4206
12. Jasanoff S (2004) States of knowledge: the co-production of science and the social order. Routledge
13. Jugend D et al (2020) Public support for innovation: a systematic review of the literature and implications for open innovation. Technol Forecast Soc Chang 156:119985
14. Sovacool BK, Axsen J, Sorrell S (2018) Promoting novelty, rigor, and style in energy social science: towards codes of practice for appropriate methods and research design. Energy Res Soc Sci 45:12–42
15. Lennon M (2021) Energy transitions in a time of intersecting precarities: from reductive environmentalism to antiracist praxis. Energy Res Soc Sci 73:101930
16. Wüstenhagen R, Wolsink M, Bürer MJ (2007) Social acceptance of renewable energy innovation: an introduction to the concept. Energy Policy 35(5):2683–2691
17. Meller C et al (2018) Acceptability of geothermal installations: a geoethical concept for GeoLaB. Geothermics 73:133–145

18. Pellizzone A et al (2015) Exploring public engagement with geothermal energy in southern Italy: a case study. Energy Policy 85:1–11
19. Dowd A-M et al (2011) Geothermal technology in Australia: investigating social acceptance. Energy Policy 39(10):6301–6307
20. Volken SP, Xexakis G, Trutnevyte E (2018) Perspectives of informed citizen panel on low-carbon electricity portfolios in Switzerland and longer-term evaluation of informational materials. Environ Sci Technol 52(20):11478–11489
21. Häring MO et al (2008) Characterisation of the Basel 1 enhanced geothermal system. Geothermics 37(5):469–495
22. Pellizzone A et al (2017) Geothermal energy and the public: a case study on deliberative citizens' engagement in central Italy. Energy Policy 101:561–570
23. Stauffacher M et al (2015) Framing deep geothermal energy in mass media: the case of Switzerland. Technol Forecast Soc Chang 98:60–70
24. Chavot P et al (2018) Social shaping of deep geothermal projects in Alsace: politics, stakeholder attitudes and local democracy. Geotherm Energy 6(1):1–21
25. Munoz-Cupa C et al (2021) An overview of microbial fuel cell usage in wastewater treatment, resource recovery and energy production. Sci Total Environ 754:142429

Application of Microbial Fuel Cell for Bioremediation of Sewage Sludge

Mohamad Farhan Mohamad Sobri and Muaz Mohd Zaini Makhtar

Abstract This chapter provides a comprehensive overview of the potential application of microbial fuel cell (MFC) technology for the bioremediation of sewage sludge. The abundance of sewage sludge and its associated contaminants presents a significant challenge to environmental sustainability, necessitating the development of innovative approaches for its treatment. The implementation of MFC technology offers a promising alternative approach that can simultaneously provide bioremediation and power generation benefits. The chapter highlights current research and studies on the utilization of MFC for sewage sludge treatment, including an overview of the mechanisms involved in the adaptation of the technology to address environmental issues associated with sewage sludge pollution. The chapter also discusses parameters for enhancing MFC performance, such as the combination of inoculum, substrate pretreatment, sludge concentration, and the effect of nitrate and sulfate. The earliest applications of MFC technology for sludge treatment are also discussed, including the configuration of the system, the use of sludge as a substrate, and the adjustment of pH to suit the system. Early MFC research also focused on nutrient recovery. Overall, the chapter aims to provide a comprehensive understanding of the potential contribution of MFC technology to sustainable wastewater management. By utilizing MFC technology for the bioremediation of sewage sludge, researchers can develop innovative solutions for addressing environmental challenges, thereby enhancing environmental sustainability.

Keywords Sewage sludge · Biomass conversion · Bioprocessing · Affordable and clean energy · Renewable energy

M. F. Mohamad Sobri · M. Mohd Zaini Makhtar (✉)
School of Industrial Technology, Bioprocess Technology Division,
Universiti Sains Malaysia, 11800 Gelugor, Penang, Malaysia
e-mail: muazzaini@usm.my

M. Mohd Zaini Makhtar
Centre for Innovation and Consultation, Universiti Sains Malaysia,
11800 Gelugor, Penang, Malaysia

M. Mohd Zaini Makhtar et al. (eds.), *Microbial Fuel Cell (MFC) Applications for Sludge Valorization*, Green Energy and Technology,
https://doi.org/10.1007/978-981-99-1083-0_5

1 Introduction

Sewage sludge may be attributed to solid, semi-solid or liquid residue generated as the bulk of the residual material removed during the wastewater treatment process from society-based liquid wastes following aerobic biological treatment, with the potential to contain unfavourable trace pollutant contaminants. Periodic removal has been imperative in preventing excessive biomass concentration in the system or possible pass through rivers and other surface waters. As such, developed countries have outlined regulations for the safety of the public and the environment via each of the following methods [1].

1. Application to the land as soil conditioner, or fertilizer in agricultural use
2. Disposal to the sea
3. Disposal on land at a surface disposal site
4. Placing on municipal solid waste landfill unit
5. Application as 'bio-soil', production for sale in marketplace, composting, land reclamation, etc.
6. Incineration
7. Sludge to energy.

Two main strategies for sewage sludge management involve disposal or reuse, such as for agricultural and for landscaping purposes [2]. However, even with reuse options available, there remain restrictions in place prior to application. With respect to current legislation, characterization, ecotoxicology and waste management routes, treatment and disposal remain the most popular approach for application. To assist in decision making, tools such as 'end-of-waste criteria' and 'Life Cycle Assessment' has been invaluable in proper assessment of the probable environmental, technical and economical evaluation between different systems. Sewage sludge treatment at wastewater treatment plant (WWTP) is seen in Fig. 1 [3].

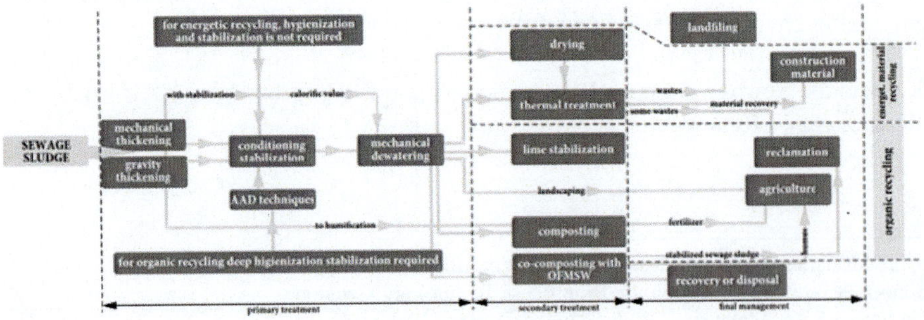

Fig. 1 Sewage sludge treatment processes at wastewater treatment plant. Reprinted with permission from [3] (CCBY)

2 Abundance of Sewage Sludge

Sewage sludge usually represents 1–2% of the treated wastewater [3]. Despite these low proportions, current production and subsequent abundance of sewage sludge cannot be understated. In excess of 10 million tonnes of dry solids (DS) of sewage sludge were produced in 26 EU Member States as of 2008 [4], while as per UN-Habitat's statistics, the existing wastewater treatment plants (WWTPs) in the United States, for example, produce over 6.5 million tonnes of dry solids annually; while it is estimated to be around 2.0 and 3.0 Mt per annum produced in Japan and, respectively. Such figures are figures are naturally anticipated to further increase as applications of WWTP in developing countries continue to grow [5]. Table 1 exemplifies the production and disposal methods of selected countries as of 2012.

Table 1 Sewage sludge production and disposal within selected countries as of 2012.

Country	Produced sewage sludge	Total disposal	Agricultural use	Compost and other applications	Landfill	Dumping at sea	Incineration
Austria	266	266	40	74	14	0	139
Belgium	157	107	19	n.d	n.d	0	89
Czech Republic	263	263	72	154	13	n.d	8
Denmark	141	115	74	n.d	1	0	34
Estonia	16	16	14	n.d	2	0	n.d
Finland	141	141	7	93	10	0	32
France	987	932	684	n.d	40	0	207
Germany	1849	1844	542	294	0	0	1009
Greece	119	119	14	0	40	0	39
Ireland	72	72	68	4	0	0	0
Israel	118,350	n.d	0	69,311	3928	45,111	0
Luxembourg	8	5	4	n.d	0	0	1
Netherlands	346	325	0	0	0	0	321
Poland	533	533	115	33	47	0	57
Portugal	339	113	102	n.d	11	0	0
Slovenia	26	26	0	2	1	0	13
Spain	2757	2577	1922	n.d	384	0	100
Sweden	207	196	48	67	7	0	1
United Kingdom	1137	1078	844	n.d	5	0	229

Source Rorat et al. [6]

3 Characteristics of Sewage Sludge

In general, sewage sludge is a heterogenous mix of microorganisms, undigested organics such as paper, plant residues, oils, faecal material, inorganic materials and moisture [6]. Depending on stabilization processes, dewatered sewage sludge (dry) may contain an average of 50–70% organic matter, 30–50% mineral components (including 1–4% incorganic carbon), 3.4–4.0% N, 0.5–2.5% P and significant amount of additional nutrients [7–9]. Organic matter present within sewage sludge mineralizes quickly owing to the small content of lignin or cellulose, with following rapid degradation potentially generating a peak in the nitrate and pollutant concentration in soil.

Presence of large concentrations of nitrogen and phosphorus makes it suitable for application as fertilizer for plants [7]. However, presence of numerous contaminants, both inorganic (including heavy metals) and organic (such as polycyclic aromatic hydrocarbons (PAHs), polychlori-nated biphenyls (PCBs), adsorbable organohalogens (AOX)), surfactants, pesticides, hormones, pharmaceuticals, nanoparticles and several others [10]. In addition to this, several pathogenic species of living organisms such as bacteria, viruses and protozoa as well as other parasitic organisms present a health hazard for humans, animals and plants [11, 12]. With appropriate sludge treatment (such as lime treatment), the number of pathogenic and parasitic organisms within sludge may be sufficiently reduced prior to application into land, thereby mitigating the health risks present [13, 14]. Different types of sewage sludge and their inherent characteristics are seen in Table 2.

4 Case for Energy Recovery

The finiteness of fossil-based energy resources continues to be a concern as globalization and localization of technologies and manufacturing capacities are coupled with a high speed of energy consumption growth [15]. Globally, rising energy consumption has been noticeable at an increasingly rapid rate, within a 10-year period, from 8,588.9 million tonnes oil equivalent (Mtoe) in 1995 to 13,147.3 Mtoe in 2015, with fossil fuels providing approximately 86.0% of the global total energy needs [16]. In BP Statistical Review of World Energy (2020), China was highlighted as the dominant country pushing global energy markets with renewables recording the biggest percentage growth yet in primary energy, with levels of power generation exceeding nuclear energy for the first time. On a more localized perspective, wastewater treatment technologies suffer from drawbacks such as energy demand, large amount of residual generation and low effectiveness in catching energy potential from wastewater [17]. Usage of inexpensive, renewable energy from sludge is expected to become more essential as the cost of sewage treatment continues to rise [15].

Evaluation of sludge as a source for energy recovery involves consideration of its respective compositions, of which the energy content of sludge is inherent in the

Table 2 Characteristics of municipal sewage sludge Kacprzak et al. [3]

Parameter	Type of sludge		
	Untreated primary sludge	Digested primary sludge	Secondary sludge
Total dry solids [TS] (%)	2.0–8.0	6.0–12.0	0.8–1.2
Volatile solids (% of TS)	60–80	30–60	59–88
Grease and fats (% of TS)	7–35	n/a	5–12
Protein (% TS)	20–30	15–20	32–41
Cellulose (% of TS)	8.0–15.0	8.0–15.0	7–9.7
Phosphorus (% of TS)	0.8–2.8	1.5–4.0	2.8–11.0
Nitrogen (% of TS)	1.5–4	1.6–6.0	2.4–5.0
Potassium (% of TS)	0–1	0–3.0	0.5–0.7
pH	5.0–8.0	6.5–7.5	6.5–8.0

volatile solids, which is subdivided into two components: readily degradable (50% in primary sludge and 25% in WAS) and not readily degradable (30% in primary sludge and 55% in WAS) [1]. Analysis has revealed the energy contained within wastewater sludge is substantial, reaching as much as 3–10 times higher than the energy required for wastewater treatment itself [18]. As per the sludge characteristics mentioned prior, considerations must be taken when deciding if sewage sludge is applicable to be salvaged for energy recovery, along with the approach to be considered.

As a conventional approach, sewage sludge is often incinerated with the exhaust gases and ashes treated, albeit its complexity and costs make it applicable only in large plants. Another established approach involves biogas production through anaerobic digestion of sewage sludge. Here, organic matter in sludge is converted by bacteria into a mixture of methane (60–60%), carbon dioxide (35–40%) and trace gases, with the resultant biogas directly applied in combined heat and power (CHP) systems for utilization in sewage treatment plants. This approach also reduces the solid content of sludge by up to 30%, further reducing the associated energy costs involved in transportation [15]. Low sludge yield may also prove to be advantageous given its accounts for approximately 25–65% of total plant operating costs [19].

5 Necessity for Bioremediation

With the potential untapped resources present in sewage sludge, many countries have recognized the value of sludge by-product for substrate in fertilization of agricultural lands as well as remediation of polluted areas. However, the potential harms must also be recognized as sewage sludge applications on agricultural land might lead to the dispersal of numerous unwanted constituents on soils possibly used for food production. Such undesirable contaminants (potentially toxic elements (PTE) such as metals, trace organic compounds (TrOC), and pathogenic organisms) may pose sanitary and environmental risks. Processes to reduce pathogens involve composting, autothermal thermophilic aerobic digestion, alkaline stabilization, pasteurization, incineration, thermal drying and wet oxidation while approaches to remove heavy metal presence include activated sludge, aerated lagoon and facultative pond to mention a few, each with differing importance in terms of implementation and operation [20].

6 Microbial Fuel Cells

As the search for renewable energy gains pace, microbial fuel cells (MFC) have emerged as a rapidly expanding field of science and technology which associate biological catalytic redox activity with abiotic electrochemical reactions and physics. By definition, MFCs form a branch of biological fuel cells, where electroactive microbes are utilized within devices, for the degradation of organics and subsequent conversion from chemical to electrical energy, by means of electrochemical reactions [21]. MFCs have been recognized for its versatility in providing direct power, as well as feedstock treatment, nutrient recovery and sensing for continuous parallel observation of processed substrates [22].

7 Mechanism of MFC

At its basic form, MFCs are made up of an anodic and a cathodic compartment separated by the presence of an ion exchange membrane (IEM), which functions in preventing the migration of electrolytes between chambers [23]. Separation of environmental conditions allows for maximization of potential differences between both anodic and cathodic electrodes, thereby optimizing power generation [24]. Organic matter is supplemented to the anode compartment to act as fuel once oxidized as substrate by anaerobic microorganisms, liberating electrons (e^-) and protons (H^+) [25]. In the cathode compartment, electrons are released to the cathode electrode surface and combined with protons and oxygen via reduction to form water [26]. Figure 2 outlines the basic configuration of an MFC.

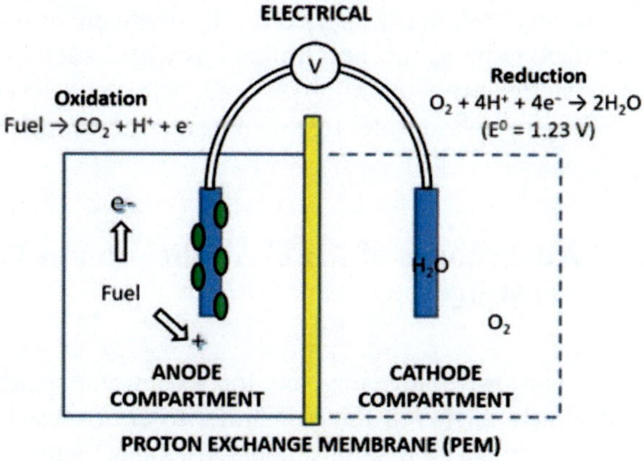

Fig. 2 Basic configuration of microbial fuel cells. Reprinted with permission [27] (CCBY)

8 Advantages of MFC

As an alternative power generator, microbial fuel cells are advantageous in comparison to conventional fuel cells by having a breadth of materials, substrates and system architectures applicable to bacteria for bioenergy production, albeit at relatively low power levels [28]. MFCs also have the potential as sustainable long-term power applications though potential issues in health and safety following bacterial usage must also be addressed prior [29].

9 MFC in Wastewater Treatment

By applying microorganisms as catalysts for energy recovery in the form of electricity, MFC has been identified as a promising anaerobic waste treatment device from a range of organic wastes such as domestic wastewater [30], industrial wastewater [31] and excess sludge [32].

10 Integration of MFC into Wastewater Treatment

Integration of MFC reactors into existing wastewater treatment plant trains can be done as a replacement to the existing biological treatment unit, such as an activated sludge reactor, for accomplishing carbon oxygen demand (COD) removal, electricity production and reduction in sludge production relative to those of an aerobic process such as activated sludge. However, a secondary process would be required to further remove COD to levels acceptable for discharge, as the current generation is minimal once the COD is less than $\sim 100 - 150$ mg/L. [33]. While the activated sludge process is currently commonly applied for the treatment of domestic and industrial wastewater, there remains inherent limitations within such as high energy consumption and waste sludge production [34]. An adaptable and sustainable technology capable of simultaneous wastewater treatment and resource recovery is therefore favoured [35].

11 Advantages of MFC Application in Wastewater Treatment

MFCs can prove advantageous for wastewater treatment by (1) direct conversion of chemical energy within substrates to electricity, (2) relatively lower amount of activated sludge production compared to other methods, (3) environmentally friendly, (4) additional gas treatment not required and (5) aeration made unnecessary. MFC application also allows for use of various types of wastewaters [18]. Direct conversion into electricity by MFC bypasses the need for separation, purification and conversion of the energy products which along with the mild operating conditions suggest as to a relatively more environmentally friendly approach [36].

Simple substrates such as sucrose, glucose and protein to undefined and complex substrates such as wastewater from domestic and municipal, brewery, dairy, pharmaceutical, food processing, agro-processing livestock, petroleum and paper recycling industry may be utilized with MFC technology. Operation of MFC differs from conventional bioremediation as organic compounds are converted to H_2O and CO_2 under aerobic conditions [37]. As the example with high strength wastewater, with the integration of MFC and bioremediation, synergy between both processes may result in enhanced energy and resource recovery with higher pollutant removal [38]. In certain cases, sludge disposal rate and treatment efficiency attained with MFC technology is comparable and even surpassed conventional technologies [39].

12 Proposed Reactions from Breakdown of Chemical Composition of Sludge

Utilization of sewage sludge as inoculum in MFCs introduces a mixed culture of microorganisms for substrate consumption and bioelectricity generation. As bioelectricity is generated with a transfer of electrons, relating the phenomenon towards the understanding of metabolism of microorganisms (in aerobic and anaerobic conditions) is considered essential.

Aerobic bacteria prevent the transfer of electrons from organic compounds to electron acceptors such as O_2, while on the other hand, anaerobic bacteria in the absence of O_2 utilize alternative electron acceptors such as nitrate and solid electrodes. Reduction of electrons occurs on the extracellular wall where the cytochromes reduce extracellular electrons from the substrate, upon which energy in the form of ATP can be produced for bacterial growth and reproduction.

In the respiration pathway of microorganisms, the metabolism involved utilizes nicotinamide adenine dinucleotide (NADH) as the energy source. By relating the biological standard potential ($E^{OJ'}[V]$) to NADH and with O_2 as an electron acceptor, a theoretical potential difference may be constructed, as seen in Eqs. 1, 2 and 3 [40].

$$\mathrm{NAD^+ + H^+ + 2e^- \rightarrow NADH}\ E^{OJ'} = -0.320\mathrm{V} \tag{1}$$

$$\mathrm{O_2 + 2H_2O + 4e^- \rightarrow 4OH^-}\ E^{OJ'} = +0.820\mathrm{V} \tag{2}$$

$$+0.820\mathrm{V} - (-0.320\mathrm{V}) = 1.14\ \mathrm{V} \tag{3}$$

In the absence of electron acceptor, electrons from the microbes are transferred directedly to the surface of the anode, which may prove complicated as several surfaces are incapable to do so.

Electricity is generated with the use of substrates as fuel for microbes in the anode chamber, upon which electrons released are passed to the cathode to reduce the electron acceptors (O_2). Protons from the anode chamber are then transferred through a membrane to the anolyte for reduction. Hence, the substrate of choice must be considered in determining the electricity output from the MFC. An example is as given by Das and Mangwani [40] on the proposed consumption of acetate and sucrose where present, as seen in Eqs. 4, 5 and 6, 7, respectively.

$$\text{Anode: } \mathrm{CH_3COO^- + H_2O \rightarrow 2CO_2 + 2H^+ + 8e^-} \tag{4}$$

$$\text{Cathode: } \mathrm{O_2 + 4e^- + 4H^+ \rightarrow 2H_2O} \tag{5}$$

$$\text{Anode: } \mathrm{C_{12}H_{22}O_{11} + 13H_2O \rightarrow 12CO_2 + 48H^+ + 48e^-} \tag{6}$$

$$\text{Cathode: } \mathrm{O_2 + 4e^- + 4H^+ \rightarrow 2H_2O} \tag{7}$$

13 Application of MFC for Bioremediation and Electricity Generation

Substrates are significant in any biological processes as carbon and energy source. Key towards determining the efficiency and economic viability of converting organic wastes as fuel to bioenergy are the characteristics, chemical composition and concentrations of the material to be used [41]. In the process of wastewater treatment and subsequent sludge stabilization, a significant amount of energy and high cost is required. However, this pales in comparison to the energy content within wastewater which is approximately 9 times higher than the energy applied for treatment [42]. Given the abundance of energy contained within sludge following treatment, it has been argued that MFCs are essential in the extraction of energy in organics for possible reduction of energy demand and expenses in a wastewater treatment plant [43]. Table 3 outlines several selected applications of sewage sludge as substrate in MFCs for bioremediation and electricity generation.

14 Earliest Sludge Treatment Using MFC Technology

Application of sludge in MFC began with work by Dentel et al. [44], which represents the earliest attempt at the conversion of organic matters in digested sludge to electricity under ambient temperature, atmospheric pressure and neutral pH upon which a maximum electrical current of approximately 0.065 mA and a maximum voltage of 0.517 V, while concluding that current is restricted to degree of degradation of organic matter. 16s rDNA analysis identified the presence of *Geobacter* as a natural constituent of microbial population in the anaerobic sludge utilized albeit not directly involved in electricity generation, by the lack of biofilm formed. Instead, it was suggested that more than one or more redox mediator may be involved in electron cycling between the anode and the solution, carried out by strains such as *Aeromonas*, *Geovibrio*, *Clostridium* and *Desulfotomacum reducens*, albeit none were positively identified in this study.

14.1 MFC Configuration

Another study by Hu [43] introduced the application of a baffle for mixing, in a membraneless, single chamber MFC with anaerobic sludge as fuel. Such a design was constructed to reduce mixing in the vicinity of the cathode and attain thick biofilm formation (>1 mm) on the cathode upon the addition of anaerobic/biomass sludge, thereby contribute to maintaining anaerobic conditions inside the reactor by minimizing oxygen diffusion through the cathode. Initial recorded electricity generated was low at 0.3 mW/m^2 under endogenous decay conditions, which suggests anaerobic sludge to be difficult for conversion into electricity. Maximum power of 161 mW/m^2 instead was only attained and attributed to glucose degradation as fuel. The baffle-chamber membraneless MFC was concluded to be effective in restricting fluid mixing within the anode chamber, with inoculation of the anaerobic sludge at a final concentration of approximately 4000 mg $CODL^{-1}$ promoted thick (>1 mm, via COD analysis) biofilm formation on the cathode. Presence of strict anaerobic conditions allows for fuel (glucose) retention that would have been consumed aerobically for prolong electricity generation.

Table 3 Selected applications of sewage sludge as substrates in microbial fuel cells for bioremediation and electricity generation

Sludge	MFC configuration	Bioremediation	Electricity generation	References
Anaerobically digested sludge	Membraneless, single chamber, with graphite foil electrodes and aerated cathode	n.a	Max. current = 0.065 mA Max. voltage = 0.517	Dentel et al. [44]
Anaerobic sludge	Membraneless, baffle-chamber, carbon anode with gas diffused Pt cathode	n.a	Anaerobic sludge: 0.3 mW/m^2 Glucose: 161 mW/m^2	Hu [43]
Desilter based sludge	Membraneless, single chamber, graphite electrodes with floating-cathode configuration	n.a	Maximum power density: 220.7 mWm^{-2}	Zhihua Liu et al. [45]
Raw sludge from second clarifier	Dual chamber, Nafion membrane, graphite fibre brush as electrodes	Maximum TCOD removal: 46.4% from initial TCOD of 10,850 mg/L	Maximum power density: 8.5 W/m^3	Jiang et al. [31]
Raw sewage sludge	GORE-TEX cloth and conductive paint catalyst as membrane, insert-type, carbon felt anode	Maximum TCOD removal: 13,167 – 6280 mgL^{-1} at pH 10.0, representing 53% removal efficiency	Maximum power density: 73 ± 5 mWm^{-2} at pH 10.0	Yuan et al. [46]
Digested sewage sludge	Dual chamber, carbon felt electrode with Nafion membrane on cathode	n.a	Maximum power: 3.1 μW	Fischer et al. [47]
Primary and digested sludge	Cation exchange membrane, tubular, carbon cloth and brush as electrodes	Primary sludge removal: 69.8 ± 24.1% TCOD removal & 68.4 ± 17.9% of volatile suspended solids (VSS) Digested sludge removal: 36.2 ± 24.4% of TCOD and 46.1 ± 19.2% of VSS	Primary sludge: 1.43 kWh/m^3 Digested sludge: 1.8 kWh/m^3	Ge et al. [48]

(continued)

Table 3 (continued)

Sludge	MFC configuration	Bioremediation	Electricity generation	References
Sludge from secondary sedimentation tank	Cloth electrode assembly, carbon anode with Pt-PTFE coated carbon cathode,	n.a	Maximum power density: 1200 mWm^{-2}	Abourached et al. [41]
Activated sludge	Cell A —non-coated carbon electrodes Cell B—Fe_2O_3 coated carbon anode with Pt-coated carbon cathode Cell C—Fe_2O_3 coated anode and cathode	Cell A—TCOD removal: 39% Cell B—TCOD removal: 60% Cell C—TCOD removal: 51%	Cell A— Power: 6.72 $mW.m^{-2}$ Cell B—Power: 73.16 $mW.m^{-2}$ Cell C—Power: 30.18 $mW.m^{-2}$	Nandy et al. [49]

n.a—not available

14.2 Sludge as Substrate and Inoculum

Work by Zhihua Liu et al. [45] focused on the application of surplus sludge as both inoculum and substrate in a single chamber floating-cathode MFC for electricity production. Maximum voltage of 440.7 mV was obtained with external resistance of 1000 Ω [44]. A periodic time of MFC was divided into 4 phases as seen in Fig. 3, beginning with a rapid fall phase due to the mass decomposition of biodegradation matter such as volatile fatty acid.

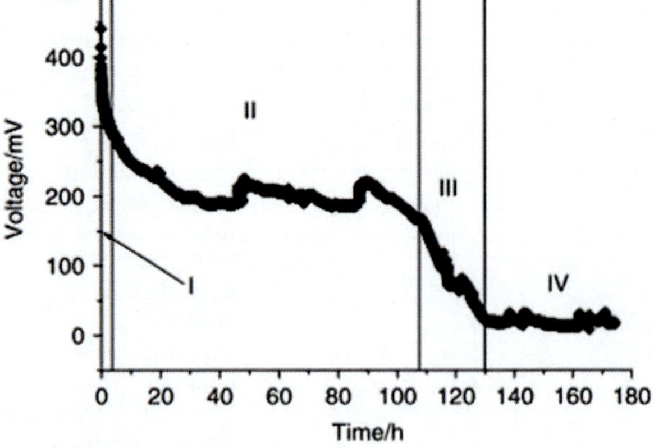

Fig. 3 Electricity generation over time in steady state from surplus sludge-based MFC. Phase I: rapid fall phase; Phase II: stationary phase; Phase III: fall phase; Phase IV: stationary phase of low voltage. Reprinted with permission [44] (CCBY)

Following this, SCOD concentration of substrate stabilizes during the stationary phase and upon reaching stationary phase of low voltage, the SCOD concentration was found to be high still, a finding contrary to that argued by Dentel et al. [44]. Across the four phases, output voltage range of 150–300 mV persisted for 107 h in stationary phase. Recorded maximum power density was 220.7 mWm^{-2} and internal resistance of 368.13 Ω in the MFC.

In another application of sewage sludge as both inoculum and substrate, a study conducted by Jiang et al. [31] constructed a two-chambered MFC with potassium ferricyanide as electron acceptor. Here, stable electrical power production during operation for 250 h was demonstrated along with TCOD removal by 46.4% from the initial TCOD of 10,850 mg/L. Several parameters such as substrate concentration, cathode catholyte concentration and anodic pH were also tested to determine their effects on bioremediation and electrical output upon which although the aforementioned parameters did not affect significantly, power production was found to be closely related to soluble chemical oxygen demand (SCOD) of sludge. Recognizing immobilization of organic matter within sewage sludge constitutes organic matter hydrolysis as the rate-limiting step [49], further ultrasonic pre-treatment of sludge applied also managed to increase TCOD removal rate albeit without additional significant enhancement on power output.

Another study by Nandy et al. [49] utilized activated sludge as inoculum and monitored for remediation with the application of Fe_2O_3 coating on carbon electrodes as an alternative. Cell C labelling MFC with coated anode and cathode (30.18 $mW.m^{-2}$) exhibited 78% higher power output in comparison to those in Cell A with non-coated carbon electrodes (6.72 $mW.m^{-2}$), albeit still lower than that of Cell B, containing Fe_2O_3 coated carbon anode with Pt coated carbon cathode (73.16 $mW.m^{-2}$). Microbial community analysis on both the biofilm formed and liquid electrolyte for all cells highlighted the presence of bacterial *Peptostreptococcaceae* as dominant in Cell A, but shifted through Cells B and C to reduced *Firmicutes* population, and enhanced *Gammaproteobacteria* and methanogens. COD removal efficiency was recorded at about 60% for cell B, 51% for cell C and 39% for cell A.

14.3 Parameter Adjustment (pH)

Work carried out by Yuan et al. [46] elucidated as to the effects of pH on electricity production from sludge-based MFC. Here, an insert-type air-cathode microbial fuel cell was constructed as pH adjustment to alkaline pH was conducted to enhance sludge substrate bioavailability and suppression of methane production. Maximum power density of 73 ± 5 mWm^{-2} attained from sludge at pH 10, which was higher than that of 33 ± 3 mWm^{-2} at pH 8 and 4 ± 0.5 mWm^{-2} at pH 6. Soluble chemical oxygen demand (SCOD) which was attributed to protein and carbohydrate content meanwhile increased with increasing pH as the alkaline condition proved favourable for enhancing the substrate bioavailability of sludge by assisting in the dissolution of insoluble organics into a soluble form. TCOD reduction from 13,167 to

6280 mg L^{-1} at pH 10.0, representing a 53% TCOD removal efficiency, was also higher than those at pH 8.0 and 6.0 (42% and 20% TCOD removal, respectively). Finally, increased coulombic efficiency from increased pH was attributed to complete methane production suppression under such conditions, thereby suggesting alkaline pH to be a successful approach for the enhancement of electricity from MFC.

14.4 Nutrient Recovery

Another example of MFC application involves the recovery of value-added elements from sewage sludge, using MFC. Work by Fischer et al. [47] focused on the recovery of orthophosphate, of which phosphor, an essential element sought globally for agricultural and industrial purposes may be extracted. Here, MFC has proven advantageous for application in ambient temperatures while serving as a power source delivering not only electrons, but also the needed protons to reduce electrochemically insoluble $FePO_4$. Liberated electrons are able to reduce the iron cations and charges are then replaced by protons, which results in orthophosphate (H_3PO_4, $H_2PO_4^-$, HPO_4^{2-}, and PO_4^{3-}) motility into the aqueous supernatant solution.

15 Long-Term Application

As a substrate to MFC, another study by Ge et al. [48] sought to investigate the long-term performance of sludge treatment by application of two MFCs, each with primary sludge and digested sludge respectively, over a period approaching 500 days. In preliminary Phase I, MFC fed with primary sludge managed to remove 69.8 ± 24.1% of total chemical oxygen demand (TCOD) and 68.4 ± 17.9% of volatile suspended solids (VSS); while the MFC with digested sludge achieved 36.2 ± 24.4% of TCOD and 46.1 ± 19.2% of VSS reduction, respectively. Further application in Phase II as a two-stage system recorded TCOD removal by 60% and VSS removal by 70% from the primary sludge. Recorded energy obtained from primary sludge MFC and digested sludge MFC were up to 1.43 kWh/m^3 from a primary sludge or 1.8 kWh/m^3, respectively. As energy analysis revealed electrical energy generated from MFC was minimal in proportion of total energy compared to energy released as methane gas, MFCs were suggested as a polishing step to effluent from anaerobic digesters, as opposed to energy recovery from primary sludge.

16 Pretreatment Method Prior MFC Usage

Application of pre-treatment methods on sludge substrate was also conducted with the aim of enhancing dissolved organic concentrations [41, 50]. Work by Jiang et al. [31] mentioned prior applied sonication of sewage sludge [31] and in work by Abourached et al. [41], fermentation of sewage sludge was applied as pre-treatment within a novel cloth electrode assembly MFC. Maximum power density of 1200 mW/m^2 was achieved after a fermentation time of 96 h. At the time, the recorded power density was an enhancement by 275% over MFCs reported prior. Even then, given known energy efficiency of MFCs treating sludge is only at 11.5%, MFCs represent an exciting modification to existing wastewater treatment infrastructure. Table 3 shows the application of MFC for bioremediation and electricity generation.

17 Sludge as Inoculum

Applications of sludge in MFC studies involve several perspectives as seen in Table 4. A popular approach is to utilize sludge as mixed culture inoculum within the anode compartment. MFCs inoculated with mixed cultures have been argued to produce significantly greater power densities than those of pure cultures [51]. Within anaerobic sludge applied as mixed culture inoculum, the simultaneous presence of electrophiles/anodophiles and groups utilizing natural mediators within the same chamber allows for wider substrate utilization [29]. Table 4 details several strains of note identified present within inoculum sludge. As an undefined inoculum, recycled activated sludge point of an aeration tank from a local municipal wastewater treatment plant was used by Liang et al. [52] in efforts to elucidate the composition and distribution of internal resistance from three separate MFC configurations. Another study by Zhidan Liu et al. [53] used mesophilic anaerobic sludge to focus on identifying the effect of operational performance and electrical response of a mediator-less MFC fed with either acetate as carbon-rich substrate or protein-rich synthetic wastewater in fed batch mode.

18 Bioremediation of Heavy Metals

For this application, it is worth noting that metal ions present in wastewater do not biodegrade into harmless end products, thereby necessitating special approaches for treatment. With several metal ions having high redox potentials, it may therefore be advantageous for reduction and precipitation in MFC [56]. If implemented properly, MFCs can therefore serve not only as remediation of wastewater but along with concomitant metal recovery too [57].

To this aim lies the works of Li et al. [58] which utilized sewage sludge inoculum as a means of disposing Cr^{6+} from real electroplating wastewater with simultaneous electricity generation, using a dual chamber MFC. Chromium removal and power density were optimum at pH 2 and application of graphite paper as cathode electrode. Maximum power density of 1600 mW/m^2 was generated at a columbic efficiency of 12% at initial Cr^{6+} concentration of 204 ppm, with a further rise in Cr^{6+} capable of

Table 4 Bacterial community composition identified from inoculum sludge

Strain	Note	Reference
Geobacter	Natural constituent of microbial population in anaerobic sludge	[29]
Pseudomonas aeruginosa	Known electrogen isolated from microbial population in anaerobic sludge. Isolated using BIOLOG gene III analysis	[54]
Actinobacillus capsulatus	Natural constituent of microbial population in anaerobic sludge. Isolated using BIOLOG gene III analysis	
Acetobacter peroxydans		
Pseudomonas mendocina		
Acinetobacter schindleri		
Escherichia coli		
Clostridium	Electrogenic, Gram-positive bacteria, proven activity without exogenous mediator necessary	[55]
Petrimonas	Fermentative bacteria, having probably symbiotic relationship with electrogenic bacteria for enhancement of MFC performance	
Rhodocyclaceae	Denitrifying bacteria, capable of bearing high dissolved oxygen concentrations in anode chamber	
Castellaniella	Capable of alleviating high levels of nitrite accumulation	
Desulfovibrio	Hydrogen consumer, capable of sulfate reduction	
Thiobacillus	Capable of sulfide oxidation	
Delftia	Related to the genus *Pseudomonas*, with possible iron reduction capabilities	[27]
Prolixibacteraceae	Reported iron (Fe^0) related redox metabolism	

enhancing power density. Removal efficiency was at 99.5% Cr^{6+} and 66.2% total Cr through the reduction of $Cr_2O_7^{2-}$ to Cr_2O_3 precipitating on the surface of cathode electrode.

Another example involves the removal of silver ions from wastewaters, as demonstrated by Choi and Cui [59]. Using sludge as inoculum, following 8 h reaction, efficiency of silver metal recovery was recorded to be as high as 99.91–98.26% across initial concentration ranges of 50 ppm to 200 ppm. Energy output was calculated to be at a rate of 69.9 kg of silver per KWh of energy output, with maximum power attained of approximately 4.25 W/m^2, maximum voltage of 0.749 V, maximum current density of 5.67 A/m^2 and a fill factor of 0.626 was achieved at 1000 ppm initial silver ion concentration.

In another study, Cu^{2+} was targeted for recovery with excess sludge as the anodic substrate within a MFC and $CuSO_4$ solution as the catholyte. Here, Cu^{2+} was proven to be a viable cathodic electron acceptor, with a stable voltage output of 0.478 V and a maximum power density of 536 mW/m^3 obtained at external resistance of 1000 Ω and Cu^{2+} concentration of 6400 mg/L. For Cu^{2+} removal from wastewater, 97.8% removal efficiency was achieved following 288 h of operation with no external resistance and initial Cu^{2+} concentration of 1000 mg/L. Final products attained were dependent on cathodic reducibility, with Cu^{2+} primarily deposited as Cu_20 and a small part as $Cu_4(OH)_6SO_4$. Excess sludge in particular was deemed crucial for supporting long-term operations, with the acclimation stage of exoelectrogenic bacteria on the anode key in determining MFC performance and cathodic reduction of Cu^{2+}.

19 Improvement Parameter for Enhancing MFC Performance

As an inoculum, numerous efforts have since been conducted for the improvement of sludge inoculum to this purpose.

19.1 Combination of Inoculums

Sun et al. [60] investigated the effect of different combinations of sludge inoculums consisting of aerobic sludge (AES), anaerobic sludge (ANS) and wetland sediment (WLS) towards overall MFC performance in wastewater treatment. To this, it was found that the application of multiple sludge inoculums was advantageous whereby MFC inoculated with ANS + WLS produced maximal power density of 373 mW/m^2 with a low internal resistance of 38 Ω. A similar observation was made with combinations of ANS + AES and ANS + AES + WLS in comparison to single sludge inoculation, with the former exhibiting the highest Coulombic efficiency. COD removal of more than 92% was recorded irrespective of the membranes and inoculums applied.

19.2 Comparison of Inoculums

In another study by Baranitharan et al. [61] a developed control inoculum (CI) was compared to that of anaerobic sludge as inoculum for use in MFC with palm oil mill effluent (POME). CI consisted of predominant microorganisms such as *Pseudomonas aeruginosa, Azospira oryzae, Acetobacter peroxydans* and *Solimonas variicoloris* isolated from palm oil anaerobic sludge in combination with a biofilm of MFC anode operated prior with anaerobic sludge. Maximum power density of MFC utilizing CI was found to be twice higher and with enhanced maximum Coulombic efficiency, albeit noticeable lower COD removal of about 32%, possibly due to the absence of necessary fermentative microorganisms within the CI for POME utilization.

19.3 Substrate Pre-Treatment

Xie et al. [62] have argued for pre-treatment of parent inoculum in anode chamber of MFC as a means of selective enrichment of a specific group of bacteria. Understanding the composition of bacterial associations to be dependent on substrate (wastewater) composition and of the symbiotic relationships within a given population, application of low-frequency ultrasonication for a short duration was proposed as stimulant, by promoting enzyme activity, cell growth and cell membrane permeability. Under stressful conditions such as high temperatures, extreme acidity and alkalinity, methanogenic bacteria without the capability of forming protective spores and with lower growth rate than hydrogen producing and electrogenic bacteria are at a disadvantage, thereby less likely to survive and thus leave the latter as dominant species in mixed culture of anaerobic sludge [63, 64]. In another study conducted by More and Ghangrekar [64], a low-frequency ultrasound pre-treatment was applied to anaerobic sludge inoculum to be used in MFC based on synthetic wastewater. Maximum power density was subsequently achieved following ultrasonication at 40 kHz, 120 W for 5 min, which was 2.5 times higher than that of untreated inoculum sludge along with enhanced COD removal by 14%.

19.4 Sludge Concentrations

Another study conducted by Khan et al. [65] applied sludge of different concentrations up to 200 ml/L to study the resultant effects on current generation and COD removal. With constant 100 Ω resistance applied, highest current generated of 314 μA was from MFC batch utilizing 200 ml/L of drainage sludge after 35 h with concomitant highest degree of COD removal of up to 78%. Under these conditions, the maximum power density was 15.12 mW/m^2 with current density at 97.34 mA/m^2. This was further tested by Amin et al. [66] as electricity generation by a membraneless MFC operating in continuous mode was modified in several parameters, including sludge concentration. Sludge feed rate of 150 mL/L showed best current generation, with a maximum value of 250 μA after 24 h of operation, albeit with less stability than that of lower concentrations. The results thereby conclude that for a given fixed amount of substrate and salt, current generation can be expected to increase with the increase in sludge concentration, attributed to higher amounts of electricity generating microorganism.

19.5 Effect of Nitrate and Sulphate

In replicating concentrations of elements commonly found in industrial wastewater for application with MFCs, Seo et al. [35] sought to study the effect of adding nitrate and sulphate independently towards the bacterial community and performance of a membraneless, single chamber microbial fuel cell. With anaerobic sludge from sewage treatment plant used as inoculum, bacteria of the genus *Clostridium*, *Petrimonas* and *Rhodocyclaceae* were found in abundance prior to addition.

Nitrate addition resulted in consistent *Clostridium* proportions, albeit with lowered *Petrimonas* (20 → 6%) and concomitant increase by *Rhodocyclaceae* (6.2 → 17.1%) and *Castellaniella* (0.4 → 4.1%) on the anode surface. On the cathode surface, nitrate addition resulted in *Rhodocyclaceae* (25.5%) being most abundant, followed by *Clostridium* (10.4%), *Castellaniella* (7.3%) and *Petrimonas* (1.5%). Known denitrifying ability of *Rhodocyclaceae* suggests for both anodic and cathodic denitrification function in MFC with its relative larger presence on the cathode biofilm as opposed to the anode and other competing denitrifying bacteria to be most suitable. *Castellaniella* instead assists in autotrophic denitrification of MFC by alleviating build-up of intermediates such as nitrite.

Addition of sulphate instead resulted in increase of *Clostridium* (23.1 → 43.2%) and decrease of *Petrimonas* (20.0 → 5.5%) proportions on the anode surface. A different proportion was noticeable on the cathode as *Desulfovibrio* (32.9%) increased significantly to abundance, followed by *Clostridium* (12.9%), *Petrimonas* (5.0%) and *Thiobacillus* (2.4%). *Desulfovibrio* and *Thiobacillus* function simultaneously as sulphate reducer and sulphide oxidizer respectively, which may explain the lower COD removal efficiencies as compared to that achieved following nitrate addition. Overall, nitrate addition resulted in higher electrical performances and higher nitrate removal efficiency (93%) while the opposite occurred following sulphate addition with concomitant lower sulphate removal efficiency (17.6%).

Acknowledgements The authors would like to thank the Universiti Sains Malaysia for the financial support of this study via APEX Era grant (1001/PTEKIND/881004). The authors have declared no conflict of interest for the manuscript.

References

1. Zsirai I (2011) Sewage sludge as renewable energy. J Residuals Sci Technol 8(4):165–179. https://doi.org/10.13140/RG.2.1.1939.0483
2. Grobelak A, Czerwińska K, Murtaś A (2019) General considerations on sludge disposal, industrial and municipal sludge. In: Industrial and municipal sludge: emerging concerns and scope for resource recovery. Butterworth-Heinemann, Oxford, pp 135–153. https://doi.org/10.1016/B978-0-12-815907-1.00007-6
3. Kacprzak M et al (2017) Sewage sludge disposal strategies for sustainable development. Environ Res 156:39–46. https://doi.org/10.1016/j.envres.2017.03.010
4. Guangyin Z, Youcai Z (2017) Sewage sludge generation and characteristics. In: Pollution control and resource recovery: sewage sludge. Butterworth-Heinemann, Oxford, pp 1–11. https://doi.org/10.1016/B978-0-12-811639-5.00001-2
5. RPA, WRC, and milieu (2008) Environmental, economic and social impacts of the use of sewage sludge on land final report Part II: report on options and impacts, p 158

6. Rorat A, Courtois P, Vandenbulcke F, Lemiere S (2019) Sanitary and environmental aspects of sewage sludge management. In: Industrial and municipal sludge: emerging concerns and scope for resource recovery. Butterworth-Heinemann, Oxford, pp 155–180. https://doi.org/10.1016/B978-0-12-815907-1.00008-8
7. Tyagi VK, Lo SL (2013) Sludge: a waste or renewable source for energy and resources recovery? Renew Sustain Energ Rev 25(71):708–728. https://doi.org/10.1016/j.rser.2013.05.029
8. Fytili D, Zabaniotou A (2008) Utilization of sewage sludge in EU application of old and new methods—a review. Renew Sustain Energy Rev 12(1):116–140
9. Samolada MC, Zabaniotou AA (2014) Comparative assessment of municipal sewage sludge incineration, gasification and pyrolysis for a sustainable sludge-to-energy management in Greece. Waste Manage 34(2):411–420. https://doi.org/10.1016/J.WASMAN.2013.11.003
10. Siebielska I (2014) Comparison of changes in selected polycyclic aromatic hydrocarbons concentrations during the composting and anaerobic digestion processes of municipal waste and sewage sludge mixtures. Water Sci Technol J Int Assoc Water Pollut Res 70(10):1617–1624. https://doi.org/10.2166/WST.2014.417
11. Fijalkowski KL, Kacprzak MJ, Rorat A (2014) Occurrence changes of Escherichia coli (including O157:H7 serotype) in wastewater and sewage sludge by quantitation method of (EMA) real time—PCR. New pub: Balaban 52(19–21):3965–3972. https://doi.org/10.1080/19443994.2014.887499
12. Kacprzak M, Stan´czyk E, Stan´czyk-Mazanek S, (2003) Changes in the structure of fungal communities of soil treated with sewage sludge. Biol Fertil Soils 38:89–95. https://doi.org/10.1007/s00374-003-0633-2
13. Farzadkia M, Bazrafshan E (2014) Lime stabilization of waste activated sludge. Health Scope 3:1–5
14. Olszewski JM, Lozano N, Haines C, Rice CP, Ramirez M, Torrents A (2013) The effect of liming on antibacterial and hormone levels in wastewater biosolids. J Environ Sci Health. Part A, Toxic/Hazard Subst Environ Eng 48(8):862–870. https://doi.org/10.1080/10934529.2013.761488
15. Dong K (2020) How renewable energy consumption lower global CO_2 emissions? evidence from countries with different income levels, vol 2017, pp 1665–1698. https://doi.org/10.1111/twec.12898.
16. BP, "BP Statistical Review of World Energy 2020," 2020
17. NACWA (2010) Renewable energy resources: banking on biosolids, p 23
18. Gude VG (2018) Wastewater treatment in microbial fuel cells—an overview. J Clean Prod 122:287–307. https://doi.org/10.1016/j.jclepro.2016.02.022
19. Katuri KP, Scott K, Head IM, Picioreanu C, Curtis TP (2011) Microbial fuel cells meet with external resistance. Bioresour Technol 102(3):2758–2766. https://doi.org/10.1016/j.biortech.2010.10.147
20. Andreoli CV, von Sperling M, Fernandes F (2007) Sludge treatment and disposal, vol 1, IWA Publishing. https://doi.org/10.2166/9781780402130
21. Santoro C, Arbizzani C, Erable B, Ieropoulos I (2017) Microbial fuel cells: from fundamentals to applications: a review. J Power Sources 356:225–244. https://doi.org/10.1016/j.jpowsour.2017.03.109
22. Gajda I, Greenman J, Ieropoulos IA (2018) Recent advancements in real-world microbial fuel cell applications. Curr Opin Electrochem 11:78–83. https://doi.org/10.1016/j.coelec.2018.09.006
23. Ho NAD, Babel S, Sombatmankhong K (2018) Bio-electrochemical system for recovery of silver coupled with power generation and wastewater treatment from silver (I) diammine complex. J Water Process Eng 23:186–194. https://doi.org/10.1016/j.jwpe.2018.04.001
24. Fornero JJ, Rosenbaum M, Angenent T (2010) Electric power generation from municipal, food, and animal wastewaters using microbial fuel cells. Electroanalysis 22(7–8):832–843. https://doi.org/10.1002/elan.200980011
25. Omeroglu S, Sanin FD (2016) Bioelectricity generation from wastewater sludge using microbial fuel cells: a critical review. Clean Soil Air Water 44(9999):1–9. https://doi.org/10.1002/clen.201500829

26. Ucar D, Zhang Y, Angelidaki I, Franks AE (2007) An overview of electron acceptors in microbial fuel cells. Front Microbiol 8:643. https://doi.org/10.3389/fmicb.2017.00643
27. Logan BE (ed) (2008). Wiley, Hoboken, NJ
28. Du Z, Li H, Gu T (2007) A state of the art review on microbial fuel cells: a promising technology for wastewater treatment and bioenergy. 25:464–482. https://doi.org/10.1016/j.biotechadv.2007.05.004
29. Ahn Y, Zhang F, Logan BE (2014) Air humidity and water pressure effects on the performance of air-cathode microbial fuel cell cathodes. J Power Sour 247:655–659. https://doi.org/10.1016/j.jpowsour.2013.08.084
30. Feng C, Huang L, Yu H, Yi X, Wei C (2015) Simultaneous phenol removal, nitrification and denitrification using microbial fuel cell technology. Water Res 76:160–170. https://doi.org/10.1016/J.WATRES.2015.03.001
31. Jiang J, Zhao Q, Zhang J, Zhang G, Lee D (2009) Electricity generation from bio-treatment of sewage sludge with microbial fuel cell. Bioresour Technol 100(23):5808–5812. https://doi.org/10.1016/j.biortech.2009.06.076
32. Logan BE, Wallack MJ, Kim K, He W, Feng Y, Saikaly PE (2015) Assessment of microbial fuel cell configurations and power densities. Environ Sci Technol Lett 2:206–214. https://doi.org/10.1021/acs.estlett.5b00180
33. Fazal U et al (2021) Microbial fuel cell: study of bioresource potential of dairy effluent and Associated Process Limitations. Research Square, pp 1–17
34. Shamsudin NA, Sabri MNIM, Tajarudin HA, Mokhtar AMA, Makhtar MMZ (2022) The future promising alternative renewable energy from microbial fuel cell BT. In: Yaser AZ, Tajarudin HA, Embrandiri A (eds) Waste management, processing and valorisation. Springer, Singapore, pp 235–263. https://doi.org/10.1007/978-981-16-7653-6_13
35. Seo Y, Kang H, Chang S, Lee Y, Cho K (2017) Toxic/hazardous substances and environmental engineering effects of nitrate and sulfate on the performance and bacterial community structure of membrane- less single-chamber air-cathode microbial fuel cells. J Environ Sci Health, Part A 53(1):13–24. https://doi.org/10.1080/10934529.2017.1366242
36. Rajasulochana P, Preethy V (2016) Comparison on efficiency of various techniques in treatment of waste and sewage water—a comprehensive review. Resource-efficient Technol 2(4):175–184. https://doi.org/10.1016/J.REFFIT.2016.09.004
37. Vu HT, Min B (2019) Integration of submersible microbial fuel cell in anaerobic digestion for enhanced production of methane and current at varying glucose levels. Int J Hydrogen Energy 44(14):7574–7582. https://doi.org/10.1016/J.IJHYDENE.2019.01.091
38. Wang Z, Mei X, Ma J, Wu Z (2012) Recent advances in microbial fuel cells integrated with sludge treatment. Chem Eng Technol 35(10):1733–1743. https://doi.org/10.1002/ceat.201200132
39. Munoz-Cupa C, Hu Y, Xu C, Bassi A (2021) An overview of microbial fuel cell usage in wastewater treatment, resource recovery and energy production. Sci Total Environ 754:142429. https://doi.org/10.1016/J.SCITOTENV.2020.142429
40. Das S, Mangwani N (2010) Recent developments in microbial fuel cells. J Sci Ind Res 69:727–731
41. Abourached C, Lesnik KL, Liu H (2014) Enhanced power generation and energy conversion of sewage sludge by CEA—microbial fuel cells. Bioresour Technol 166:229–234. https://doi.org/10.1016/j.biortech.2014.05.027
42. Ting CH, Lee DJ (2007) Production of hydrogen and methane from wastewater sludge using anaerobic fermentation. Int J Hydrogen Energy 32(6):677–682. https://doi.org/10.1016/J.IJHYDENE.2006.06.063
43. Hu Z (2008) Electricity generation by a baffle-chamber membraneless microbial fuel cell. J Power Sour 179(1):27–33. https://doi.org/10.1016/j.jpowsour.2007.12.094
44. Dentel SK, Strogen B, Chiu P (2004) Direct generation of electricity from sludges and other liquid wastes. Water Sci Technol 50(9):161–168. https://doi.org/10.2166/wst.2004.0561
45. Liu Z et al (2009) Production of electricity from surplus sludge using a single chamber floating-cathode microbial fuel cell. Water Sci Technol 60(9):2399–2404. https://doi.org/10.2166/wst.2009.313

46. Yuan Y, Chen Q, Zhou S, Zhuang L, Hu P (2011) Improved electricity production from sewage sludge under alkaline conditions in an insert-type air-cathode microbial fuel cell. J Chem Technol Biotechnol 87(1):80–86. https://doi.org/10.1002/jctb.2686
47. Fischer F, Bastian C, Happe M, Mabillard E, Schmidt N (2011) Microbial fuel cell enables phosphate recovery from digested sewage sludge as struvite. Bioresour Technol 102(10):5824–5830. https://doi.org/10.1016/j.biortech.2011.02.089
48. Ge Z, Zhang F, Grimaud J, Hurst J, He Z (2013) Long-term investigation of microbial fuel cells treating primary sludge or digested sludge. Bioresour Technol 136:509–514. https://doi.org/10.1016/j.biortech.2013.03.016
49. Nandy A, Sharma M, Venkatesan SV, Taylor N, Gieg L, Thangadurai V (2019) Comparative evaluation of coated and non-coated carbon electrodes in a microbial fuel cell for treatment of municipal sludge. Energies 12(6):1034. https://doi.org/10.3390/en12061034
50. Mohd Zaini Makhtar M, Tajarudin HA, Samsudin MDM, Vadivelu VM, Shoparwe NF, 'Izzah Zainuddin N (2021) Membrane-less microbial fuel cell: Monte Carlo simulation and sensitivity analysis for COD removal in dewatered sludge. AIP Adv 11(6):65016. https://doi.org/10.1063/5.0039014
51. Pant D, Van Bogaert G, Diels L, Vanbroekhoven K (2010) A review of the substrates used in microbial fuel cells (MFCs) for sustainable energy production. Bioresour Technol 101(6):1533–1543. https://doi.org/10.1016/j.biortech.2009.10.017
52. Liang P, Huang X, Fan MZ, Cao XX, Wang C (2007) Composition and distribution of internal resistance in three types of microbial fuel cells. Appl Microbiol Biotechnol 77(3):551–558. https://doi.org/10.1007/s00253-007-1193-4
53. Liu Z, Liu J, Zhang S, Su Z (2009) Study of operational performance and electrical response on mediator-less microbial fuel cells fed with carbon-and protein-rich substrates. Biochem Eng J 45(3):185–191. https://doi.org/10.1016/j.bej.2009.03.011
54. Chabert N, Amin Ali O, Achouak W (2015) All ecosystems potentially host electrogenic bacteria. Bioelectrochem 106:88–96. https://doi.org/10.1016/j.bioelechem.2015.07.004
55. Logan BE, Regan JM (2006) Electricity-producing bacterial communities in microbial fuel cells. Trends in Microbiol 14(12):512–518. https://doi.org/10.1016/j.tim.2006.10.003
56. Wang H, Ren ZJ (2014) Bioelectrochemical metal recovery from wastewater: a review. Water Res 66:219–232. https://doi.org/10.1016/J.WATRES.2014.08.013
57. Mathuriya AS, Yakhmi JV (2014) Microbial fuel cells to recover heavy metals. Environ Chem Lett 12(4):483–494. https://doi.org/10.1007/S10311-014-0474-2/FIGURES/3
58. Li Z, Zhang X, Lei L (December 2008) Electricity production during the treatment of real electroplating wastewater containing Cr^{6+} using microbial fuel cell. Process Biochem 43(12):1352–1358. https://doi.org/10.1016/J.PROCBIO.2008.08.005
59. Choi C, Cui Y (March 2012) Recovery of silver from wastewater coupled with power generation using a microbial fuel cell. Biores Technol 107:522–525. https://doi.org/10.1016/J.BIORTECH.2011.12.058
60. Sun J, Hu Y, Bi Z, Cao Y (2009) Improved performance of air-cathode single-chamber microbial fuel cell for wastewater treatment using microfiltration membranes and multiple sludge inoculation. J Power Sour 187(2):471–479. https://doi.org/10.1016/j.jpowsour.2008.11.022
61. Baranitharan E et al (2015) Enhanced power generation using controlled inoculum from palm oil mill effluent fed microbial fuel cell. Fuel 143:72–79. https://doi.org/10.1016/j.fuel.2014.11.030
62. Xie B, Liu H, Yan Y (2009) Improvement of the activity of anaerobic sludge by low-intensity ultrasound. J Environ Manag. https://doi.org/10.1016/j.jenvman.2007.09.004
63. Zhu H, Béland M (2006) Evaluation of alternative methods of preparing hydrogen producing seeds from digested wastewater sludge. Int J Hydrogen Energy 14(31):1980–1988. https://doi.org/10.1016/J.IJHYDENE.2006.01.019
64. More TT, Ghangrekar MM (2010) Improving performance of microbial fuel cell with ultrasonication pre-treatment of mixed anaerobic inoculum sludge. Bioresour Technol 101(2):562–567. https://doi.org/10.1016/j.biortech.2009.08.045

65. Khan MR, Amin MSA, Sarker S, Ferdaus K (2012) Design and fabrication of membrane less microbial fuel cell (ML-MFC) using food industries wastewater for power generation. J Chem Eng 27(2):55–59
66. Amin MSA, Haque T, Tarannum R, Khan MR (2014) Wastewater treatment and electricity generation by membrane less microbial fuel. Int J Environ Eng 6(3):314. https://doi.org/10.1504/ijee.2014.064306
67. Palanisamy G, Jung HY, Sadhasivam T, Kurkuri MD, Kim SC, Roh SH (2019) A comprehensive review on microbial fuel cell technologies: processes, utilization, and advanced developments in electrodes and membranes. J Clean Prod 221:598–621. https://doi.org/10.1016/j.jclepro.2019.02.172
68. Vavilin VA, Rytov SV, Lokshina LY (1996) A description of hydrolysis kinetics in anaerobic degradation of particulate organic matter. Bioresour Technol 56(2–3):229–237. https://doi.org/10.1016/0960-8524(96)00034-X

Microbial Fuel Cell Technology as Advanced Sewage Sludge Treatment

Muhammad Najib Ikmal Mohd Sabri, Nur Atiqah Mohd Abdul Rasik, Kavita Pusphanathan, Muaz Mohd Zaini Makhtar, and Hafiza Shukor

Abstract The microbial fuel cell (MFC) has emerged as an innovative and sustainable renewable energy technology, offering a potential alternative to address the global energy crisis. Operating through electrochemical processes, MFCs harness the power of electrogenic bacteria (EB) as biocatalysts to generate electricity. This chapter highlights the untapped potential of sewage sludge, derived from wastewater treatment, as a valuable fuel source within the MFC system. Extensive research has demonstrated the abundance of organic components present in sewage sludge, making it highly amenable to degradation through microbiological pathways within the MFC. Despite the lack of large-scale commercial utilization of MFC technology in wastewater treatment plants, the significant progress and promising findings indicate its effectiveness in addressing the challenges associated with sewage sludge management. The MFC system not only facilitates the simultaneous generation of energy but also contributes to bioremediation efforts. The redox potential inherent in MFCs enables this dual functionality, effectively integrating energy production with the treatment of sewage sludge. This chapter sheds light on the potential of MFC technology as an advanced approach for sewage sludge treatment. By harnessing the capabilities of electrogenic bacteria and capitalizing on the rich organic composition of sewage sludge, MFCs offer a sustainable solution that can simultaneously address energy needs and promote efficient waste management in wastewater treatment plants. The abundant and promising data accumulated thus far underscore the viability and potential of MFCs in mitigating the challenges associated with sewage sludge waste.

M. N. I. Mohd Sabri · N. A. Mohd Abdul Rasik · K. Pusphanathan ·
M. Mohd Zaini Makhtar (✉) · H. Shukor
School of Industrial Technology, Bioprocess Technology Division,
Universiti Sains Malaysia, Penang, Malaysia

M. Mohd Zaini Makhtar
Centre for Innovation and Consultation (CIC), Universiti Sains Malaysia,
Penang, Malaysia

H. Shukor
Faculty of Chemical Engineering & Technology, Universiti Malaysia Perlis, Kangar, Malaysia

M. Mohd Zaini Makhtar et al. (eds.), *Microbial Fuel Cell (MFC) Applications for Sludge Valorization*, Green Energy and Technology,
https://doi.org/10.1007/978-981-99-1083-0_6

Keywords Sewage sludge · Biomass conversion · Microbial fuel cell · Affordable and clean energy · Renewable energy

1 Introduction

Dewatering process is the middle process that happens in sewage sludge management after being pre-treated by digestion or conditioning, thickening before any further processing occurs such as incineration, landfill and composting. Generally, centrifuge involves in dewatering process hence about 20%–25% commonly a good dewatering performance practice in industrial scale [1]. In addition, total cost for managing through the process was estimated about US$ 0.33 billion per year [2]. Commonly incineration considered as a major solver for the lack of new space for sludge to be dumped in landfill. However, side effect of incineration has been argued as approximately 30% of solids from the sludge remain as ash [3]. There are several losses could be found from sludge process in which possible for power generation through biogas production. Data from Bunus Centralized Sewage Treatment Plant (Kuala Lumpur) highlighted the possible of biogas generation up to 2500 m^3/days hence 15,000 kWh/m^3 of power losses statistically [4].

Various types of energy recovery technique that had been through by recent researchers such as thermal hydrolysis, pyrolysis and water extraction through hydrate formation [5]. An attempt of several research to apply thermal hydrolysis onto dewatered sludge for improvement in anaerobic digestion performance which solid content from 10 to 20 wt% unfortunately has drawbacks which are high operation cost and large energy consumption [6, 7]. Next combustion–based technologies named as pyrolysis can be applied to dewatered sludge as well for energy recovery [8, 9]. Specifically, research that focuses on the effect of conditioning operation on pyrolysis performance is limited thus study by [10] on minimizing the bio-crude yield of sludge pyrolysis through attempted of $Fe_2(SO_4)_3$ for producing Fe—amended char.

2 Sludge Disposal Techniques

The diverse disposal method of sewage sludge been implemented throughout the years for example thermal drying, wet air oxidation, incineration and landfill. Both thermal drying and incineration are included in the thermal processing of sewage sludge [11]. Thermal drying is an efficient stabilization, dewatering and pathogen removal process, involving short–term high temperatures, through which sewage sludge may reach sterilization. Incineration is a stabilization process that destroys organic substances and pathogenic organisms through combustion obtained in the presence of excess oxygen [12]. These two methods required enormous energy inputs moreover it has been assumed by Oladejo (2019) that the main purpose of incineration

Table 1 Sewage sludge disposal methods

Methods	Remarks	References
Thermal drying	• Produce pellets for agricultural reuse, sanitary landfills disposal and fuel for boilers and industrial heater	[11]
Wet air oxidation	• Suitable for high diluted of effluent • Improve sludge characteristic for anaerobic digestion	[13]
Incineration	• Produce sludge ash; greatest volume reduction • Destroys organic substances and pathogenic organisms through combustion	[13, 17]
Landfill	• Safe disposal and no damage to public health • Emit methane which 20 times more active than carbon dioxide	[15, 16]

is for burning off harmful elements from waste before final disposal not for electric power generation as conventional used. Wet oxidation is a method that is implemented when the effluent is too diluted to be incinerated thus can improve the sewage sludge characteristics for the anaerobic digestion [13]. Meanwhile, landfills are still retained by metropolitan areas directly after mechanical biological treatment of the unsorted waste [14]. The treated organic fraction when buried continues to emit methane, which is known to be a greenhouse gas effect [15, 16]. Lastly, disposal through landfills may become unsustainable in terms of environmental perspective due to a shortage of land in addition to rising health concerns with regards to the suitability of sludge constituents. Table 1 sums up techniques for sludge disposal.

3 Current Usage of Sewage Sludge

Generally,the sewage sludges are used widely as a source of nutrient for vegetables and plants. These sludges play a pivotal role as soil conditioner such reduces toxic levels and amendment raises pH without a doubt to improve the adsorption of nutrients and stimulates microbial activity. The sewage sludge also can be innovated through advance material inform of biochar. Biochar is a versatile carbonaceous porous product derived from various biomass which came with an appealing alternative for environment remediation as it has an enormous surface area and high porosity that empower it to utilize as an catalyst and adsorbent material in the removal of a wide range of organic and inorganic contaminants in wastewaters [18, 19]. Several researchers reported that the derivation of lipids from sewage sludge may use as biofuel production which lipids are converted to biodiesel by means of transesterification or to bio-oil by pyrolysis [20]. Table 2 summarizes the benefit for each usage of sewage sludge.

Microbial fuel cell (MFC) has been considered as one of the efficient alternative renewable bioenergy technologies since 1911 electrical generated by bacteria [24]. Moreover, through degradation of organic matter includes abundant of biomass in dewatered sludge can directly produce electricity [25].

Table 2 Current usage of sewage sludge

Type of usage	Remarks	References
Biochar	• Applicable as a nutrient source in refining soil fertility • Role as restoration of contaminated soils	[21, 22]
Soil conditioner	• Suitable for high diluted of effluent • Suitable for agricultural purposes; increases the shoots and roots	[13, 23]
Biofuel	• Potential feedstock; low cost and abundant accessibility	[20]

4 Microbial Fuel Cell

Microbial fuel cell (MFC) converts chemical energy to electrical energy by certain microorganisms. Regarding to this, MFC is a bio-electrical device that harness the natural metabolism of electrogenic bacteria (EB) to produce electrical energy (Ren, 2014). Involvement of electrochemical interactions between the microorganisms called as electrogenic microbes which the electrons are transferred from the substrate to the anode electrode. Electron transportation phenomenon known as extracellular electron transport (EET) [26]. Later this MFC innovated with non-mediated where both chemical and electron shuttles mediators had been terminated. This MFC device generally configurates through two compartments; anode and cathode separated by an ion-permeable material. The configuration must innovate accordingly as adaptation through times is hence suitable for all different applications. As shown in Fig. 1, both single chambered and double chambered MFC known as basic configuration and been used broadly in various applications [27].

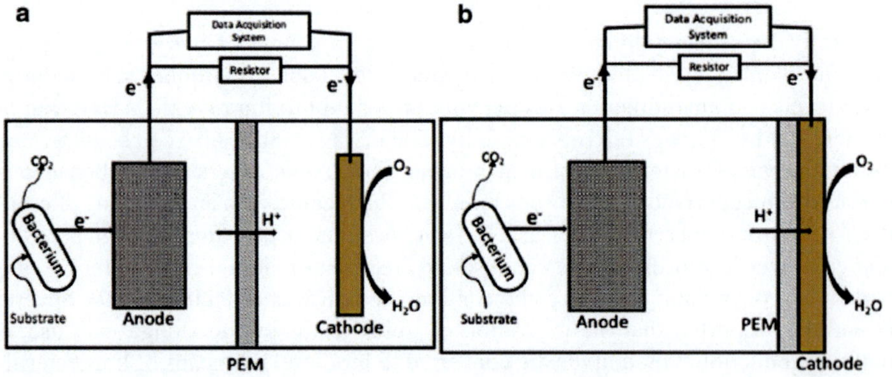

Fig. 1 Basic configuration of MFC; (A) double chambered MFC, (B) single chambered MFC. Reprinted with permission [27] (CC-BY)

4.1 History of Microbial Fuel Cell

Early 1791, researcher named Luigi Galvani examined that current gave a twitching effect on dead frog legs hence there are relation between electric and biological [28]. Back in 1911 which idea for the implementation of microbial fuel cells to produce electricity was first conceived by Potter & Waller (1911). The idea showed that it is possible to generate electricity from cultures of the bacterium *Escherichia coli*. Later 1931, Cohen (1931) added some significant knowledge when he reported that the creation of an assembly of microbial half fuel cells stack connected in series capable of producing over 35 V but the current generated through this stack was only 2 mA. A great impact factor held on early 1980s in the area of MFC research was received by the work of Allen and Peter Bennetto [29], power density improves by the usage of chemical mediators as electron shuttle which meant fuel cell would only function till the mediators were present.

Most recently, MFCs contained various mediators in an oxidized state. These mediators can easily shorten by capturing the electrons from within membrane and released the electrons to the anode and became oxidized again in the bulk solution in the anodic chamber [30]. MFC capacity to clean wastewater and deliver clean drinking water while simultaneously generating electricity, would allow developing countries towards sustainable water treatment [31, 32]. Table 3 summarizes preliminary history of MFC started on 1791 until early 2000.

Table 3 Preliminary history of MFC

Year	Description	References
1791	• Luigi Galvani • Applied current to dead frog legs, the legs twitching • Biological reactions and electric current are closely related	[28]
1911	• Publication by Potter about MFC report on the ability of microorganisms to transform organic substrate (chemical energy) into electricity • The production of electrical energy from living cultures either *Escherichia coli* or *Saccharomyces* • The first MFC proving that biological process produces bioelectricity	[24, 33, 34]
1980	• H. Peter Benneto • Succeed in extracting electric power from MFCs • Employed pure cultures to catalyze the oxidation of organics and utilizing artificial electron mediators; to facilitate electron transfer	[29, 35]
Early 2000s	• Two robots were developed: Chew–Chew and EcoBot I • These two robots are powered by MFCs	[36, 37]

5 Dewatered Sludge as MFC Feedstock

It should be noted that most of the studies on MFC have been focused on wastewater or activated sludge in municipal wastewater treatment plant (MWTP) but have overlooked the end product of MWTP, dewatered sludge. The sewage sludge is commonly taken to landfills or burned in incinerators. Handling sewage sludge is one of the largest contributors to the operational cost of MWTP and it indirectly elevates local environmental problems [4]. Furthermore, there are a few reports on the use of dewatered sludge from MWTPs for energy generation using MFC. There was a huge energy reserve in the sewage sludge without being recognized [38].

There were in the form of biodegradable organic matter and the energy could be recovered. It is reported that there is a conventional wastewater treatment plant in Toronto, Canada, which contained energy about 9.3 times more energy than was used to treat the wastewater. While the study by Logan (2006) stated that the processing of wastewater for domestic, animal and food approximately consist of 17 GW. This amount was equivalent to the energy needed to supply for the whole water infrastructure in the U.S. It is a promising energy and if the energy managed to be recovered that means the treatment plant could be run using its own energy supply [39].

Sewage sludge generated daily from the wastewater in Indah Water Konsortium (IWK) treatment plant were analyzed for its capability to support growth of the EB for the electricity generation. They can be used as value-added substrate instead of polluting the environment. Approximately 3 million cubic meters of sludge was produced annually, and it is also estimated that IWK will be producing 10 million cubic meters of sludge by the year 2035 [4] and make it the most favourable substrate for bioconversion as they are renewable and abundantly available. Their efficiency and economic viability of converting to bioenergy depend on their characteristic and components in it [40].

6 Advantages Compared to Anaerobic Digester

Since the dawn of the twenty-first century, the awareness on the protection environment come out for alternative fuels around the world thus focusing on the MFCs due to greener and bioenergy production. Variety of biomass, rich in carbohydrates, protein, hydrocarbons, alcohols and organic acids, moreover polymeric carbohydrates such as cellulose and starch also could be used as a fuel for the MFC. Accordingly, there also aid to continuous monitoring of quality wastes and minimal investment on the fuels Click or tap here to enter text. Table 4 summarizes the advantages of MFC compared to anaerobic digesters in which MFC directly converted organic substrate into electrical energy and able to treat on low concentration of substrate [41].

Table 4 Advantages of MFC compared to anaerobic digesters

Advantages	Remarks	References
Efficient conversion of substrates to electricity	Roles of electrogenic bacteria (EB) inefficiently convert and consume substrate for electricity generation	[26]
Powerful exoelectrogens oxidize organic matter	Utilization of a wide range of organic and inorganic matter into direct current (DC) electricity	[42, 43]

7 Electrogenic Microorganisms in MFC

Electroactive or electrogenic microorganisms are the core of the MFC technology. Additionally there are various mechanisms of electron transfer such as mediated electron transfer and interspecies electron transfer besides direct electron transfer itself [44]. Electrogenic microorganism can be fractionating into anaerobic and aerobic as described below.

7.1 Aerobic Electrogenic Bacteria

Fundamentally aerobic bacteria can form biocathode which catalyze the reduction of oxygen at the cathodes [45]. Research conducted by Qu (2012) highlighted that bacterial diversity and operating environment affect the biodegraded products generation. This phenomenon can be seen on the reductive breakdown of azo bonds been further degraded through aerobic condition by the presence of several oxidoreductases also called as oxidative degradation [46].

7.2 Anaerobic Electrogenic Bacteria

Generation of electricity through microorganism by exchanging electrons with electrodes while oxidizing organic also called as bacterial exocellular electron transfer principle plays a vital role in anaerobic microbial communities that degrade both inorganic electron acceptors; iron- and manganese- oxide and organic matter for growth [45, 47]. Hence these exocellular bacterial can be isolated from anaerobic sludge and municipal effluent thus can be categorized into various functional groups based on types of anaerobic respiration [48]. Via anaerobic respiration, both purple non-sulphur bacteria photosynthetic *Rhodopseudomonas palustris* DX-1 and *Rhodoferax ferrireducens*, non-photosynthetic found to generate electricity in MFC [48, 49].

7.3 Fungi

Fungi belonging to the group of white-rot—well known wood degraders—were found to have an extracellular oxidative ligninolytic enzymatic system; degrade lignin. For the last 10 years, fungi-based MFCs have appeared, their results have revealed the electrogenic potential. In addition enzyme of fungi has proven that been the best catalyst for oxidative reduction that can assure electrogenic activity in MFC activity meanwhile degradation of various xenobiotic compounds and dyes through enzymes of this system [50–51]. Yeast-based is the most intensively studied systems for fungi-based MFCs where direct electron transfer was proved via cytochrome c [52]. MFC that complied with fungi species for treating waste waters from the distillery industries, for example, are *Aspergillusawamori*, *Trichodermaviride* and *Trichodermaatroviride* while *Pleurotusostreatus* used as decolourisation of dye from textile industries effluent [53].

7.4 Algae

Organisms that contain chlorophyll size range from unicellular to multicellular are called as algae. There are two different algal growth types, autotrophic and heterotrophic. Autotrophic termed by the growth system of algae that use carbon dioxide as a carbon source in the presence of light energy while heterotrophic termed as algae that grow in the absence of light, in photobioreactors (PBRs), by utilizing a carbon dioxide source from provided substrates. These two modes can be combined to form a mixed culture (mixotrophic) growth mode, photosynthesis and respiration metabolism simultaneously function to assimilate organic carbon and carbon dioxide [53, 54]

Research outcome by Kruzic reported that metabolism of algae on bicarbonate and oxygen using solar energy may be integrated with aeration system to replace a sustainable photosynthetic one [55]. Subsequently, algae grown in cathode chamber of an MFC, produced electricity by a photosynthetic process [56]. The overall biochemical reaction that happened in both anode and cathode chamber where mediator was used had stated by Zhou (2012) below:

$$C_6H_{12}O_6 + 6H_2O \rightarrow 6CO_2 + 24H^+ + 24e \quad (1)$$

$$6CO_2 + 12H^+ + 12e \rightarrow C_6H_{12}O_6 \text{ (biomass)} + 3O_2 \quad (2)$$

7.5 Bacteria

Wastewater is the popular power source of MFC thus at anode chamber should have similar functions to methanogenic anaerobic digesters microbial communities except for microorganisms that are capable of transferring electrons to the electrode. Moreover, there are two mechanisms for electron transport in MFC, firstly direct electron transfer: (a) c-cytochromes, (b) nanowire and (c) electron shuttle. Availability of c-cytochrome in most archaea and eubacteria so a usefully role for electricity generation through electron transfer by electrogenic bacteria. Nanowire by bacteria was studied as a new way of transferring electrons to the electrode by electrically conductive pili (Das 2018). Electron shuttle secreted by most gram-negative bacteria, for example, flavin secretion that can be utilized by the organism as carbon resource although may be limited in field-applied MFC [57]. Secondly mediator electronic transfer; is essential for bacteria that cannot transfer electrons due to enabling of electron shuttle from cell membrane to electrode for example ferricyanide and benzoquinone usage to facilitate electron transfer from bacteria to electrodes [58–60].

8 Microbial Fuel Cell Concept

The concept is a linkage between negative terminal (anode) and positive terminal (cathode). Anode terminal oxidized organic matter such as fuel and released CO_2, electrons and protons while the cathode terminal received the electrons that produced via an external circuit as the result of electrophilic attraction at the cathode electrode. The migration of protons from the anode to the cathode through the separator or called mediated [61]. This mediated generally must possess the quality of high proton transfer rate, low gas permeability, good thermal stability and resistance against biofouling.

8.1 Biological Concept

The basic of molecular diversities are made up of chemical based which mostly of the element carbon. Carbon is unparalleled in its ability to form molecules that are large, varied and complex, making possible the diversity of organisms that evolved on Earth.

Basically, there are seven chemical groups that are most important in biological processes which are carbonyl, hydroxyl, carboxyl, amino, phosphate, sulfhydryl and methyl groups. Major sources of energy in cellular processes are the phosphate group, it complicated name is adenosine triphosphate (ATP) [62]. This ATP will be split off when reacted with water and later becomes adenosine diphosphate (ADP), the reaction released energy then can be used by the cell. In MFCs energy production

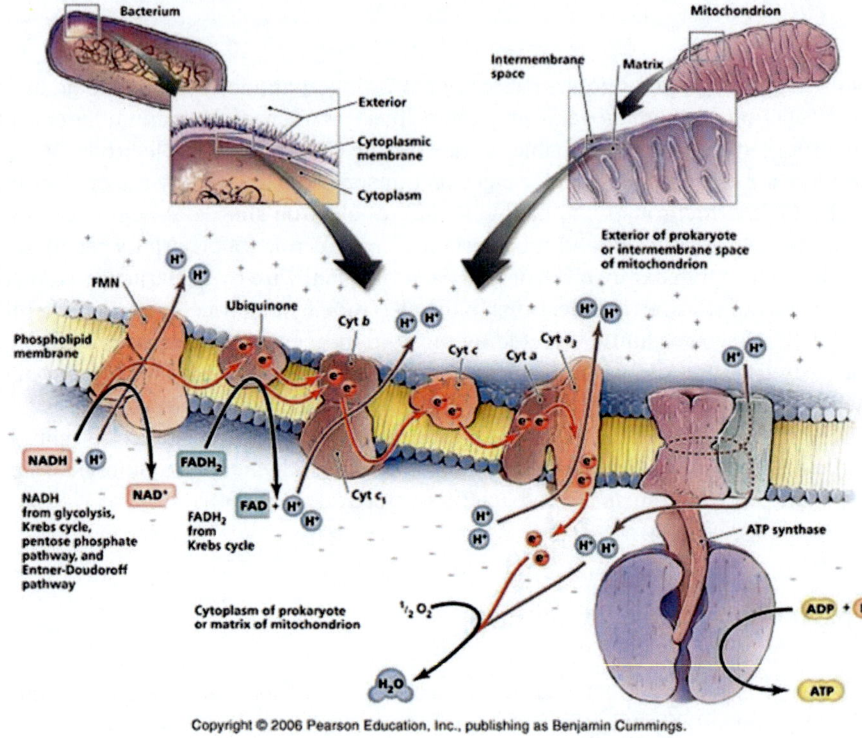

Fig. 2 The schematic diagram of electron transport chain (Reprinted with permission [63] (CC-BY))

occurs when electrons were passed through an electron transport chain (ETC) and protons are translocated across the cell membrane to generate energy in the form of adenosine triphosphate (ATP). Roughly, this ETC mechanism had been illustrated in Fig. 2.

8.2 *Chemical Concept*

Chemical reaction that usually occurs in MFC system is reduction and oxidation popularly called as redox reaction which reduction of oxygen takes place at the cathode resulting in water molecule while oxidation for example hydrogen at the anode was helped by a conductive catalyst; platinum (Pt) [43, 59]. MFC developed as an anode catalyst where microorganism is used as a biocatalyst for the redox reaction. Capability of electrogenic bacteria for generating and transferring electrons through nanowires (*Geobacter sulfurreducens*) and electron shuttle (*Pseudomonas aeruginosa*) [59, 64].

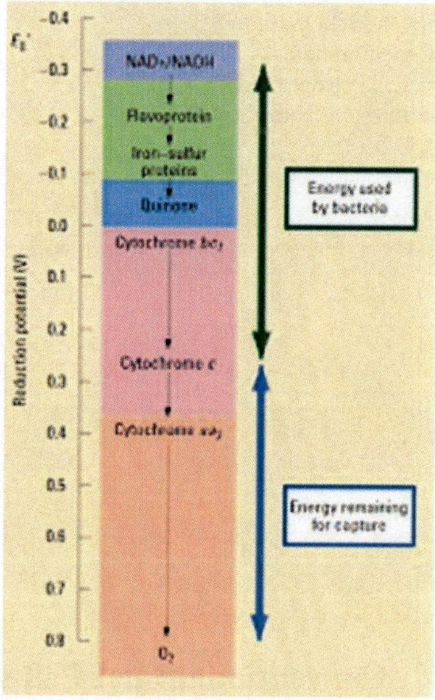

Fig. 3 The schematic diagram of standard redox potential (Reprinted with permission [66] (CC-BY))

Present of lowest redox of mediator theoretically lowest anodic redox thus maximize the redox difference between anode and cathode, affect the voltage different but it would not necessarily be the most efficient at pulling electrons away. A mediator with a higher E^0 redox would give a higher overall power than a mediator with the lowest redox [65]. The schematic of standard redox potential is shown in Fig. 3.

8.3 *Electrical Concept*

In MFCs, bacteria as a living catalyst is used to decompose organic substrates into electricity. Electrical energy yield happened or occurred in MFC when biomass-based materials oxidize by resulting in the generation of free electrons which pass through external circuit [67]. These phenomena occur during microbial metabolism which involved redox reaction. The electrons pass through an electron transport chain (ETC) and protons are translocated across the cell membrane to generate energy in the form of adenosine triphosphate (ATP) (Fig. 4). Production electrical power (W) based on the rate of electrons moving through the circuit; current (amps) besides electrochemical potential difference (V) across the electrodes [68]. Table 5 shows the MFCs with different substrate and the maximum current produced.

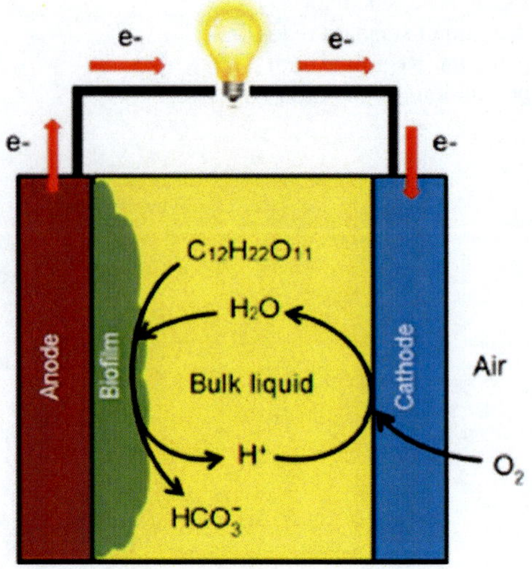

Fig. 4 Schematic diagram of membraneless microbial fuel cell (Reprinted with permission from [69] (CC-BY))

9 Conventional Fuel Cell vs Microbial Fuel Cell

Conventional or typical fuel cell provided with greater control for the designer over operating conditions. The realm of conventional fuel cell is packed with a variety of well-understood technologies that delivers high performance with respect to efficiency and power density. There are reasons where conventional fuel cell may incompetent which were the demand for chemical selectivity and high-cost production correspond with technology and high performance. For example, current densities that produced by Direct Methanol Fuel Cells (DMFCs) are often lower by about 100–300 mA/cm^2. Methanol oxidation owing to the high kinetic resistance as compared to hydrogen oxidation besides performance limited caused by methanol crossover from anode to cathode [67]. Table 6 tabulated the difference between conventional fuel cell and microbial fuel cell.

10 Microbial Fuel Cell Design

Common microbial fuel cell designs consist of an anodic chamber and cathodic chamber separated by proton exchange membrane (PEM) chamber, fundamental for the construction of MFC in a diversity of architecture to produce high power density and coulumbic efficiencies. Power output, coulombic efficiency, stability and longevity are usually evaluated in MFC. Not only the cost of materials but feasibility scaling up also been considered in the real application of MFC. Popular MFCs designs are single chamber, double chamber, tubular membrane, stack design and lastly flat-plat.

Table 5 MFCs with different substrate and the maximum current produced [70]

Types of substrate	Concentration	Source inoculums	Type of MFC	Current density (mA/cm^2)
Starch processing wastewater	10 g/L	Starch processing wastewater	One chamber air cathode MFC with carbon paper anode (25 cm^2)	0.09
Starch	10 g/L	Pure culture of *Clastridiumbutyricum*	Two chambered MFC with woven graphite anode (7 cm^2) and ferricyanide catholyte	1.3
Acetate	1 g/L	Pre-acclimated bacteria from MFC	Cube shape one chamber MFC with graphite fibre brush anode (7170 m^2/m^3 brush volume)	0.8
Corn stover biomass	1 g/L	Domestic wastewater	One chamber membrane-less air cathode MFC with carbon paper anode (7.1 cm^2) and carbon cloth electrode	0.15
Landfill leachate	6000 mg/L	Leachate and sludge	Two chambered MFC with carbon veil electrode (30 cm^2)	0.0004
Domestic wastewater	600 mg/L	Anaerobic sludge	Two chambered mediator-less MFC with plain graphite electrode (50 cm^2)	0.06

The architecture of the optimizations of MFCs aimed to reduce the internal resistance and increase the cell power output. Roughly all of the designs stated above had an addition either PEM or assisted chemical for electron transportation through media. Along the appropriate optimization of architecture, these microbial fuel cells are able to power a wide range of devices such as power sensors for environmental parameters monitoring at various intervals, store energy in external storage device; capacitor and power devices placed under water environment [71]

Table 6 Conventional fuel cell and Microbial fuel cell

Subject	Conventional fuel cell	Microbial fuel cell
Mediator	Artificial Abiotic fuel cells Comprise of inorganic catalyst	Natural Biotic fuel cells Assist of microorganism such as *Geobacter sulfurreducens;* nanowires and *Pseudomonas aeruginosa;* electron shuttle
Advantage	Aeration consumes more energy Enhance power generation, short half–life and instability limit	Higher affinity for oxygen with cathode Enhance the chemical oxygen demand removal Inexpensive catalyst and cheap substrate Can operate at ambient temperature and atmospheric pressure
Disadvantage	Expensive catalyst substances	Systematic configuration undetermined
Example	Solid oxide fuel cell and proton exchange membrane fuel cell	Up-flow reactor and stacked MFC *Pseudomonas aeruginosa* and *Geobacter sulfurreducens*

10.1 Single Chamber MFC

This type of MFC design is purposely to solve scale-up problems on two chambers MFCs due to complex design and cost even can be operated in either batch or continuous mode [72]. The design was used to characterize the performance of either anodic or cathodic chambers separately. A common single chamber possessed aeration on an anodic chamber without including a cathodic chamber. The reduction of internal resistance of MFCs thus enhances electricity production [73]. Schematic diagram of the single chamber that is provided with proton exchange membrane (PEM) layered on the cathode (Fig. 5).

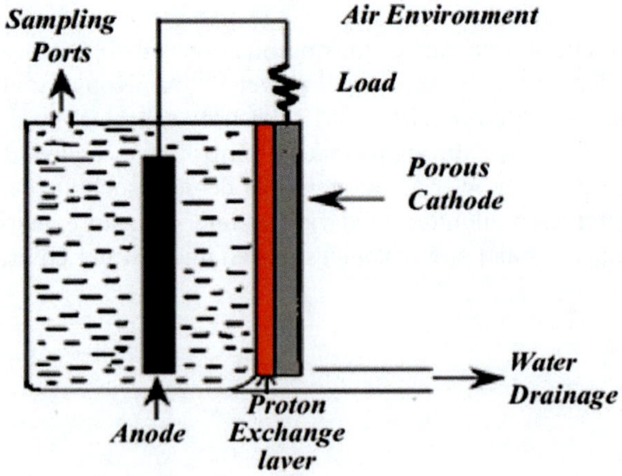

Fig. 5 The schematic diagram of single chamber microbial fuel cells (Reprinted with permission from [73] (CCBY))

10.2 Double Chamber MFC

Fundamental or conventional design for microbial fuel cells often run and investigated in batch mode with a defined medium such as acetate or glucose solution to generate electricity. They were built by one cathode chamber and one anode chamber, connected by a bridge and separated by a proton or cation exchange membrane to allow protons to move across to the cathode while blocking the diffusion of oxygen into the anode. Chemically the plain carbon cathode was catalyst and coated in ferricyanide due to platinum expensively [72, 73].

Plain carbon electrode immersed in ferricyanide solution as the electron acceptor and the cathodic reaction is $Fe(CN)_6^{3-} + e = Fe(CN)_6^{4-}$. Reaction in the cathode chamber reduced ferricyanide to ferrocyanide, addition of chemical compulsory after it is depleted. Because of that ferricyanide are not environmentally friendly and not economic to use on cathodes. That is the reason some researcher stated that power densities in two-chamber MFCs are possible to be increased by enhancements of cathode such as concentration increment of dissolved oxygen.

According to He (2005), dual chamber and cylindrical shaped of microbial fuel cells suitable and useful in powering autonomous sensors for long-term because they are relatively easy to scale up. Maximum power generated about 1530 kWh/day of electricity by 24-hour operation perpendicular with 0.204 kWh/m^3 closed to aerobic trickling filter consumed. Schematic diagram of various architecture of dual chambers that provided with PEM as bridge layered on different shapes such as cylindrical, rectangular and miniature (Fig. 6).

10.3 Tubular Membrane MFC

Architecture working of tubular membrane commonly designed in continuous flow mode, initial flow moving through anode chamber and directly up into the cathode chamber in the same column. Although this design has high possibility to be scaled up but there is drawbacks for this design. Based on the implementation of tubular design in wetlands by Wetser (2017) indicated that electricity generation was not optimal due to complication of oxygen crossover from cathode to anode. Practically orientation of anode and cathode tube that been placed as closely as possible inside the reactor developed 112–240 mV higher than outside the reactor. Schematic diagram of tubular membrane built in granular anode that also provided with PEM (Fig. 7).

10.4 Stack MFC

Both parallel and series circuits can be investigated using stack microbial fuel cell. The usage of copper wires in this system is interconnected in series or parallel to

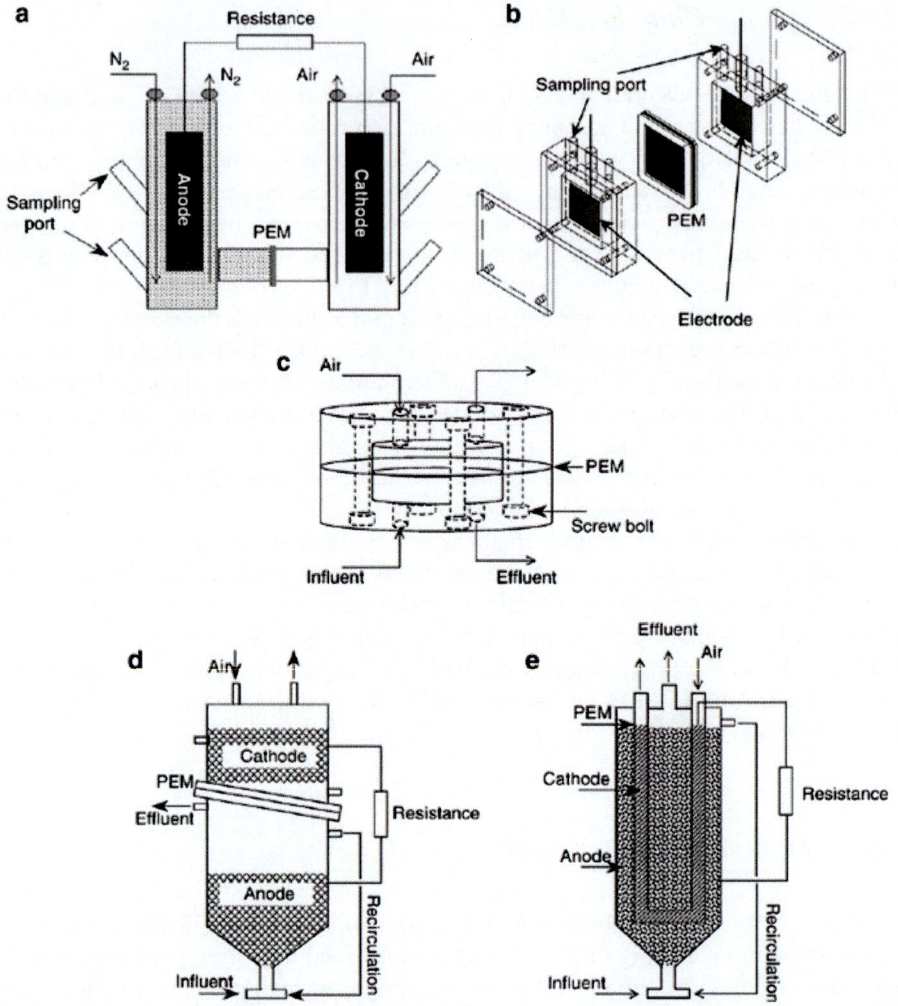

Fig. 6 The schematic diagram of various double chamber microbial fuel cell (Reprinted with permission from [72, 74] (CCBY))

the electrodes and held each other by screw bolts. Both Aelterman (2006) and Li (2008), observed from their research that the effect of maximum power output per MFC unit was no visible adverse which Coulombic efficiency diverged greatly in two arrangements with parallel connection giving about six times efficiency more when both the series were operated at the same volumetric flow rate. Again both research by [75] and [76] supported previous research thus highlighting that the performance of stacked MFC is low caused by voltage reversal in individual cells, increase ohmic, inactive surface area on the cathode, kinetic and transportation resistances. Schematic

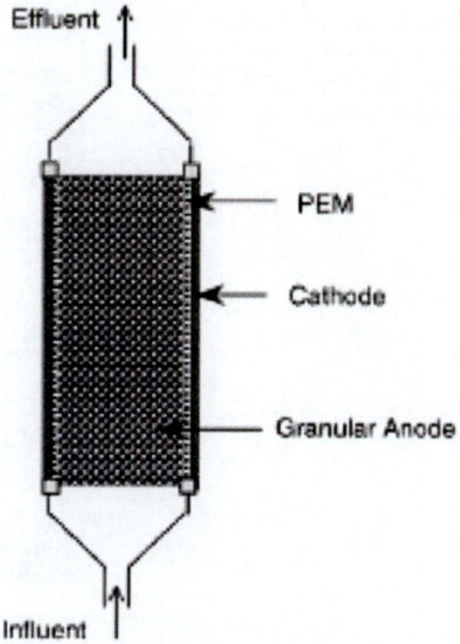

Fig. 7 The schematic diagram of tubular membrane microbial fuel cell (Reprinted with permission from [72] (CCBY))

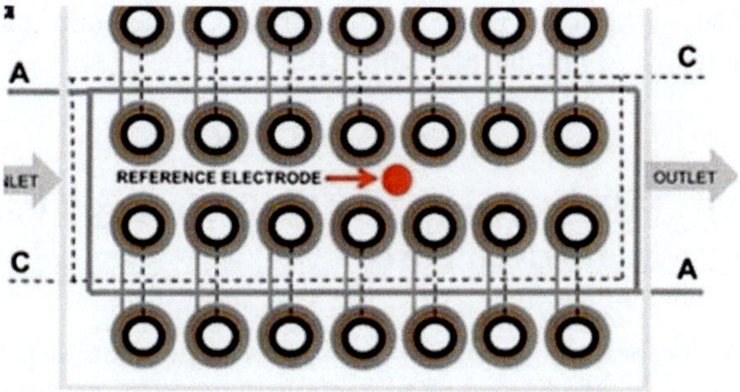

Fig. 8 The schematic diagram of stack microbial fuel cell (Reprinted with permission from [77] (CC-BY))

diagram of stack MFC built in six individual units of granular graphite anode that also provided with PEM (Fig. 8).

10.4.1 Flat–Plat

Basically the structure of flat-plat similar to chemical fuel cell whose designed by [78]. Hot pressed method on cathode with PEM sandwiched between two non-conductive (polycarbonate) plates and placed on top of anode. Coulombic efficiency diverged greatly in two arrangements with parallel connection giving about six times efficiency more when both the series were operated at the same volumetric flow rate. Maximum power density for domestic wastewater obtained was about 72 mW/m^2

Fig. 9 The schematic diagram of flat-plat microbial fuel cell (Reprinted with permission from [80] (CC-BY))

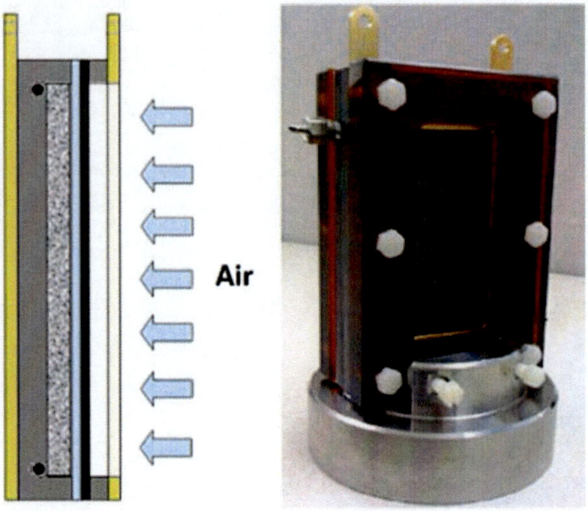

increment, 2.8 times compared to single chambered MFC. However this design drawback is high in anodic resistance [79]. Schematic diagram (upper, side view; top, lower view) of flat-plat MFC (Fig. 9).

11 MFC Technology: World Energy's Paradigm as a Driven Force

Non-renewable energy sources such as oil, natural gas and coal account for 85 per cent of global energy. Oil provides over 40% of the world's energy [81]. Linearly from 2012, rising transportation fuel consumption and robust industrial demand have resulted in an increase in non-renewable energy usage [81]. About 1.3 billion people in the globe do not have access to electricity, and another three billion rely on traditional fuels, which can have negative consequences for their health, ecosystems and development. According to what is known about global energy demand, it is expected to grow at a pace of 1.6 per cent per year on average from 2008 to 2030 [82].

Non-renewable energy is the most frequently stated problems with constant increment of prices and CO_2 emission which both coal and natural gas is the major impact. Along with this statement is evidence that oil prices are expected to remain between US$ 50.0 and US$ 80.0/barrel until 2030 as stated. Moreover, the increase in the market is due to structural changes and energy efficiency gains in the market. Besides European oil consumption will be reduced by 3.0% over the next 15 years. In addition, oil supply globally is increasing by 14.0 mb/d to 104.0 mb/d by 2040 even though the drift timely expenditure specifically [83].

There is a large volume of published studies describing on greenhouse gases (GHG) which cause by reradiated infrared radiations by CO_2, CH_4, O_3, NO_2 and NO slightly by water vapours thus significantly to maintain Earth's temperature by 33°C (Kumar et al. 2018). Between 2000 and 2010 annual anthropogenic GHG emissions have increased directly coming from energy supply (47%), industry (30%), building (3%) and transport (11%) [83, 84]. Emission of greenhouse gases continuously will cause further warming called as global warming which is irreversible and gives pervasive impacts for people and ecosystems. It has been reported that major drivers of increment in CO_2 emissions are from both fossil fuel combustion and coal impacted by economic and population growth globally [83, 84]. Furthermore, according to Ahmad (2011), Malaysia's petroleum resources are very limited compared to other international areas, at roughly 5.5 billion barrels, with petroleum output peaking in 2004 at roughly 861.8 thousand barrels per day. Nevertheless, these resources will be consumed and become more expensive in the long run.

Effect of oil and natural gas usage can be shown in Fig. 10, CO_2 emission has doubled since early 1970s, accelerating environmental change and climate degradation [83, 84]. Most countries occur a sustained increase due to global economic shift perpendicularly with power consumption clearly world economy is booming which gross domestic product (GDP) growing 2.5 times over the past three decades [85].

Figure 11 explained about the statistic of world's total electricity generation since 1990 and overview of the percentage increase/decrease in world energy, oil, gas, coal, CO_2 emissions and the share of renewable energy in electricity generation respectively. The natural gas, oil and coal cover up to 84% of world's primary energy consumption in 2019 [87]. The increment came from both public and private utilities hence China; Asia significantly contributing almost half of the increase in 2017 due to

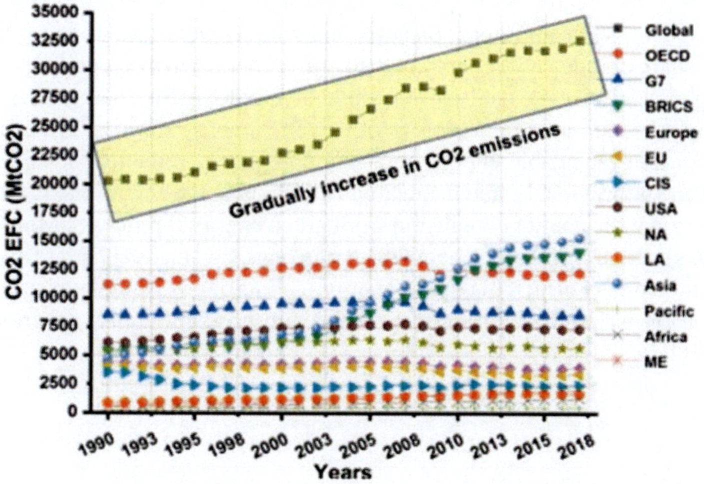

Fig. 10 CO_2 emissions from fuel combustion ($MtCO_2$) (Reprinted with permission from [86] (CC-BY))

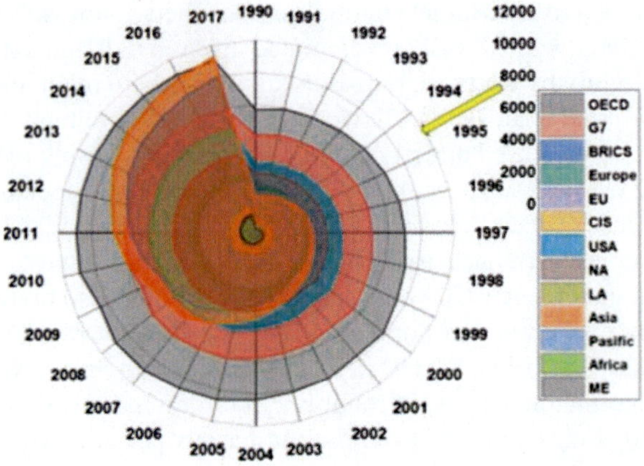

Fig. 11 World's total electricity generation (TWh) (Reprinted with permission from [86] (CCBY))

the high demand for electricity combined to accelerate the development of production capacity (Ahmad and Zhang 2020). According to the New Policy Scenario (NPS), global primary energy requirements increased by 37% between 2012 and 2040 which considers existing government initiatives. As a result, by 2040, oil, coal and natural gas were expected to account for around a third of total demand. Various states declared new initiatives to decrease CO_2 emissions at a long-term climate summit named the Paris Agreement in 2015; nonetheless, emissions continue to rise by 20% (Ahmad and Zhang 2020). Meanwhile, Climate Change Conference 2021 (COP 26) that had been conducted by United Nation (UN) also highlighted on several circumstances that are better than Paris Agreement in which almost 90% had been covered by a net–zero target (Lord, et al. 2021).

In the EU, the United States, India, Japan and China, stringent environmental policies have a significant impact on solar and wind energy development. Without a doubt, the share of renewable energy in the entire generation of electricity is gradually increasing daily [86]. As well as that bulk of employment sources is increasing perpendicularly with a large number of renewable energy sources. The major issue arises when non-renewable energy becomes completely reliable, causing prices to skyrocket and as previously stated, negatively impacting the environment. As a result, a new solution with a better conclusion from green energy is required. Microbial fuel cells (MFCs) which convert biochemical energy consisted in the substrate to electrical energy can be a part of it. This green energy technology is capable of utilizing any type of carbon waste that is seen to be impactful on community, government and environment.

12 Conclusion

This chapter summarizes the utilization of sewage sludge in MFCs for electricity production and waste treatment. While simple substrates like acetate and glucose were commonly used in the early years, recent research has focused on utilizing more unconventional substrates, such as sewage sludge, with the aim of both waste utilization and enhancing MFC output. The generation of bioenergy, in the form of electricity, from renewable sources like sewage sludge through MFCs holds significant development potential. It not only contributes to energy self-sufficiency but also addresses concerns about competition with food production that are associated with conventional biofuels. The findings presented in this chapter highlight the evolving landscape of MFC technology and its potential for sustainable sewage sludge treatment and renewable energy generation.

Acknowledgements The authors would like to thank the Universiti Sains Malaysia for the financial support of this study via APEX Era grant (1001/PTEKIND/881004) The authors have declared no conflict of interest for the manuscript.

References

1. To VHP, Nguyen TV, Vigneswaran S, Ngo HH (2016) A review on sludge dewatering indices. Water Sci Technol 74(1):1–16. https://doi.org/10.2166/wst.2016.102
2. Hanum F et al. (2019) Treatment of sewage sludge using anaerobic digestion in Malaysia: current state and challenges. Front Energy Res 7(MAR):1–7. https://doi.org/10.3389/fenrg.2019.00019
3. Kasina M, Kowalski PR, Michalik M (2016) Metals accumulation during thermal processing of sewage sludge—characterization of fly ash and Air Pollution Control (APC) residues. Energy Procedia 97:23–30. https://doi.org/10.1016/j.egypro.2016.10.012
4. Indah Water Konsortium (2013) Indah water cleaning the unseen
5. Wu B, Dai X, Chai X (2020) Critical review on dewatering of sewage sludge: influential mechanism, conditioning technologies and implications to sludge re-utilizations. Water Res 180:115912. https://doi.org/10.1016/j.watres.2020.115912
6. Duan N, Dong B, Wu B, Dai X (2012) High-solid anaerobic digestion of sewage sludge under mesophilic conditions: feasibility study. Biores Technol 104:150–156. https://doi.org/10.1016/j.biortech.2011.10.090
7. Dai X, Duan N, Dong B, Dai L (2013) High-solids anaerobic co-digestion of sewage sludge and food waste in comparison with mono digestions: stability and performance. Waste Manage 33(2):308–316. https://doi.org/10.1016/j.wasman.2012.10.018
8. Tian K, Liu W-J, Qian T-T, Jiang H, Yu H-Q (Sep.2014) Investigation on the evolution of n-containing organic compounds during pyrolysis of sewage sludge. Environ Sci Technol 48(18):10888–10896. https://doi.org/10.1021/es5022137
9. Zhao M, Wang F, Fan Y, Raheem A, Zhou H (2019) Low-temperature alkaline pyrolysis of sewage sludge for enhanced H_2 production with in-situ carbon capture. Int J Hydrogen Energy 44(16):8020–8027. https://doi.org/10.1016/j.ijhydene.2019.02.040
10. Yu G, Chen D, Arena U, Huang Z, Dai X (2018) Reforming sewage sludge pyrolysis volatile with Fe-embedded char: minimization of liquid product yield. Waste Manage 73:464–475. https://doi.org/10.1016/j.wasman.2017.08.004

11. Baru PA, Hassan S (2018) Akademia Baru characterization of Malaysian sewage sludge dried using thermal dryer. J Adv Res Fluid Mech Therm Sci 5(January):24–29
12. Tambo N, Kobayashi M, Thebault P, Haubry A (1982) Sludge treatment and disposal. Water Supply 1(2/3). https://doi.org/10.1142/9781848160798_0015
13. Hii K, Baroutian S, Parthasarathy R, Gapes DJ, Eshtiaghi N (2014) Bioresource Technology: a review of wet air oxidation and thermal hydrolysis technologies in sludge treatment. Biores Technol 155:289–299. https://doi.org/10.1016/j.biortech.2013.12.066
14. Technical EEA (2020) Horizon 2020 Mediterranean report, no. 6
15. Lou XF, Nair J (2009) The impact of landfilling and composting on greenhouse gas emissions—a review. Biores Technol 100(16):3792–3798. https://doi.org/10.1016/j.biortech.2008.12.006
16. Mancini G, Luciano A, Bolzonella D, Fatone F, Viotti P, Fino D (2021) A water-waste-energy nexus approach to bridge the sustainability gap in landfill-based waste management regions. Renew Sustain Energy Rev 137(October 2020):110441. https://doi.org/10.1016/j.rser.2020.110441
17. Oladejo J, Shi K, Luo X, Yang G, Wu T (2019) A review of sludge-to-energy recovery methods. Energies 12(1):1–38. https://doi.org/10.3390/en12010060
18. Sharma M, Singh J, Baskar C, Kumar A (2018) A comprehensive review on biochar formation and its utilization for wastewater treatment. Pollut Res 37:1–18
19. Varjani S, Kumar G, Rene ER (2019) Developments in biochar application for pesticide remediation: current knowledge and future research directions. J Environ Manage 232:505–513. https://doi.org/10.1016/j.jenvman.2018.11.043
20. Bora AP, Gupta DP, Durbha KS (2020) Sewage sludge to bio-fuel: a review on the sustainable approach of transforming sewage waste to alternative fuel. Fuel 259:116262. https://doi.org/10.1016/j.fuel.2019.116262
21. Vijayaraghavan K (Jan 2019) Recent advancements in biochar preparation, feedstocks, modification, characterization and future applications. Environ Tech Rev 8(1):47–64. https://doi.org/10.1080/21622515.2019.1631393
22. Agegnehu G, Srivastava AK, Bird MI (2017) The role of biochar and biochar-compost in improving soil quality and crop performance: a review. Appl Soil Ecol 119:156–170. https://doi.org/10.1016/j.apsoil.2017.06.008
23. Gubišová M et al (2020) Sewage sludge as a soil amendment for growing biomass plant Arundo donax L. *Agronomy* 10(5). https://doi.org/10.3390/agronomy10050678
24. Potter MC, Waller AD (Sep 1911) Electrical effects accompanying the decomposition of organic compounds. *Proc R Soc London Ser B, Contain Pap Biol Character* 84(571):260–276. https://doi.org/10.1098/rspb.1911.0073
25. Shamsuddin NA, Mohd Sabri MNI, Tajarudin HA, Shoparwe NF, Makhtar MMZ (May 2021) Effect of thermal pre-treatments method on sludge degradation process prior usage in membrane-less microbial fuel cell for electricity generation. IOP Conf Ser Earth Environ Sci 76(1):012092. https://doi.org/10.1088/1755-1315/765/1/012092
26. Khoo KS, Chia WY, Tang DYY, Show PL, Chew KW, Chen WH (2020) Nanomaterials utilization in biomass for biofuel and bioenergy production. Energies 13(4):1–20. https://doi.org/10.3390/en13040892
27. Tamboli E, Eswari JS (2019) Chapter 3.2—microbial fuel cell configurations: an overview. Elsevier B.V. https://doi.org/10.1016/B978-0-444-64052-9.00016-9
28. Piccolino M (Jul 1998) Animal electricity and the birth of electrophysiology: the legacy of Luigi Galvani. Brain Res Bull 46(5):381–407. https://doi.org/10.1016/s0361-9230(98)00026-4
29. Bennetto HP, Stirling JL, Tanaka K, Vega CA (Feb1983) Anodic reactions in microbial fuel cells. Biotechnol Bioeng 25(2):559–568. https://doi.org/10.1002/bit.260250219
30. Rezaei F, Richard TL, Logan BE (2008) Enzymatic hydrolysis of cellulose coupled with electricity generation in a microbial fuel cell. Biotechnol Bioeng 101(6):1163–1169. https://doi.org/10.1002/bit.22015
31. Khera J, Chandra A (2012) Microbial fuel cells: recent trends. Proc Natl Acad Sci India Sect A Phys Sci 82(1):31–41. https://doi.org/10.1007/s40010-012-0003-2

32. Logan BE (2012) Essential data and techniques for conducting microbial fuel cell and other types of bioelectrochemical system experiments. Chemsuschem 5(6):988–994. https://doi.org/10.1002/cssc.201100604
33. Bullen RA, Arnot TC, Lakeman JB, Walsh FC (May 2006) Biofuel cells and their development. Biosens Bioelectron 21(11):2015–2045. https://doi.org/10.1016/j.bios.2006.01.030
34. Shukla A, Suresh P, Berchmans S, Rajendran A (Aug 2004) Biological fuel cells and their applications. Curr Sci 87
35. Roller SD, Bennetto HP, Delaney GM, Mason JR, Stirling JL, Thurston CF (Mar 1984) Electron-transfer coupling in microbial fuel cells: 1. comparison of redox-mediator reduction rates and respiratory rates of bacteria. J Chem Technol Biotechnol. Biotechnol 34(1):3–12. https://doi.org/10.1002/jctb.280340103
36. Wilkinson S (2000) 'Gastrobots'—benefits and challenges of microbial fuel cells in food powered robot applications. Auton Robot 9(2):99–111. https://doi.org/10.1023/A:1008984516499
37. Ieropoulos I, Melhuish C, Greenman J (2003) Artificial metabolism: towards true energetic autonomy in artificial life. Adv Artif Life: 792–799
38. Mohd Zaini Makhtar M, Tajarudin HA, Samsudin MDM, Vadivelu VM, Shoparwe NF, Izzah Zainuddin N (Jun 2021). Membrane-less microbial fuel cell: Monte Carlo simulation and sensitivity analysis for COD removal in dewatered sludge. *AIP Adv* 11(6):65016. https://doi.org/10.1063/5.0039014
39. Logan BE et al (2006) Microbial fuel cells: methodology and technology. Environ Sci Technol 40(17):5181–5192. https://doi.org/10.1021/es0605016
40. Pant D, Van Bogaert G, Diels L, Vanbroekhoven K (2010) A review of the substrates used in microbial fuel cells (MFCs) for sustainable energy production. Bioresour Technol 101(6):1533–1543. https://doi.org/10.1016/j.biortech.2009.10.017
41. Pham TH et al (2006) Microbial fuel cells in relation to conventional anaerobic digestion technology. Eng Life Sci 6(3):285–292. https://doi.org/10.1002/elsc.200620121
42. Li M et al (2018) Microbial fuel cell (MFC) power performance improvement through enhanced microbial electrogenicity. Biotechnol Adv 36(4):1316–1327. https://doi.org/10.1016/j.biotechadv.2018.04.010
43. Debabrata D (2018) Microbial fuel cell: a bioelectrochemical system that converts waste to watts. Springer International, Switzerland
44. Akiba T, Bennetto HP, Stirling JL, Tanaka K (1987) Electricity production from alkalophilic organisms. Biotech Lett 9(9):611–616. https://doi.org/10.1007/BF01033196
45. Chabert N, Amin Ali O, Achouak W (2015) All ecosystems potentially host electrogenic bacteria. Bioelectrochemistry 106:88–96. https://doi.org/10.1016/j.bioelechem.2015.07.004
46. Danish M et al (2021) Science of the total environment integrated air cathode microbial fuel cell-aerobic bioreactor set-up for enhanced bioelectrodegradation of azo dye Acid Blue 29. Sci Total Environ 756:0–10. https://doi.org/10.1016/j.scitotenv.2020.143752
47. Pous N, Koch C, Colprim J, Puig S, Harnisch F (2014) Extracellular electron transfer of biocathodes: revealing the potentials for nitrate and nitrite reduction of denitrifying microbiomes dominated by Thiobacillus sp. Electrochem Commun 49:93–97. https://doi.org/10.1016/j.elecom.2014.10.011
48. Guang L, Koomson DA, Jingyu H, Ewusi-Mensah D, Miwornunyuie N (2020) Performance of exoelectrogenic bacteria used in microbial desalination cell technology. Int J Environ Res Public Health 17(3):10–12. https://doi.org/10.3390/ijerph17031121
49. Xing D, Zuo Y, Cheng S, Regan JM, Logan BE (Jun.2008) Electricity generation by Rhodopseudomonas palustris DX-1. Environ Sci Technol 42(11):4146–4151. https://doi.org/10.1021/es800312v
50. Martínez ÁT et al (2005) Biodegradation of lignocellulosics: microbial, chemical, and enzymatic aspects of the fungal attack of lignin. Int Microbiol 8(3):195–204. https://doi.org/10.2436/im.v8i3.9526
51. Wesenberg D, Kyriakides I, Agathos SN (2003) White-rot fungi and their enzymes for the treatment of industrial dye effluents. Biotechnol Adv 22(1):161–187. https://doi.org/10.1016/j.biotechadv.2003.08.011

52. Wilkinson S, Klar J, Applegarth S (Oct 2006) Optimizing biofuel cell performance using a targeted mixed mediator combination. Electroanalysis 18(19–20):2001–2007. https://doi.org/10.1002/elan.200603621
53. Sivasankar V, Mylsamy P, Omine K (2018) Microbial fuel cell technology for bioelectricity. Springer International Publishing, Cham. https://doi.org/10.1007/978-3-319-92904-0
54. Lee Y-K (Nov 2003) Algal nutrition—heterotrophic carbon nutrition. Handb Microalgal Cult :116–124. https://doi.org/10.1002/9780470995280.ch7
55. Kruzic AP, Kreissl JF (May2009) Natural treatment and onsite systems. Water Environ Res 81(10):1346–1360
56. Juang DF, Lee CH, Hsueh SC (2012) Comparison of electrogenic capabilities of microbial fuel cell with different light power on algae grown cathode. Biores Technol 123:23–29. https://doi.org/10.1016/j.biortech.2012.07.041
57. Marsili E, Baron DB, Shikhare ID, Coursolle D, Gralnick JA, Bond DR (2008) Shewanella secretes flavins that mediate extracellular electron transfer. Proc Natl Acad Sci USA 105(10):3968–3973. https://doi.org/10.1073/pnas.0710525105
58. Konovalova EY et al (May 2018) The microorganisms used for working in microbial fuel cells. In: AIP Conference Proceedings, vol 1952. https://doi.org/10.1063/1.5031979
59. Logan BE (2009) Exoelectrogenic bacteria that power microbial fuel cells. Nat Rev Microbiol 7(5):375–381. https://doi.org/10.1038/nrmicro2113
60. Schröder U (2007) Anodic electron transfer mechanisms in microbial fuel cells and their energy efficiency. Phys Chem Chem Phys 9(21):2619–2629. https://doi.org/10.1039/b703627m
61. Feng C et al. (2014) Characterization of exoelectrogenic bacteria enterobacter strains isolated from a microbial fuel cell exposed to copper shock load. PLoS One 9(11). https://doi.org/10.1371/journal.pone.0113379
62. Wu Y et al (2020) Enhanced current production by exogenous electron mediators via synergy of promoting biofilm formation and the electron shuttling process. Environ Sci Technol 54(12):7217–7225. https://doi.org/10.1021/acs.est.0c00141
63. Miriam R, Federico Aulenta, Marianna V, Largus T Angenent (2011) Cathodes as electron donors for microbial metabolism: which extracellular electron transfer mechanisms are involved? Bioresour Technol 102(1):324–333, ISSN 0960-8524. https://doi.org/10.1016/j.biortech.2010.07.008
64. Bond D, Lovley D (2003) Electricity production by geobacter sulfurreducens attached to electrodes. Appl Environ Microbiol 69(3):1548–1555. https://doi.org/10.1128/AEM.69.3.1548
65. Ieropoulos IA, Greenman J, Melhuish C, Hart J (2005) Comparative study of three types of microbial fuel cell. Enzyme Microb Technol 37(2):238–245. https://doi.org/10.1016/j.enzmictec.2005.03.006
66. Kumari S (2012) Studies on marine microbial fuel cell, Doctoral dissertation
67. Barton SC, Gallaway J, Atanassov P (2004) Enzymatic biofuel cells for implantable and microscale devices. Chem Rev 104(10):4867–4886. https://doi.org/10.1021/cr020719k
68. Zhao F, Slade RCT, Varcoe JR (2009) Techniques for the study and development of microbial fuel cells: an electrochemical perspective. Chem Soc Rev 38(7):1926–1939. https://doi.org/10.1039/b819866g
69. Sirinutsomboon B (2014) Modeling of a membraneless single-chamber microbial fuel cell with molasses as an energy source. Int J Energy Environ Eng 5(2–3):1–9. https://doi.org/10.1007/s40095-014-0093-5
70. Roy S, Marzorati S, Schievano A, Pant D (2017) Microbial fuel cells, vol 3. Elsevier. https://doi.org/10.1016/B978-0-12-409548-9.10122-8
71. Logan BE, Regan JM (2006) Electricity-producing bacterial communities in microbial fuel cells. Trends Microbiol 14(12):512–518. https://doi.org/10.1016/j.tim.2006.10.003
72. Du Z, Li H, Gu T (2007) A state of the art review on microbial fuel cells: a promising technology for wastewater treatment and bioenergy. Biotechnol Adv 25:464–482. https://doi.org/10.1016/j.biotechadv.2007.05.004
73. Parkash A (2016 July) Microbial fuel cells: a source of bioenergy. Microbial and Biochemical Technology (July). https://doi.org/10.4172/1948-5948.1000293

74. He Z, Minteer SD, Angenent LT (2005) Electricity generation from artificial wastewater using an upflow microbial fuel cell. Environ Sci Technol 39(14):5262–5267. https://doi.org/10.1021/es0502876
75. Feng C, Tsai C-C, Ma C-Y, Yu C-P, Hou C-H (2017) Integrating cost-effective microbial fuel cells and energy-efficient capacitive deionization for advanced domestic wastewater treatment. Chem Eng J 330:1–10. https://doi.org/10.1016/j.cej.2017.07.122
76. Feng Y, He W, Liu J, Wang X, Qu Y, Ren N (2014) A horizontal plug flow and stackable pilot microbial fuel cell for municipal wastewater treatment. Biores Technol 156:132–138. https://doi.org/10.1016/j.biortech.2013.12.104
77. Santoro C et al. (2018 Feb) Ceramic microbial fuel cells stack: power generation in standard and supercapacitive mode. Sci Rep 8. https://doi.org/10.1038/s41598-018-21404-y
78. Min B, Logan BE (2004) Continuous electricity generation from domestic wastewater and organic substrates in a flat plate microbial fuel cell. Environ Sci Technol 38(21):5809–5814. https://doi.org/10.1021/es0491026
79. Helder M, Strik DP, Hamelers HVM, Buisman CJN (2012) The flat-plate plant-microbial fuel cell: the effect of a new design on internal resistances. Biotechnol Biofuels 5(1):70. https://doi.org/10.1186/1754-6834-5-70
80. Ding HH, Chang S, Liu Y (Nov 2017) Biological hydrolysis pretreatment on secondary sludge: enhancement of anaerobic digestion and mechanism study. Biores Technol 244(Pt 1):989–995. https://doi.org/10.1016/j.biortech.2017.08.064
81. Calderone L (2019) What is the future of non-renewable resources? Other Energy Topics.
82. Kumar A, Ogita S, Yau YY (2018) Biofuels: greenhouse gas mitigation and global warming. Springer, India. https://doi.org/10.1007/978-81-322-3763-1
83. IPCC (2014) Climate change 2014, mitigation of climate change. In: Contribution of working group III to the fifth assessment report of the intergovernmental panel on climate change. Cambridge, UK
84. IPCC (2014) Climate change. Synthesis report. Contribution of working groups I, II and III to the fifth assessment report of the intergovernmental panel on climate change. Geneva
85. Kan S, Chen B, Chen G (2019) Worldwide energy use across global supply chains: Decoupled from economic growth? Appl Energy 250:1235–1245. https://doi.org/10.1016/j.apenergy.2019.05.104
86. Ahmad T, Zhang D (2020) A critical review of comparative global historical energy consumption and future demand: the story told so far. Energy Rep 6:1973–1991. https://doi.org/10.1016/j.egyr.2020.07.020
87. Robert R (2020) Fossil fuels still supply 84 % of world energy—and other eye openers from BP's annual review

Utilization of Electrogenic Bacteria Consortium for Sewage Sludge Treatment via Organic Compound Degradation

Amira Suriaty Yaakop, Ooi Kah Hong, and S. M. Salman

Abstract Electrogenic bacteria (EB), also known as electrigens and exoelectrogens, can transmit electrons extracellularly beyond the cell membrane to or from electron acceptors such as electrodes, oxide minerals, and other bacteria under anaerobic or microaerobic conditions. These bacteria catalyze electrochemical oxidations or reductions at an anode or cathode, respectively, to generate an electric current in microbial fuel cells. Originally, these fuel cells were inefficient and could only be used as a battery in isolated locations. Microbial fuel cell (MFC) has been used as one of the green technology sources and one of the promising approaches for green energy production. Currently, few studies highlight the electron production ability of bacteria found in MFC and the application also is underexplored. Hence, this chapter will discuss the EB ability, characteristic, and mechanism in producing electrons. How EB can be utilized in sludge degradation and what are the secondary metabolites produced by the EB that can benefit the environment.

Keywords Electrogenic bacteria (EB) · Renewable energy · Microbial fuel cell · Bacteria enzymes · Secondary metabolites

1 Introduction

In recent decades, chemical use has increased due to increased industrial, agricultural, and human activities. In addition, human social growth and a false sense of personal safety produce a large number of chemicals in both the aquatic and terrestrial environments. For example, river water is used for agriculture, bathing, and drinking in many areas worldwide. Furthermore, sewage water is used for cultivation in many impoverished nations without being treated. As a result, untreated agricultural wastewater

A. S. Yaakop (✉) · O. K. Hong · S. M. Salman
School of Biological Sciences, Universiti Sains Malaysia, 11800 Minden, Penang, Malaysia
e-mail: amirasuriaty@usm.my

M. Mohd Zaini Makhtar et al. (eds.), *Microbial Fuel Cell (MFC) Applications for Sludge Valorization*, Green Energy and Technology,
https://doi.org/10.1007/978-981-99-1083-0_7

introduced a slew of contaminants into the environment and required urgent solutions for their removal. Chemical and physical methods for removing pollutants are extensively used and successful. However, these technologies have certain drawbacks, including cost, efficiency, and the creation of secondary pollutants. As a result, biological therapy appears to be more favourable and is regarded as a green removal method. One of the methods for wastewater treatment is using microbial fuel cells that involve the utilization of Electrogenic Bacteria (EB). These bacteria are also reported to have various hydrolytic enzymes that can work well in degrading complex carbon compounds, including sludge.

2 Mechanisms of Electrogenic Bacteria

Electrogenic bacteria (EB) are a group of bacteria capable of converting chemical energy into electrical energy by carrying out the respiration process [1]. The oxidation of organic substrates would thereby generate electrons that could be received by external electron acceptors such as metal oxides, other bacteria, or even electrodes for electricity production [2, 3]. EB could be broadly classified into two categories, namely electricigens and electrotrophs [4]. Electricigens contain a subset of bacteria with the ability to completely oxidize organic compounds to carbon dioxide (CO_2) and donate electrons directly to an electron acceptor whereas microorganisms that can act to reduce terminal electron acceptors are known as electrotrophs [4]. Electron transfer from the electrogenic bacteria to the electron acceptors primarily utilized two main mechanisms (Fig. 1), namely (i) Direct or non-mediated transfer as well as (ii) mediated transfer [2].

Direct electron transfer (DET) requires a direct contact between the anode surface and the microorganism's outer membrane, whether through direct cell attachment or nanowires (pili). Pili are nanowires that connect the membrane of the microorganism to the anode surface. The benefits of pili development include the ability for many layers of biofilm microorganisms to participate in electron transfer while bulk microorganisms do not [4].

Indirect electron transfer through external or internal mediators. In this type, a redox-active material (mediator) is responsible for the electron transfer between the microorganism and the anode surface. This redox can either be exerted naturally by the microorganisms (internal) or can be added from outside (external). These mediators whether internal or external will be responsible for the electron transfer from the bulk microorganisms to the anode surface. The electron transfer in the mediated electron transfer is higher than that in the DET [4].

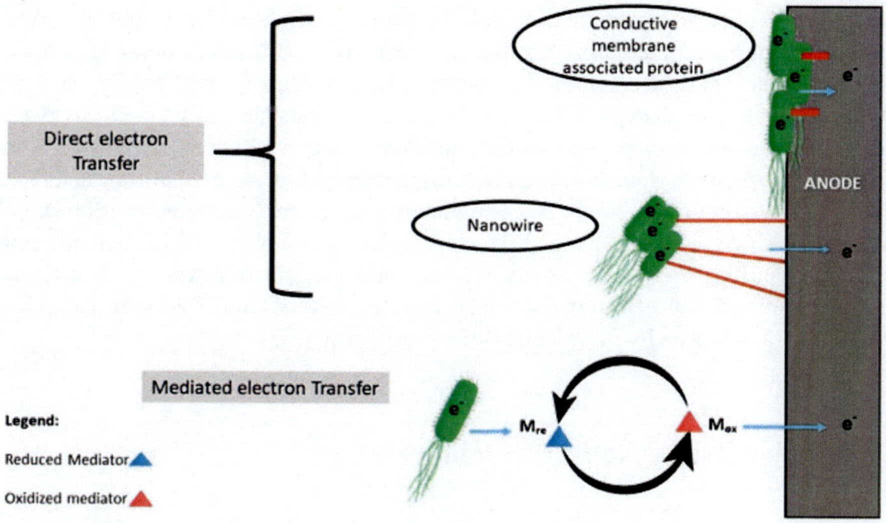

Fig. 1 Mechanism of electron transfer from the electrogenic bacteria (EB) to the electron acceptors

3 Flagella

Direct transfer of electrons would involve the establishment of physical contact between the electron acceptors and the electrogenic bacteria themselves. Micro-sized, protein-based biological extensions of electrically conductive flagella or pili known as bacterial/microbial nanowires which are found on the outer membrane of electrogenic bacteria play the vital role of transporting the electrons [2]. The nanowires are typically found in the genus of *Geobacter sp.* and *Shewanella sp.* for long-range electron transfer into the cytosol [2]. One type of nanowire known as Type IV pili are crucial in the secretion systems for effectors, microbial adherence, and locomotion in addition to making contact between the bacterial species and electron acceptors [2]. Bacteria such as *Geobacter sulfurreducens* were a model organism studied showing that Type IV pili were involved in the attachment of the bacteria and subsequently the translocation of the electrons during the reduction of iron (III) ions (Fe^{3+}) [2].

4 Extracellular/Mediated Transfer

Mediated transfer, on the other hand, is a condition where the electrogenic bacteria forfeits physical contact with the electrodes and rely on endogenous redox mediators/electron shuttles instead, commonly in the forms of secondary metabolites produced by the bacteria, that shuttles the electrons produced to an external electron acceptor [2]. Electrons in mediated-transfer-utilizing bacteria are first transported to the cell surface as an aftermath of metabolic pathways. Subsequently, they are transferred into potential mediators residing in either the periplasm or outer membrane that's capable of diffusing into the extracellular environment and finally received by external electron acceptors [2, 5]. Several compounds currently known to effectively mediate electrons would include thionine, methyl viologen, 2-hydroxy-1,4-naphtoquinone, methylene blue, humic acids, anthraquinone-2,6-disulfonic acid, and flavin [2]. Among them, flavin has been demonstrated by Marsili et al. (2008) and Velasquez-Orta et al. (2010) as a class of secondary metabolite vital in the mediation of electrons using *Shewanella oneidensis* as a model organism where higher concentrations and removal of flavin increased peak current by four-fold and decreased electrons transfer rate by more than 70% respectively [2].

5 Biofilm/Redox-Active Proteins

Membrane-bound redox-active proteins known as cytochromes are another component that helped in facilitating the short-range direct transfer of electrons generated by electrogenic bacteria to the electron acceptors [2]. Cytochromes, particularly cytochrome C (cyt-C) are among the key components in the adenosine triphosphate (ATP) synthesizing electron transport chain (ETC) process. Cyt-C acts as an intermediatory protein that passes down electrons in a serial redox reaction until they encounter a terminal electron acceptor. Cyt-C such as OmcS, OmcE, and OmcB found on the surface of *Geobacter sulfurreducens* are some common manifestations of the role of redox-active protein in transferring electrons which are crucial in the reduction of iron (III) oxides, an example of naturally occurring terminal electron acceptors [2]. Electron transfer occurs through biofilm formation by electrogenic bacteria where the electrons travel through the biofilm to the electron acceptors using the nanowires linked to the cytochromes.

6 Light-Dependent Direct Transfer

Photosynthetic bacteria, a group of solar energy harnessing bacteria through the use of chlorophyll in order to produce sugars [6], were also found to have electrogenic properties, with particular attention given to cyanobacteria. Cyanobacteria, a group of microorganisms having the oxygenic photosynthetic pathways of photosystems I and II (PSI and PSII) [7], were discovered to possess conserved light-dependent electrogenic activity where they own the innate ability to transfer electrons to their surroundings in response to illumination [8]. They are able to utilize the electrons from the photolysis of water by PSII which would be transmitted to extracellular electron acceptors via plastoquinone and cytochrome bd quinol oxidase during the occurrences of photosynthetic electron transport chain (P-ETC) [9]. Such an activity is driven by red and blue but not green light which shows consistencies in the predominant role of phycobilisomes, pigment-protein complexes comprising *phycocyanins* and *allophycocyanins* that absorb red light in cyanobacteria [8].

Photosynthesis in cyanobacteria begins with photolysis, a reaction where water is split into hydrogen ions, oxygen, and electrons, aided by the catalysis of proteins in PSII and driven by light energy [9]. Electrons acquired from the photolysis reaction are transported along the P-ETC from PSII to plastoquinone (P.Q.), then to the cytochrome b_6f (cyt-b_6f) complex, plastocyanin (P.C.), and eventually to PSI [9]. However, under a high light condition where bd quinol oxidases are maximally active to prevent the reduction of quinol pool, electrons at P.Q. would instead divert to bd quinol oxidase before being released into the surrounding medium [9]. The mechanism of the movement of electrons from the bd quinol oxidases to the environment remains unknown despite the conclusion of the study by Pisciotta et al. (2011) [9].

7 Bacteria Genera Involve as Electrogenic Bacteria (EB)

Electrogenic bacteria are organisms that can transfer electrons to extracellular electron acceptors and have the potential to be used in devices such as bioelectrochemical systems. In 2009, Logan BE wrote a review and list some of the well-known electrogenic bacteria (Table 1) [10]. Besides the reported bacterium, this subtopic highlighted some other bacteria genera that were also used extensively in microbial fuel cells (MFC).

7.1 Bacillus

Bacillus is a Gram-positive, sporulating, rod-shaped bacterium genus belonging to the Firmicutes phylum. It has around 266 named species. The plural *Bacilli* is the name of the bacterial class to which this genus belongs, and the term is also used to describe the form (rod/cylindrical) of specific bacteria. *Bacillus* species can either be obligatory aerobes, meaning they require oxygen to survive, or facultative anaerobes, meaning they can survive without it. If oxygen has been utilized or is present, cultured *Bacillus* species will produce the enzyme catalase [1]. Only one spore is produced by each bacterium, and it is resistant to heat, cold, radiation, desiccation, and disinfectants. Most *bacilli* are saprophytes meaning that they live on dead or decaying organic matter. *Bacilli* have a diverse array of physiologic properties that allow them to thrive in a wide range of environments, including desert sands, hot springs, and Arctic soils. *Bacillus* species can be thermophilic, psychrophilic, acidophilic, alkaliphilic, halotolerant, or halophilic, meaning they can thrive at pH values, temperatures, and salt concentrations that few other organisms can. *Bacillus insolitus* may thrive at temperatures as low as 0° Celsius. The *Bacillus* cell wall is a structure on the outside of the cell that serves as a second barrier between the bacteria and the environment while maintaining the rod shape and withstanding the

Table 1 Exoelectrogenic bacteria that power microbial fuel cells [10]

Year	Microorganism	Comment
1999	*Shewanella putrefaciens* IR-1	Direct proof of electrical current generation in an MFC by a dissimilatory metal-reducing bacterium (Gammaproteobacteria)
2001	*Clostridium butyricum* EG3	First Gram-positive bacterium was shown to produce electrical current in an MFC (phylum Firmicutes)
2002	*Desulfuromonas acetoxidans*	Identified in a sediment MFC community and shown to produce power (Deltaproteobacteria)
	Geobacter metallireducens	Shown to generate electricity in a poised potential system (Deltaproteobacteria)
2003	*Geobacter sulfurreducens*	Generated current without poised electrode (Deltaproteobacteria)
	Rhodoferax ferrireducens	Used glucose (Betaproteobacteria)
	A3 (Aeromonas hydrophila)	Deltaproteobacteria
2004	*Pseudomonas aeruginosa*	Produced low amounts of power through mediators such as pyocyanin (Gammaproteobacteria)
	Desulfobulbus propionicus	Deltaproteobacteria
2005	*Geopsychrobacter electrodiphilus*	Psychrotolerant (Deltaproteobacteria)
	Geothrix fermentans	Produced an unidentified mediator (phylum Acidobacteria)
2006	*Shewanella oneidensis* DSP10	Achieved a high power density (2 W per m^2 or 500 W per m^3) by pumping cells grown in a flask into a small (1.2 mL) MFC (Gammaproteobacteria)
	S. oneidensis MR-1	Various mutants identified that increase current or lose the ability for current generation (Gammaproteobacteria)
	Escherichia coli	Found to produce current after long acclimation times (Gammaproteobacteria)
2008	*Rhodopseudomonas palustris* DX-1	Produced high power densities of 2.72 W per m^2 compared with an acclimated wastewater inoculum (1.74 W per m^2) (Alphaproteobacteria)
	Ochrobactrum anthropi YZ-1	An opportunistic pathogen, such as *P. aeruginosa* (Alphaproteobacteria)
	Desulfovibrio desulfuricans 56	Reduced sulphate when growing on lactate; resazurin in the medium was not thought to be a factor in power production (Deltaproteobacteria)
	Acidiphilium sp. 3.2	Current at low pH and in the presence of oxygen in a poised potential system (Alphaproteobacteria)

(continued)

Table 1 (continued)

Year	Microorganism	Comment
	Klebsiella pneumoniae L17[b]	The first time this species produced current without a mediator (Gammaproteobacteria)
	Thermincola sp. strain JR59	Phylum Firmicutes
	Pichia anomala[b]	Current generation by a yeast (kingdom Fungi)

[a]Air cathode microbial fuel cells (MFCs), except where noted [10]
[b]Ferricyanide cathode

pressure created by the cell's turgor. Teichoic and teichuronic acids make up the cell wall. *B. subtilis* is the first bacterium to discover the involvement of an actin-like cytoskeleton in cell shape determination and peptidoglycan synthesis, as well as the complete collection of peptidoglycan-synthesizing enzymes. The cytoskeleton plays a crucial role in the formation and preservation of shape. *Bacillus* species like *Bacillus cereus, Bacillus subtilis, and Bacillus megaterium* have the ability to produce electricity in microbial fuel cell (MFC). Islam et al. prove that by using cyclic voltammetry (CV) and electrochemical impedance spectroscopy (EIS), the CV of *B. cereus* MFCs displayed a significant redox peak, indicating that *B. cereus* exhibits electrogenetic characteristics. Furthermore, the addition of *B. cereus* to activated sludge (AS) increased the MFC's power generation (4.83 W/m^3) and C.E. (22%) as compared to the comparable values for an MFC infected simply with AS (1.82 W/m^3, 12%) [11].

7.2 *Geobacter*

Geobacter is a genus of Proteobacteria. It is rod-shaped bacteria with flagella. *Geobacter* is known for its remarkable electron transfer and environmental restoration abilities, which lends itself to a variety of industrial applications. The sequencing of *Geobacter sulfurreducens* has revealed additional information about the bacteria's capabilities. *Geobacter* has been discovered to be able to transition towards metallic substances because of this sequencing. *Geobacter* also contains genes that allow it to operate in the presence of oxygen, according to previous studies *Geobacter* appears to have approximately 100 genes that code for distinct c-type cytochrome forms. Geobacter's ability to reduce metals and generate power is due to its wide range of c-type cytochromes [2]. *Geobacter* was initially discovered in the Potomac River in 1987 [3]. *G. sulfurreducens* was later discovered in a hydrocarbon-contaminated soil sample in Oklahoma [2]. *Geobacter* species have also been discovered to be able to breathe through a graphite electrode. Geobacter species are anaerobic respiration bacterial species with bioremediation properties. *Geobacter* was discovered to be the first organism capable of oxidizing organic chemicals and metals, such as iron, radioactive metals, and petroleum compounds, into ecologically friendly carbon dioxide utilizing iron oxide or other readily accessible metals as electron acceptors.

7.3 *Klebsiella*

Klebsiella is a Gram-negative, oxidase-negative genus of rod-shaped bacteria with a polysaccharide-based capsule. *Klebsiella* species may be found all over the world. This is assumed to be owing to different sublineages generating niche adaptations and biochemical adaptations that make them more suited to a certain environment. They can be found in various environments, including water, soil, plants, insects, and people. Edwin Klebs, a German-Swiss scientist, and was given the name Klebsiella. For many years, Klebsiella bacillus was known as Friedlander bacillus named after Carl Friedlander, who characterized it. Klebsiella bacteria are found in the typical flora of humans and animals' noses, mouths, and intestines. Klebsiella is a Gram-negative bacterium that is normally non-motile. When compared to other members of the Enterobacteriaceae family, they are shorter and thicker. The cells are rod-shaped and range in size from 0.3 to 1.5 μm width by 0.5–5.0 μm long. They come in singles, pairs, chains, and end-to-end links. Like the other members of the Enterobacteriaceae family, *Klebsiella* may grow on conventional lab media and has no particular growth needs. The species is aerobic but anaerobic in nature. Their optimal growing temperature is from 35° to 37° Celsius, with a pH of around 7.2 [4].

Although certain species create a visible capsule or slime layer that may be utilized for serologic identification, molecular serotyping may eventually replace this approach. On their cell surfaces, certain species of this genus often express two kinds of antigens. The first, O antigen, is a component of lipopolysaccharide (LPS), which comes in nine different types. The second is K antigen, a capsular polysaccharide with over 80 variants [5]. Both have a role in pathogenicity and serve as the foundation for serogrouping. Several vaccines have been developed based on these two key antigenic determinants [6]. Some *Klebsiella* species like *Klebsiella pneumoniae* and *Klebsiella variicola* are used in microbial fuel cells (MFC); however, there was not much explanation on its electrogenic mechanism can be found.

7.4 *Pseudomonas*

Pseudomonas is a Gram-negative Gammaproteobacteria genus that belongs to the Pseudomonadaceae family and has 191 validly recognized species [7]. Species of this genus display some characteristics: Rod-shaped, Gram-negative, flagellum one or more providing motility, aerobic, non-spore-forming, catalase-positive, and oxidase-positive. Members of the genus exhibit a wide variety of metabolic diversity, allowing them to colonize a wide range of habitats. *Pseudomonas aeruginosa* in its role as an opportunistic human pathogen, *Pseudomonas syringae as a plant pathogen, Pseudomonas putida* as a soil bacterium, and plant growth-promoting *Pseudomonas fluorescens, Pseudomonas lini, Pseudomonas migulae,* and *P. graminis* are among the best-studied species [8, 8].

The *pseudomonads* were discovered early in the history of microbiology due to their extensive prevalence in water and plant seeds such as dicots. Walter Migula classified *Pseudomonas* as a genus of Gram-negative, rod-shaped, polar-flagellated bacteria with some sporulating species in 1894 and 1900 in hazy terms [10, 11]. The latter assertion was eventually proven to be erroneous due to the reserve materials' refractive grains. *Pseudomonas pyocyanea* (*basonym* of *Pseudomonas aeruginosa*) proved to be the most accurate descriptor [12].

With a few exceptions, the production of pyoverdine, a luminous yellow-green siderophore under iron-limiting circumstances, is another feature associated with *Pseudomonas* species. Certain *Pseudomonas* species, such as *Pseudomonas aeruginosa* and *Pseudomonas fluorescens*, can create other forms of siderophores, such as pyocyanin and thioquinolobactin. *Pseudomonas* species also typically give a positive result to the oxidase test, which indicates the absence of gas formation from glucose.

One extensively use Pseudomonas species in MFC is *P.aeruginosa* [1]. It is exoelectrogenic bacteria that can generate secondary metabolites like pyocyanin (PYO) and 1-hydroxy-phenazine (OHPHZ) and use them as redox-active metabolites for facilitating the extracellular electron transfer between bacterial cells and electrodes in MFCs [12, 13].

7.5 *Lysinibacillus*

Lysinibacillus are motile, Gram-positive bacteria with rod-shaped cells that generate ellipsoidal or spherical endospores [13]. Previously, *Lysinibacillus* species were categorized as Bacillus species. Two previously listed species, *Bacillus sphaericus* and *Bacillus fusiformis*, were reclassified to the new genus *Lysinibacillus*, together with a unique species (*Lysinibacillus boronitolerans*). In 2007, Ahmed, Yokota A, Yamazoe, and Fujiwara suggested *Lysinibacillus* as a separate genus from *Bacillus* based on the discovery of lysine, aspartic acid, alanine, and glutamic acid in its peptidoglycan [14]. Since then, other *Bacillus* species have been proposed to be reclassified as *Lysinibacillus*. Tests for oxidase and catalase are positive, while indole and H2S generation are negative. A4 is the peptidoglycan type found on the cell wall (LyseAsp). Iso-C 15: 0 is the most abundant fatty acid in cells. MK-7 is the most abundant menaquinone, whereas diphosphatidylglycerol, phosphatidylglycerol, and ninhydrin-positive phosphoglycolipid are the most common polar lipids. The amount of G C in the sample is 35e38 mol% [14].

Lysinibacillus is well-known for its insecticidal effect against a variety of insects, including mosquitos, which are known to transmit several human illnesses. Furthermore, some *Lysinibacillus* species can remove heavy metals from the environment. *Lysinibacillus spp.* have recently grabbed researchers' interest as plant growth promoters and disease control agents that might be employed instead of agrochemicals.

The use of *Lysinibacillus* spp. in MFCs is a relatively new addition to the long list of biocatalysts that have been studied for their potential in power generation. *Lysinibacillus sphaericus* has been shown to generate a maximum power density of 85 mW/m^2 and current density of $\approx$270 mA/m^2 using graphite felt as an electrode. The species has also been found to be efficient in utilizing proteinaceous material which is useful to treat a specific type of wastewaters like wastewater from slaughterhouses or from meat packaging industry [14]. *Lysinibacillus macroides* is another *Lysinibacillus* spp. that has been found to be electrogenic bacteria. Uma et al. (2017) investigated the adhesion of bacteria as a biofilm on pencil graphite lead using a fluorescence microscope and a scanning electron microscope (SEM) [15].

8 Sludge as Carbon-Rich Sources for EB

Global water contamination issues have stemmed from rapid industrialization and urbanization worldwide. Sludge is produced in large quantities by traditional wastewater treatment plants. In 2005, the United States produced 7.6 million tonnes of dry sludge; by 2010, this figure is expected to rise to 8.2 million tonnes [16]. In 2005, the EU generated 10 million tonnes of dry sludge [17]; from 2007 to 2013, China significantly produced 6.25 million tonnes of dry solids (DS) sludge, with output expected to rise to 3.6 million tonnes in 2010 [18]. Malaysia is no exception to the global trend of sewage sludge volume increasing annually as a developing country. Malaysia produces roughly 3 million metric tonnes of sewage sludge each year, with that number predicted to rise to 7 million metric tonnes by 2020. (Indah Water Konsortium Sdn Bhd, 2010). Due to various pollutants such as organic contaminants, pathogens, and a portion of heavy metals [19, 20], sewage sludge is considered a potential source of secondary environmental contamination. These pollutants can enter the natural ecosystem through improper disposal methods, posing a risk to human health and the environment [21]. For example, when agricultural land is treated with sewage sludge, the concentration of organic contaminants (OCs) rises above that of reference soil, and some OCs enter inhabited plant and animal tissues [22]. On the other hand, the high levels of heavy metals in sludge ash, which is utilized as a construction material, pose a significant risk of contamination. Such sludge cannot be disposed of until it has undergone the necessary treatments. Therefore, it will benefit sludge to be converted into a simpler organic compound that does not give a thread to the environment.

8.1 Composition of Sludge

During wastewater treatment, a considerable number of contaminants are removed from the liquid phase and transported to the semisolid phase, which is known as sludge. Sludge disposal standards are governed by European legislation, namely, the Urban Wastewater Treatment Directive (91/271/EEC). However, the definition varies in every country. Table 2 shows the definitions of sewage sludge according to various legal acts [23].

As a result of expanding population, industrialization, and wastewater treatment requirements, sludge generation in wastewater treatment plants (WWTP) and landfill accumulation has increased in many countries [13]. The long-term management of this hazardous waste has become a major environmental challenge due to the potential for detrimental ecological repercussions. [14]. The end product of the sludge treatment is the dewatering process to reduce the waste volume to prepare it for disposal. Before disposal, drying the sludge with a dewatering filter press reduces its weight and volume.

The organic concentration of dewatered sewage sludge (about 84% moisture content, organic content: 33.4% protein, 6.6% lipid, and 3.3% carbohydrate on an organic basis) is high, and its volume is about one-tenth that of sewage sludge. The consistency of dewatered sewage sludge is solid. As a result, anaerobic digestion is not possible. Compared to dewatered sewage sludge, the volume of the liquidized sludge remains unchanged. The number of proteins in the liquidized sludge (27.6%) decreased through liquefaction, but the ammonium content increased. The amount of lipids in the liquidized sludge rose (13.8%). According to these findings, proteins were hydrolysed and degraded to volatile acids and ammonia via amino acids. The liquid phase (6.9% VS concentration) contains adequate anaerobic digesting substrates (15.5% lipid and 9.2% carbohydrate) [24]. These findings suggest that anaerobic liquid phase digestion could result in a higher digestion ratio and higher methane generation. The solid phase, which had a low moisture content (77.0%), may be composted and used for various purposes, including pipeline transportation and anaerobic digestion (Chapter 'Microbial Fuel Cells (MFC) as an Alternative Energy Source: The Perceptions and Attitudes Towards Sustainable and Renewable Energy in Malaysia', Table 3) [25].

9 Sludge Utilization by EB in MFC

Using biological decomposition of organic matter to produce energy, wastewater treatment with MFCs has been researched for the removal and recovery of contaminants such as Chemical Oxygen Demand (COD), heavy metals, and ammonia (NH^3) [26]. MFCs can be used to recover high-value products such as silver (Ag) or chromium (Cr), which can improve the sustainability of the process for large-scale operations [27]. The anaerobic method is used to remediate wastewater and relies

Table 2 Definitions of sewage sludge according to various legal acts

Legal act	Definition	Comment
Council Directive 91/271/EEC concerning urban wastewater treatment (Council Directive 91/271/EEC, 1991)	Sewage sludge includes domestic septage, scum, or particles removed during primary, secondary, or advanced wastewater treatment methods and material created from sewage sludge. Solid, semisolid, or liquid sludge can form during domestic sewage treatment at treatment works. Sewage sludge does not contain the ash formed during the combustion of sewage sludge in a sewage sludge incinerator, nor the grit and screens produced during the preliminary treatment of residential sewage in a treatment plant	It concerns only urban wastewater, not industrial. It does not include faecal sludge
Council Directive 86/278/EEC (Council Directive 86/278/EEC, 1986)	Residual sludge from: • Sewage plants treating domestic or urban wastewater and from other sewage plants treating wastewater of a composition similar to domestic and urban wastewater • Residual sludge from septic tanks or other similar installations for the treatment of sewage • Residual sludge from sewage plants other than that mentioned above	It is a wide definition that includes all types of sludge considered in the chapter

(continued)

Table 2 (continued)

Legal act	Definition	Comment
USA Part 503 (40 CFR Part 503)	Sludge formed during the treatment of household sewage in treatment works might be solid, semisolid, or liquid. Domestic septage, scum, or particles removed in primary, secondary, or advanced wastewater treatment procedures and material generated from sewage sludge are all examples of sewage sludge. The ash produced during the combustion of sewage sludge in a sewage sludge incinerator, as well as the grit and screens produced during the preliminary treatment of household sewage in a treatment plant, are not included in sewage sludge	It concerns only the sludge produced during domestic wastewater treatment and septage sludge
EPA definition 40 CFR 260 (40 CFR Part 260, n.d.)	Sludge is defined as any solid, semisolid, or liquid waste generated by a municipal, commercial, or industrial wastewater treatment plant, water supply treatment plant, or air pollution control facility, excluding treated effluent	The definition is broad, encompassing both industrial and municipal sludge, as well as electroplating sludge

Table 3 Properties of dewatered sewage sludge, liquidized sludge, and its centrifuged phases [24]

	Moisture content (%)	Organic content (%)	Ash content (%)
Dewatered sewage sludge	83.8	12.4	3.8
Liquidized sludge	84.8	11.4	3.8
Liquid phase	92.5	6.9	0.6
Solid phase	77.0	18.0	5.0

on bacteria to transport the electrons [28]. The high organic concentration shown in Table 3 of dewatered sewage sludge serve as good carbon source for the electrogenic bacteria.

9.1 Enzyme Utilization in Wastewater Treatment

Wastewater is the source of growth of several anaerobic and facultative anaerobic bacteria, which have the ability to transfer electrons to an anode as a terminal electron acceptor and are thus classified as electrogenic bacteria [2]. At the anode of an MFC, the organic substrate is oxidized by electrochemically active microorganisms. Subsequently, the microorganisms transfer the electrons resulting from this oxidation of the anode, which then passes through an external circuit to the cathode, thus generating electricity.

Traditional activated sludge can remediate most pollutants effectively, while resistant contaminants like oil, grease, pharmaceuticals, pesticides, and plastics are difficult to be eradicated. Pharmaceuticals, insecticides, plastics, and personal care items, all of which are micropollutants, are referred to as emerging pollutants or emerging concern contaminants [29, 30]. These pollutants mainly include oil, grease, and organic micropollutants (Fig. 2). Oil and grease-containing wastewater are usually produced by dairies, oil mills, slaughterhouses, and food waste [20, 21]. The transfer rate of substrates, products, and oxygen will be harmed if oil and grease float on water surfaces [31]. The floating oil and grease may induce a filamentous microbe bloom, resulting in floating sludge, poor sedimentation, and sludge biomass loss, as well as poor performance of activated sludge [31].

Enzyme is a powerful biocatalyst that can destroy compounds in a controlled environment [32, 33]. Enzymes contain active sites that bind to specific substrates, lowering the activation energy during enzymatic operations. As a result, the reaction kinetics and specificity of these processes are quite high. There are six enzymes, and the most used ones in wastewater treatment are hydrolases and oxidoreductases [34, 35]. These two enzymes can biocatalyst most pollutants in wastewater due to their wide range of substances. Currently, commercially available enzymes include lipase,

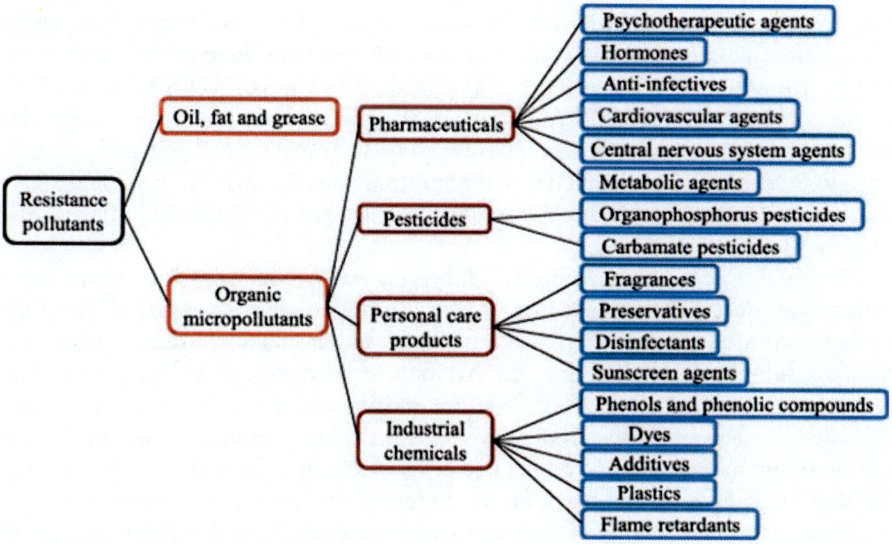

Fig. 2 Typical resistant pollutants in wastewater and their classification. Reprinted with permission from [38] (CC-BY)

laccase, and peroxidase. Laccase and peroxidase are commonly used to remove organic micropollutants [36, 37].

9.2 *Enzyme Production by Bacteria*

Extracellular hydrolytic enzymes such as amylases, proteases, lipases, DNases, pullulanases, and xylanases have a wide range of potential applications in the food, feed, biomedical, and chemical industries [39, 40]. Concerns over the accumulation of micropollutants in the aquatic environment inspired a flurry of studies into biological micropollutant degradation in wastewater treatment systems. [41]. Some organic micropollutants may be hazardous and bioaccumulative, posing serious risks to human health and the environment. This bioaccumulation is usually linked to a compound with high lipid solubility and the capacity to accumulate in living creatures' fatty tissues for an extended time. These persistent compounds migrate up the food chain, and their concentration rises as they're absorbed and degraded in certain organs, increasing their environmental toxicity [42].

Physiochemical and biological approaches are the most common wastewater treatment technologies nowadays. Physiochemical procedures such as chemical oxidation, distillation, membrane-based separation techniques, and adsorption have been used for wastewater treatment [34]. These are treatment-oriented options, but they are exceedingly expensive and may cause more pollution and harm [35]. The three common microbial enzymes use in wastewater treatment are oxygenases,

Laccases, and Cellulases. Microbial oxygenases belong to the oxidoreductase family of enzymes and are involved in the oxidation of reduced substrates by transferring oxygen from atomic oxygen (O^2) to a cosubstrate such as FAD/NADH/NADPH. These enzymes play an important role in the metabolism of organic molecules by increasing their reactivity or dissolvability in water or causing the aromatic rings to be cleaved. In conjunction with multifunctional enzymes, oxygenases also intervene in the dehalogenation reactions of halogenated methanes, ethanes, and ethylenes [43].

Microbial Laccases (p-diphenol: dioxygen oxidoreductase) are multicopper oxidases produced by plants, fungi, insects, and microorganisms that catalyze the oxidation of a wide range of reduced phenolic and aromatic substrates while also reducing sub-atomic oxygen to water. Aromatic compounds, including phenols and aromatic amines, are among the most common types of poisons and are heavily regulated in many countries. Coal conversion, oil refining, resins and plastics, wood safeguarding, metal coating, colours and other synthetic chemicals, textiles, mining and dressing, and pulp and paper are just a few of the industries where they can be found in wastewater. During enzymatic hydrolysis, cellulose is debased by the cellulases to reducing sugars that yeasts or microorganisms can ferment to ethanol. In addition, Cellulases cause the removal of cellulose microfibrils, which are produced during washing and the utilization of cotton-based materials. In the paper and pulp industry, cellulase is utilized to remove ink during the recycling of paper. For as far back as a decade, there has been an interest in the enzymatic hydrolysis of cellulose. This intrigue comes from the benefits that such a procedure would offer, in particular, the conversion of lignocellulosic and cellulosic waste to a helpful energy source through the creation of sugars, ethanol, biogas, or other vigorous end products [43]. Beside the three common enzymes used in wastewater treatment, selection of electrogenic bacteria can be based on substrate specificity. For example, Nandy et al. (2013) used *Lysinibacillus sphaericus* in its MFCs using high protein waste due to its enzymes ability to utilize substrates mainly rich in protein components like beef extract and produce good electricity [14].

Incorporation of EB that produce hydrolytic microbial enzymes will help to treat wastewater. Industrial chemicals, pesticides, and petroleum hydrocarbons pollute water and are discovered as environmental pollutants in various aquatic and terrestrial environments due to their widespread use. The main process in the hydrolysis of organic contaminants is bacterial activity (Table 4). Only substances with a molecular mass of less than 600 Daltons can pass through cell pores. Therefore, extracellular enzyme activity is an important step in the breakdown and use of the organic compound [46].

In sludge degradation, bacteria tend to aggregate and create sludge flocs in the activated sludge process, which are made up of microbial, prokaryotic (bacteria, archaea), and eukaryotic (algae, fungal) microorganisms held together by extracellular polymeric molecules (EPS). Various studies reported that sludge flocs constitute 60%–70% of the organic fraction [45] and are further utilized by the bacteria.

Cell fractionation begins with killing microbial cells, which is done by a hydrolytic enzyme (mainly protease). An improvement in lysis efficiency can thus contribute to

Table 4 Applications of microbial enzymes [44]

S.I. No	Enzyme	Substrate	Reaction
1	Oxidoreductase		
1.1	Oxygenase		
1.1.1	Monooxygenase	Alkane, steroids, fatty acids, and aromatic compounds	Incorporation of oxygen atom to substrate and utilize substrate as reducing agent. Desulfurization, dehalogenation, denitrification, ammonification, and hydroxylation of substrate
1.1.2	Dioxygenase	Aromatic compounds	Introduction of two oxygen atoms to the substrate results in intradiol cleaving and extradiol cleaving with the formation of aliphatic product
1.2	Laccase	Ortho and paradiphenols, aminophenols, polyphenols, polyamines, lignins, and aryldiamines	Oxidation, decarboxylation, and demethylation of a substrate
1.3	Peroxidase		
1.3.1	Lignin peroxidase	Halogenated phenolic compounds, polycyclic aromatic compounds, and other aromatic compounds	Oxidation of substrate in the presence of cosubstrate H_2O_2 and mediator-like veratryl alcohol
1.3.2	Manganese peroxidase	Lignin and other phenolic compounds	In the presence of Mn^{2+} and H_2O_2 the cosubstrate catalyzes the oxidation of Mn^{2+} to Mn^{3+} which results in an Mn^{3+} chelateoxalate, which in turn oxidizes the phenolic substrates
1.3.3	Versatile peroxidase	Methoxybenzenes and phenolic aromatic	The enzyme catalyzes the electron transfer from an oxidizable substrate, with the formation and reduction of compound I and compound II intermediates
2	Hydrolase		
2.1	Lipase	Organic pollutants such as oil spill	The hydrolysis of triacylglycerols to glycerols and free-fatty acids
2.2	Cellulase	Cellulosic substance	Hydrolyses the substrate to simple carbohydrates

(continued)

Table 4 (continued)

S.I. No	Enzyme	Substrate	Reaction
2.3	Protease	Proteins	Enzymes that hydrolyze peptide bonds in an aqueous environment

a reduction in overall sludge output and play a key role in lowering investment and operational costs and optimizing the current sewage treatment system. Sludge hydrolysis can be aided by hydrolase. Previous research has revealed that several microbial strains with extracellular hydrolytic enzyme secretions and other commercial hydrolytic enzymes are frequently utilized directly in reactors to accelerate sludge lysis [31]. Hydrolysis enzymes, including protease, amylase, and lipase enzymes, are the primary agents of deflocculation, hydrolysis, and oxidation of sludge-activated sludge flocs to be broken down into simpler carbohydrate molecules. The first step in protein degradation involves breaking the protein down into peptides, or into two peptides and amino acids. Next, amino acids can be transformed into organic acids with low molecular weight, ammonia, and carbon dioxide [31].

10 EB Consortium Potential for Sludge Degradation

Currently, although the key principles of MFC design and operation are well understood, the microbiological aspects remain unclear. In an MFC system, the electron transfer from the bacteria to the anode can proceed directly from their cell membrane to the anode or indirectly by means of a mediator [4]. The direct transfer method or MFC inoculated with the mixed culture exhibited significant potential for harvesting energy from organic matter and degrading organic matter [1]. Thus, the selection of a highly performing microbial consortium (either pure or mixed culture) is crucial.

Consortium or co-culture electrogenic bacteria have been shown to improve the performance in MFCs. The ecological networks between the microorganisms in a specific co-culture system containing *P. aeruginosa* and *Klebsiella variicola* were studied by Islam et al. (2018) [46]. Compared to these two bacteria alone, the co-culture showed a 3 times higher current density in MFCs. Metabolite study revealed that the fermentative metabolite (1,3-propanediol) produced by *K. variicola* stimulated *P. aeruginosa* to create more pyocyanin, resulting in improved performance of co-culture MFCs fed with palm oil mill effluent through synergistic interactions (POME). This research shows that "interspecies ecological communication" based on metabolites can improve MFC electrochemical activity [46].

Other studies also reported some outstanding performance of co-culture bacteria in MFC. For instance, Venkataraman et al. (2011) [47] reported that the fermentation product (2,3-butanediol) produced by *Enterobacter aerogens* stimulated the production of mediator by *P. aeruginosa* that boosted up the current density by 14-fold in co-culture MFC compared to their monocultures. In another study, Wang et al.

(2015) [48] reported that the metabolite-enabled mutualistic interaction between *Shawanella oneidensis* and *E. coli* helped to achieve higher power generation in co-culture MFC compared to the monocultures in MFC. Thus, there is huge potential that the utilization of a perfect EB consortium could help the MFC performance in degrading the sludge and at the same time produce good current generation.

11 Conclusion

The chapter represents how electrogenic bacteria (EB) play essential parts in MFC, the potential of this microbe in converting the organic sources of any pollutants could benefit the world in handling the continuous waste production. The optimization of EB as a single or consortium bacteria should be explored especially in managing sludge problems. This is due to the versatility of the MFCs in converting any organic source to the current generation which is very suitable for sludge valorization. More detailed analyses should also be done in exploring the potential of the EBs as well, there were still undiscovered mechanisms that need to be understood despite the significant advancements. Up until now, there are still many EB that has not yet been identified and characterized. More discoveries are likely to be made due to an accurate analysis of the microbial diversity within EB that emerged in various habitats and advancements in culture conditions and analytical equipment. Additionally, employing inocula from harsh settings will help scientists find new electrogenic bacteria and possibly new ways to transmit electrons to electrodes. Future tools such as genomic and transcriptomic analyses could help to understand the amount and type of genetic changes that accumulate in evolving populations on electrodes over several generations. Over evolution, mutation rates can fluctuate, favourably accumulating genetic variations that are better suited to the MFC environment. Genetic differences between derived and ancestor organisms can be detected on a whole-genome level. This will absolutely help the researcher to set up the MFC more efficiently.

Acknowledgements Universiti Sains Malaysia (Malaysia) financially supported this research article under the APEX ERA Grant: 1001/PBIOLOGI/881005. The author (Amira Suriaty Yaakop) would like to thank Universiti Sains Malaysia for funding this project.

References

1. Agrawal K et al (2019) Microbial fuel cell: a boon in bioremediation of wastes. Microbial wastewater treatment. Elsevier, pp 175–194
2. Slate AJ et al (2019) Microbial fuel cells: an overview of current technology. Renew Sustain Energy Rev 101:60–81
3. Tahernia M et al (2020) Characterization of electrogenic gut bacteria. ACS Omega 5(45):29439–29446

4. Drendel G et al (2018) Microbial fuel cells, related technologies, and their applications. Appl Sci 8(12)
5. Uria N, Ferrera I, Mas J (2017) Electrochemical performance and microbial community profiles in microbial fuel cells in relation to electron transfer mechanisms. BMC Microbiol 17(1):1–12
6. Pollard TD et al (ed) (2017) Chapter 19—Mitochondria, chloroplasts, peroxisomes, in cell biology, 3rd edn. Elsevier, pp 317–329
7. Sánchez-Baracaldo P, Cardona T (2020) On the origin of oxygenic photosynthesis and Cyanobacteria. New Phytol 225(4):1440–1446
8. Pisciotta JM, Zou Y, Baskakov IV (2010) Light-dependent electrogenic activity of Cyanobacteria. PLoS One 5(5):e10821
9. Pisciotta JM, Zou Y, Baskakov IV (2011) Role of the photosynthetic electron transfer chain in electrogenic activity of cyanobacteria. Appl Microbiol Biotechnol 91(2):377–385
10. Logan BE (2009) Exoelectrogenic bacteria that power microbial fuel cells. Nat Rev Microbiol 7(5):375–381
11. Islam MA et al (2017) Electrogenic and antimethanogenic properties of *Bacillus cereus* for enhanced power generation in anaerobic sludge-driven microbial fuel cells. Energy Fuels 31(6):6132–6139
12. Bellin DL et al (2014) Integrated circuit-based electrochemical sensor for spatially resolved detection of redox-active metabolites in biofilms. Nat Commun 5(1):1–10
13. Chen W et al (2015) An U.V.—vis spectroelectrochemical approach for rapid detection of phenazines and exploration of their redox characteristics. Biosens Bioelectron 64:25–29
14. Nandy A, Kumar V, Kundu PP (2013) Utilization of proteinaceous materials for power generation in a mediatorless microbial fuel cell by a new electrogenic bacteria *Lysinibacillus sphaericus* VA5. Enzyme Microb Technol 53(5):339–344
15. Uma Vanitha M et al (2017) Microbial fuel cell characterization and evaluation of *Lysinibacillus macroides* MFC02 electrigenic capability. World J Microbiol Biotechnol 33(5):91
16. Rorat A et al (2019) Sanitary and environmental aspects of sewage sludge management. Ind Munic Sludge pp 155–180
17. Kelessidis A, Stasinakis AS (2012) Comparative study of the methods used for treatment and final disposal of sewage sludge in European countries. Waste Manage 32(6):1186–1195
18. Yang G, Zhang G, Wang H (2015) Current state of sludge production, management, treatment and disposal in China. Water Res 78:60–73
19. Raheem A et al (2018) Opportunities and challenges in sustainable treatment and resource reuse of sewage sludge: a review. Chem Eng J 337:616–641
20. Feng L, Luo J, Chen Y (2015) Dilemma of sewage sludge treatment and disposal in China. Environ Sci Technol 49(8):4781–4782
21. Ruiz-Gómez N et al (2017) Co-pyrolysis of sewage sludge and manure. Waste Manage 59:211–221
22. Meng X-Z et al (2016) Organic contaminants in Chinese sewage sludge: a meta-analysis of the literature of the past 30 years. Environ Sci Technol 50(11):5454–5466
23. Wiśniowska E et al (2019) 10—Sludge legislation-comparison between different countries. In: Prasad MNV et al (eds) Industrial and municipal sludge. Butterworth-Heinemann, pp 201–224
24. Inoue S et al (1996) Organic composition of liquidized sewage sludge. Biomass Bioenerg 10(1):37–40
25. Xu Y, Gong H, Dai X (2020) High-solid anaerobic digestion of sewage sludge: achievements and perspectives. Front Environ Sci Eng 15(4):71
26. Jadhav DA, Ghosh Ray S, Ghangrekar MM (2017) Third generation in bio-electrochemical system research—a systematic review on mechanisms for recovery of valuable by-products from wastewater. Renew Sust Energy Rev 76:1022–1031
27. Munoz-Cupa C et al (2021) An overview of microbial fuel cell usage in wastewater treatment, resource recovery and energy production. Sci Total Environ 754:142429
28. Logan BE et al (2019) Electroactive microorganisms in bioelectrochemical systems. Nat Rev Microbiol 17(5):307–319

29. Sauvé S, Desrosiers M (2014) A review of what is an emerging contaminant. Chem Cent J 8(1):1–7
30. Teodosiu C et al (2018) Emerging pollutants removal through advanced drinking water treatment: a review on processes and environmental performances assessment. J Clean Prod 197:1210–1221
31. Cammarota M, Freire D (2006) A review on hydrolytic enzymes in the treatment of wastewater with high oil and grease content. Biores Technol 97(17):2195–2210
32. Brandelli A, Sala L, Kalil SJ (2015) Microbial enzymes for bioconversion of poultry waste into added-value products. Food Res Int 73:3–12
33. Yao Y et al (2020) Insights into the improvement of the enzymatic hydrolysis of bovine bone protein using lipase pretreatment. Food Chem 302:125199
34. Mishra B et al (2020) Engineering biocatalytic material for the remediation of pollutants: a comprehensive review. Environ Technol Innov, pp 101063
35. Varga B et al (2019) Enzymatic treatment and subsequent toxicity of organic micropollutants using oxidoreductases—A review. J Clean Prod 221:306–322
36. Ji C et al (2016) Biocatalytic degradation of carbamazepine with immobilized laccase-mediator membrane hybrid reactor. J Membr Sci 502:11–20
37. Stadlmair LF et al (2017) Mass spectrometry based in vitro assay investigations on the transformation of pharmaceutical compounds by oxidative enzymes. Chemosphere 174:466–477
38. Feng S et al (2021) Roles and applications of enzymes for resistant pollutants removal in wastewater treatment. Biores Technol 335:125278
39. Sánchez-Porro C et al (2003) Diversity of moderately halophilic bacteria producing extracellular hydrolytic enzymes. J Appl Microbiol 94(2):295–300
40. Sonune N, Garode A (2018) Isolation, characterization and identification of extracellular enzyme producer *Bacillus licheniformis* from municipal wastewater and evaluation of their biodegradability. Biotechnol Res Innov 2(1):37–44
41. Grandclément C et al (2017) From the conventional biological wastewater treatment to hybrid processes, the evaluation of organic micropollutant removal: a review. Water Res 111:297–317
42. Burkhardt-Holm P (2011) Linking water quality to human health and environment: the fate of micropollutants. Institute of Water Policy Working Paper, vol 3, p 2011
43. Pandey K et al (2017) Application of microbial enzymes in industrial waste water treatment. Int. J. Curr. Microbiol. Appl. Sci 6(8):1243–1254
44. Karigar CS, Rao SS (2011) Role of microbial enzymes in the bioremediation of pollutants: a review. Enzyme Res 2011:805187
45. Xu Q (2003) New equipment, techniques and technology of sludge treatment
46. Islam MA et al (2018) An insight of synergy between *Pseudomonas aeruginosa* and *Klebsiella variicola* in a microbial fuel cell. ACS Sust Chem Eng 6(3):4130–4137
47. Venkataraman A et al (2011) Metabolite-based mutualism between *Pseudomonas aeruginosa* PA14 and *Enterobacter aerogenes* enhances current generation in bioelectrochemical systems. Energy Environ Sci 4(11):4550–4559
48. Wang VB et al (2015) Metabolite-enabled mutualistic interaction between *Shewanella oneidensis* and *Escherichia coli* in a co-culture using an electrode as electron acceptor. Sci Rep 5(1):1–11

Assessment of Sludge-Based Microbial Fuel Cell Performance via Electrochemical Impedance Spectroscopy

Mohamad Farhan Mohamad Sobri, Muhammad Hazwan Hussin, and Muaz Mohd Zaini Makhtar

Abstract The effective, non-destructive method known as electrochemical impedance spectroscopy (EIS) can be a useful supplement to the methods currently being used to examine microbial fuel cells (MFCs). Its use in MFC research should be investigated further, such as in the examination of MFC internal resistance, electrode materials, catalyst coatings on electrodes and growth of biofilms. Specifically, this chapter details the application of EIS in comprehending electro-chemical processes involved in sludge-based MFC, the biological and/or abiotic variables that restrict power production, with the goal of comprehending and overcoming them.

Keywords Internal resistance · Impedance spectroscopy · Microbial fuel cell · Sludge

---

M. F. Mohamad Sobri (✉) · M. Mohd Zaini Makhtar
School of Industrial Technology, Universiti Sains Malaysia, 11800 Gelugor, Penang, Malaysia
e-mail: mfarhansobri@student.usm.my

M. Mohd Zaini Makhtar
e-mail: muazzaini@usm.my

M. F. Mohamad Sobri
Faculty of Chemical Engineering and Technology, Universiti Malaysia Perlis, 02600 Arau, Perlis, Malaysia

M. H. Hussin
School of Chemical Sciences, Universiti Sains Malaysia, 11800 Gelugor, Penang, Malaysia
e-mail: mhh@usm.my

M. Mohd Zaini Makhtar
Centre for Innovation and Consultation, Universiti Sains Malaysia, 11800 Gelugor, Penang, Malaysia

M. Mohd Zaini Makhtar et al. (eds.), *Microbial Fuel Cell (MFC) Applications for Sludge Valorization*, Green Energy and Technology,
https://doi.org/10.1007/978-981-99-1083-0_8

1 Introduction

EIS is an effective technique for researching chemical and physical processes in both solids and liquids. For many years, EIS has been used to investigate various corrosion processes, batteries, and fuel cells. However, application of EIS by MFC researchers to investigate resistance in the microbial fuel cell is relatively recent. Using a device known as a potentiostat, a MFC can be linked either in a three-electrode mode or a two-electrode mode, depending on the goal of the measurement required [1]. An individual electrode is examined using the three-electrode mode. The working electrode is either the anode or the cathode, while the counter electrode is the other electrode. The third lead is connected to a reference electrode that has been inserted into either the cathode or anode compartment, such as Ag/AgCl. When a cell voltage is applied, the Rin of the entire cell is measured using the two-electrode mode [2]. The working electrode is one electrode, while the reference and counter electrodes are on the other electrode. In the two-electrode mode, no additional reference electrode is required. The opposite electrode serves as a counter electrode to the working electrode. A reference is connected to the third lead. The potentiostat may be set to determine impedance spectra across a large frequency range, such as from 100 kHz to 1 MHz, using the EIS function [3]. The lowest frequency limit for MFC investigations should be 1 or 5 MHz in order to obtain valid results.

The analysis of electrochemical reactions on electrodes, bacterial metabolism, as well as the surface and material properties of electrodes are all made possible by the EIS measurement of a single electrode. These analyses are crucial to understanding the electricity-generating process and enhancing the power output of MFCs. A few studies have used the EIS approach to look at some of these MFC elements, but further study is unquestionably required.

2 Resistances Within MFC

a. Internal Resistance and Constituents

MFC performance and efficiency are generally viewed in terms of electrical parameters (power density, current density, potential difference, and internal resistance) as well as biodegradation efficiency (removal of organics). Both parameters are combined as the Coulombic efficiency (CE) defined as the ratio of amount of electrons harvested as currents and total equivalents present in organic matter available to microbial oxidative metabolism (recognized as chemical oxygen demand (COD)) [4]. In a given MFC, the power generated is quantified in terms of power output, with voltage (V) and current (I) as per the following equation:

$$\mathrm{P} = V_{cell} \times \mathrm{I}$$

In the absence of current generation, the voltage of an MFC is known as open circuit voltage (OCV) while the thermodynamic value between the cathode and anode, represented as potential difference, is known as overall cell electromotive force (E_{emf}) [5, 6]. In theory, OCV value should be close to the cell emf; however, this is inapplicable in practice with OCV value often significantly lower, owing to E_{emf} not considering the internal potential losses of the system. As per Logan et al. [5] the difference in value, known as overpotential, is quantified as the sum of respective overpotentials at the anode ($\Sigma\eta_a$) and cathode ($\Sigma\eta_c$) and ohmic losses (IR_Ω) of the system, as seen in the equation below.

$$E_{cell} = E_{emf} - (\Sigma\eta_a + |\Sigma\eta_c| + IR_\Omega)$$

Polarization is the change of a given electrode potential (or MFC voltage) from its equilibrium state upon application of a current [6]. $\Sigma\eta_a + |\Sigma\eta_c|$ represents polarization of a cell at the anode and cathode, respectively [4]. For efficient MFC performance, minimization of the magnitude of polarization element is crucial.

Internal resistance (R_{int}) of electrochemical cells such as microbial fuel cells primarily consist of three components; (i) charge-transfer/polarization/activation resistance (R_{ct}), (ii) ohmic/solution resistance (R_Ω), and (iii) concentration/diffusion resistance (R_m) [6]. In general, the internal resistance of MFCs originates from electrode materials, catalyst coatings on electrodes, undeveloped biofilm and electrochemical reactions on both anode and cathode [4].

b. **Charge-Transfer Resistance**

Polarization resistance, particularly at the anode, represents one of the major components in the internal resistance of MFC [7]. Activation polarization may be attributed to the activation energy required to be overcome by the reacting species and may prove to be the rate-limiting step when the rate of an electrochemical reaction on an electrode surface is dictated by slow reaction kinetics. The processes involved the adsorption of reactant species, transfer of electrons across the double-layer cell membrane, desorption of product species, and physical nature of the electrode surface [8]. R_{ct} occurs on the interfaces between the electrode and surrounding electrolyte [9, 10]. On the anode, the reduction of charge-transfer resistance has been managed by the proper development of biofilm on the anode surface [7] as well as the addition of mediators for microbes that do not readily release electrons to the anode. Activation polarization on the cathode has instead been managed by the preference of certain electrodes such as platinum (Pt) over graphite, owing to the lower energy barrier required to overcome in the cathodic oxygen reaction, producing water. In general, activation polarization is dominant at low current density with the electronic barriers necessary to be overcome prior to the flow of current and ions [8].

c. **Ohmic Polarization**

Ohmic losses or ohmic polarization account for the resistances from the flow of electrons across the electrodes and interconnections and flow of ions through a cation exchange membrane (CEM) including anodic and cathodic electrolytes [11, 12]. Reduction of such losses focuses on minimizing electrode spacing, application of membrane with low resistivity, ensuring proper contacts are achieved and where practical, increasing solution conductivity to the limits tolerable to the bacteria applied [5]. In this context, proton exchange membranes (PEM) generate a transmembrane potential difference that forms a major resistance [8].

D. **Concentration/Diffusion Resistance**

Concentration polarization can be defined as loss of potential attributed to the inability to retain the initial substrate concentration in the bulk fluid, of which slow mass transfer rates for reactants and products are often the cause [8]. Cathodic overpotential attributed to a lack of dissolved oxygen gradient for the cathodic reaction has limited the power density output of several MFCs [13]. Hence, for mitigation of concentration polarization, mass transfer should be improved by measures such as stirring and bubbling for the reduction of concentration gradient in an MFC. Nevertheless, given the energy investment necessary may outweigh the power outputs of the MFC, a balance must be struck should this approach be required [8].

3 Electrochemical Impedance Spectroscopy as Method of Assessment of Resistance

Compared to conventional fuel cells, MFCs power outputs are relatively of lower production with maximum power output largely dependent and attributed to their high R_{int} [14, 15]. Thus, for improvement of power output from MFCs, reduction of R_{int} is considered key, particularly by affecting anodic and cathodic charge-transfer resistances, electrolyte resistance, and membrane resistance (if present) [9, 16]. Application of a suitable method for measuring and assessment of R_{int} would provide researchers with a better comprehension of the limitations within a particular MFC [9].

For measurement of internal resistance (R_{int}) of an MFC, either a resistance calculation method applying the slope of the V-I curve collected through a discharge test, known as electrode potential slope analysis (EPS) or electrochemical impedance spectroscopy (EIS) may be used [1].

In general, polarization (voltage vs. current) curves represent the standard form of demonstrating MFC performances. In addition to showing the maximum current and open circuit voltage of the cell, characteristics of voltage–current behaviour also allow for major indications of internal resistances [4]. Alternatively known as electrode potential slope analysis (EPS), R_{int} is calculated from the slope of the V-I curve and has proven reliable for MFCs having ohmic resistance (R_{Ω}) which involves solution/contact resistance and other primary resistances within the cell [1].

It must be highlighted however that EPS is limited in distinguishing key resistances within, such as R_{ct} of each electrode, mass transfer resistance, solution resistance, and membrane resistance along with difficulties in determining the values of each R_{ct}, R_{Ω}, and R_{m} using the peak power point calculation approach [17]. Thus, the application of electrochemical impedance spectroscopy (EIS) using alternating current (AC) and with an appropriate equivalent circuit constructed, represents a complementary method for quantification of the different components constituting internal resistance [16, 18].

Similar to EPS, EIS has been a significant non-intrusive tool in the study of physical and chemical processes in both solutions and solids, applied for stability assessment, characterization, and performance diagnosis of fuel cells [3, 19]. Its popularity can be attributed to its capacity of separating influences of different cell components, and in particular its high surface sensitivity, which allows for the detection of changes unforeseen in other techniques (www.palmsens.com). Known applications include identification of electrochemical phenomena present on electrodes, recording electrode transfer resistance, detection of the role of mediators in anodophilic electron transfer, and conduct real time monitoring of biofilm formation [7]. As an electrochemical method, EIS is advantageous in that it is a steady-state technique, utilizes small signal analysis, able to probe signal relaxations over a large breadth of applied frequency, all while using commercially available electrochemical working stations (potentiostat) [3].

4 Basic Concept of Electrochemical Impedance Spectroscopy

For measurement, the MFC is first connected to the potentiostat coupled with Frequency Response Analyzer. Application of EIS involves the introduction of a small current as a perturbation signal, enough to not create a large over-potential to disrupt operation of the system. The AC signal introduced is of amplitude 5–10 mV over a frequency range between 100 kHz and 1 MHz, sufficient to initiate a current response from the MFC. Connection of the MFC to the potentiostat can be done in either three-electrode mode or two-electrode mode. For analysis of individual impedance present at each electrode, the three-electrode mode is to be applied of which given an example of anodic impedance, anode acts as the 'working' electrode, cathode as 'counter' electrode, and a 'reference' electrode (Ag/AgCl) is placed near the anode. In turn, the positions of working and counter electrodes are switched, with the reference electrode placed near the cathode instead for measurement of cathodic impedance. The two-electrode configuration instead measures impedance of the cell as a whole. Here, one of the electrodes serves as a working electrode while the other electrode serves as both a counter and reference electrode. Nevertheless, care must be applied in this mode as such measurements assume that the potential of the counter/reference electrode remains the same throughout, which may be an impractical assumption to make when working with MFCs.

4.1 Impedance and Representation via Nyquist and Bode Plots

In response to the perturbation signal, the current, I, and the voltage, U, of the cell are measured and exemplified as impedance, Z, via the relationship $Z = \mathrm{d}U/\mathrm{d}I$ [20]. Measurement of impedance spectra over a range of voltages in turn allows for the determination of R_{int} [21]. Expression of impedance is divided into a real and imaginary parts. Plotting of 'real impedance' (Z_{real}) on X-axis against 'imaginary impedance' (Z_{imag}) on Y-axis forms a 'Nyquist Plot', as seen in Fig. 1, as a semicircle [21]. Each point on the plot represents an impedance value at a given frequency, with Z_{imag} being negative. Low-frequency impedance is located on the right of the X-axis while impedances of high frequency are generated on the left. [3]. Impedance at the high frequency limit may be attributed to R_{Ω} with the diameter of the semicircle representative of R_{ct} [21].

On the Nyquist plot, impedance may also be represented as a vector (arrow) of length |Z| with the angle between the arrow and the X-axis known as the phase angle, Φ [3]. However, the Nyquist plot is limited by the inability to ascertain the frequency directly by looking at the plot [23]. The Bode plot represents an alternative form of representing impedance where two separate logarithmic plots consisting of phase angle vs. frequency as well as impedance vs. frequency, respectively [21], are included, as seen in Fig. 2.

Measurement of impedance involves application of a potential wave to the working electrode and recording the resultant current wave. From the two waves, Z, Φ, Z_{real}, and Z_{imag} are obtained and sketched with the spectrum obtained plotting impedance

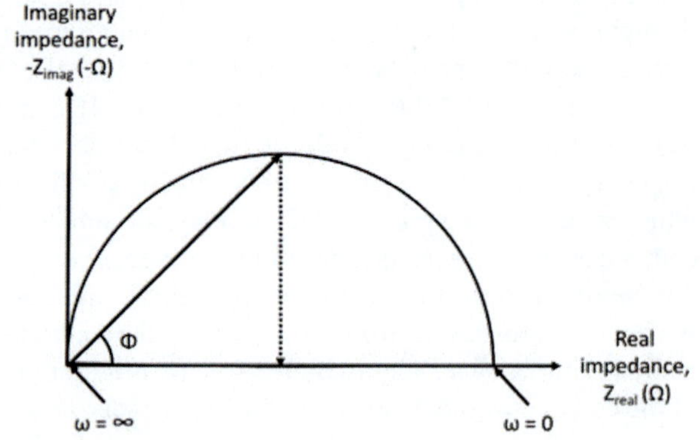

Fig. 1 Nyquist plot with phase angle (Φ) and radial frequency (ω) illustrated. Figure adapted from [21]

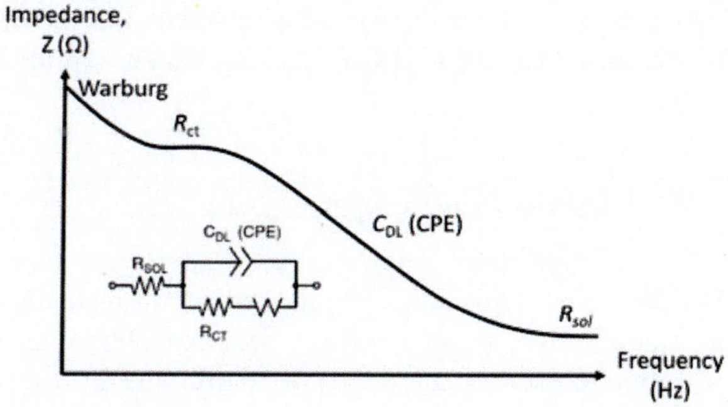

Fig. 2 Bode plot with Randles equivalent model of an electrochemical interface comprising both non-Faradaic and Faradaic phenomena. Equivalent model adapted from Magar et al. [3]

across different frequencies [3]. Generally, Nyquist plot allows for analysing the resistive processes while plotting frequency against phase angle on the Bode plot allows for the determination of capacitance within the electrochemical systems [24].

4.2 Equivalent Circuit

Fitting EIS spectra with an electrical equivalent circuit give the values of resistance, capacitance, and impedance associated with physical and chemical processes [25]. As seen in Fig. 2, most of the circuit elements in the model are typical electrical elements such as resistors, inductors, and capacitors, of which a Randles circuit is one of the most often used cell models and a starting point for other complex models. For an ideal resistor, the impedance will be purely real and in phase with the voltage across it whereas for an ideal inductor and capacitor, the impedance will be purely imaginary. As the frequency is raised, an inductor's impedance increases while the capacitor's impedance decreases. Connection between elements can be done in either series or parallel combinations, collectively forming a nested circuit allowing for the calculation of equivalent impedance [23].

5 Electrochemical Parameters of Effect to Impedance

5.1 Electrolyte Resistance

Electrolyte resistance represented by the ionic solution (electrolyte) and microorganisms present within form a significant factor contributing to overall cell impedance. Resistance of ionic solution is dependent on the type of ions, ionic concentration, temperature, and geometry of the projected area with conductivity largely requiring the determination of the current flow path and geometry of the electrolyte solution. As such, both ionic solution resistance and the resistance from microorganisms may be represented as two resistances in parallel at the electrolyte level, and calculated by the fitting of EIS analysis to an equivalent circuit mode [23].

5.2 Double-layer Capacitance

Double-layer capacitance is indicative of instances in charge separation within a system, such as an interface between electrode and electrolyte, ion exchange membranes, across the bacterial membrane in the maintenance of proton motive forces, and others, representative of a capacitor in an electrical circuit. Its value is dependent on factors such as electrode polarization, ionic concentration, types of ions, temperature, electrode roughness, oxide layers, and others. In particular, biofilm formation on the electrode surface may affect the thickness of the double layer, thereby resulting in concomitant significant effect on double-layer capacitance of a given system [23].

5.3 Polarization Resistance

For a given electrode, should an applied electrode potential be shifted from equilibrium potential, the electrode is said to be polarized and thus leading to oxidation/reduction of a species on the electrode's surface [23].

5.4 Charge-transfer Resistance

Transfer of electrons from the ionic species within the solution to solid metal is defined as charge-transfer reaction, in turn dependent on the type of reaction, temperature, potential, and concentration of the reactants [23].

5.5 Diffusion

Diffusion forms a significant part of mass transfer processes from bulk electrolyte through to the biofilm. Specifically, substrates diffuse through the biofilm, are oxidized and the resultant products diffuse back to the bulk electrolyte. Nevertheless, only at lower AC frequency is such an impedance appreciated as at higher AC frequency, the impedance is considered negligible as the reactants will not have enough time to move from bulk to the surface for electrochemical reactions. Once diffusion thickness is infinite, the diffusion element is simplified to Warburg impedance, appearing as a 45° slope diagonal on the Nyquist plot and a plane shift of 45° on the Bode plot [23].

6 Concept and Assumptions of EIS

In contrast to voltage vs. current polarization curves, EIS represents a superior approach where separation and identification of different loss mechanisms in a fuel cell are made by the association of each to particular frequency ranges [26]. At very high frequencies, the impedance is attributed to ohmic losses, primarily the membrane ionic resistance. Within intermediate frequencies, charge-transfer resistance coupled with double-layer capacitor, contributes to the impedance while at low frequencies, diffusion/mass-transport losses contribute to the fuel cell impedance [27]. Ohmic losses in particular can be singled out from other internal MFC resistances using EIS, with charge-transfer resistance and diffusion resistance of a greater challenge, owing to the three-dimensional nature of an anode in an MFC, particularly with the presence of a biofilm and the relatively larger electrode spacing [16].

7 Applications of EIS for Assessment of Sludge-Based Microbial Fuel Cell Performance

7.1 Inoculum Sludge Choice

Applications of EIS with sludge-based MFCs have been demonstrated across several studies. In a study by Cetinkaya et al. [28], using different sludges as inoculums, EIS was utilized to elucidate the electrochemical properties of each anode. From this study, gum industry sludge was found to be the best in comparison to anaerobic brewery sludge and anaerobic phosphate sludge, with a maximum power density of 406 mW m^{-2}, by having the best electron transfer efficiency and subsequently, the smallest charge-transfer resistance. Variation in power production was attributed to the diverse biochemical and electrochemical interactions present in microbial communities in the inoculum, where an electrogenic bacteria in isolated culture inoculum present in the biofilm may prove to be non-electrogenic in the mixed-culture inoculum. In modelling, double-layer capacitance of the biofilm (CPE) was used in place of a capacitor, accounting for surface roughness and speed of reactions across the electrode surface, affecting the overall kinetics of the reaction. The equivalent circuit model and fitting parameters obtained are shown in Fig. 3 and Table 1(a), respectively.

EIS was also further applied with a choice of anaerobic gum industry sludge as inoculum following application of different industrial wastewaters (slaughterhouse industry, gum industry, and chocolate industry wastewaters) on MFC performance.

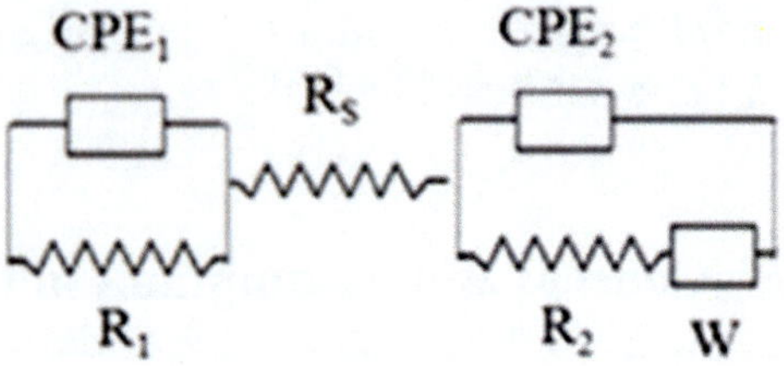

Fig. 3 Equivalent circuit diagrams of the MFC anodes. Adapted from [28] (CCBY)

Here, slaughterhouse industry wastewater was found to have relatively higher double-layer capacitance, suggesting the presence of a thick biofilm on the surface of the electrode. Coupled with lower solution resistance, electron transfer is suggested to be easier thereby resulting in a maximum power density of 267 mW m^{-2}.

In another study by Baranitharan et al. [29], a developed control inoculum (CI) was compared to that of anaerobic sludge as inoculum for use in MFC with palm oil mill effluent (POME). While the maximum power density of MFC utilizing CI was twice higher and with enhanced maximum Coulombic efficiency, there was noticeable lower COD removal of about 32%, possibly due to the absence of necessary fermentative microorganisms within the CI for POME utilization. EIS analysis further revealed a significant reduction of charge-transfer resistance by as much as 40% during operation with CI, indicating substantial improvements in electron transfer between the microorganisms and anode from CI use.

7.2 *Bioremediation Studies*

In another study, Huang et al. [30] focused on the utilization of anaerobic sludge from anaerobic tank in a wastewater treatment plant as inoculum in MFC for remediation of cadmium (Cd) contaminated soil, coupled with studies on degree of remediation across differing Cd concentrations and electrode spacing. While results showed that Cd concentrations had no effect on the output voltage and internal resistance, EIS analysis carried out in two-electrode configuration with AC signal set at an open circuit potential of $\pm$ 10 mV and frequency range from 100 to 10 MHz further managed to support Cd concentration as having no effect on the charge-transfer on the anodic biofilm. However, the same cannot be said of electrode spacing, of which the resultant R_{ct} and internal resistance has been found to be proportional to, given the larger ion migration path necessary.

Table 1 Fitting parameters obtained for different industry sludges according to the Nyquist curves

(a)

Substrate	Sludge	CPE_1 (µF)	n_1	R_1 (Ω)	R_S (Ω)	CPE_2 (µF)	n_2	R_2 (Ω)	W	Reference
Dairy wastewater	Anaerobic gum	0.03555	0.433	30.19	52.2	0.001188	0.4034	10.15	0.004113	[28]
	Anaerobic brewery	0.000415	0.2171	28.8	4.932	0.00375	0.5287	59.93	0.004095	
	Anaerobic phosphate	0.006727	0.191	15.92	11.89	0.04092	0.6447	19.58	0.003982	

(b)

Sludge	Wastewater	CPE_1 (µF)	n_1	R_1 (Ω)	R_S (Ω)	CPE_2 (µF)	n_2	R_2 (Ω)	W	Reference
Anaerobic gum	Chocolate	1448	0.3585	24.9	34.41	5294	0.6691	26.91	0.007582	[28]
	Gum	1936	0.3125	31.32	33.01	6856	0.6494	43.21	0.004971	
	Slaughterhouse	5681	0.2088	35.58	25.46	6021	0.6167	44.89	0.004335	

7.3 MFC Variant Assessment

Another study by Meng et al. [31] has focused on the application of a new variant of MFC designed for low-cost, energy-saving desalination from resultant bioelectricity generated from dewatered sludge, known as microbial desalination cell (MDC). The introduction of a middle chamber between the anode and cathode allows for the migration of anions and cations within to the adjacent chambers upon oxidation of organics in the anode chamber subsequently promoting electron transfer. Dewatered sludge acted as an anodic inocula and substrate with EIS applied at three typical open circuit voltages in the frequency range of 10^5 to 0.01 Hz with sinusoidal perturbation of 5 mV amplitude in a three-electrode configuration. Ohmic resistance between anode and cathode electrodes was found to be unchanged. R_{ct} for both electrodes on the other hand reduced significantly with an increase of OCV with R_{ct} of cathode slightly higher and decreased slowly upon startup suggesting the cathode reaction to be the rate-limiting step for reactor startup. Warburg impedance changed slightly with OCV change while for double-layer capacitance (C_{dl}), a decreased trend was noticed in the anode with an increased trend present in the cathode.

Meng et al. [32] further developed a microbial capacitive desalination cell (MCDC) to address the difficulties current MDC systems encounter with salt accumulation in anodic wastewater and sludge. Under similar analysis conditions, the R_s, C_{dl}, $R_{ct,}$ and W constitute the main energy loss in the MCDC, of which R_s has been argued to be the key factor. An increase in OCV resulted in an increase in Rs between the anode and cathode electrodes, possibly due to the rise in membrane resistance from the growth of biofilm on ion exchange membranes. On the other hand, R_{ct} of each individual electrode was substantially reduced with increased OCV, indicative of the rise in electrode activity, and enhanced ease of electrochemical reactions proceeding, thereby increasing electricity generation. R_{ct} of cathode was as prior, slightly higher and decreased slowly upon startup which suggests the cathode reaction to be the rate-limiting step for reactor startup. Warburg impedance of the anode reduced slightly while that of the cathode remained relatively consistent with increasing OCV. For double-layer capacitance (C_{dl}), a decreased trend was noticed in the anode, attributed to thickening of the electric double layer by the multiplication of adsorbed organic on the anode surface with an increased trend present in the cathode suggestive of the increased active cathode surface area. Taşkan et al. [33] evaluated the capability of utilizing sewage sludge with sediment MFC (SMFC) using Ti-TiO_2 electrode. Along with a maximum power density of 187 mW/m^2, the application of EIS in three-electrode configuration within the frequency range of 0.1 Hz to 100 kHz at the applied potential of 5 mV resulted in R_{ct} and R_s of 84.7 Ω and 338.1 Ω, respectively. High value of R_s was predicted prior due to the low ionic conductivity of the anode while low charge-transfer resistance suggests as to the rapid electron transfer from the bacteria to the anode surface, allowing for enhanced power generation.

7.4 Optimization Studies

Work by He et al. [34] focused on discovering the effect of electrolyte pH on the rate of anodic and cathodic reactions in an air-cathode MFC. Using wastewater sludge as inocula, it was found that the MFC could tolerate initial pH of up to 10 with optimal pH for power generation between 8–10. Bacterial metabolism caused a buffer effect and a change in electrolyte pH. Specifically, EIS analysis revealed power output to be restricted most by polarization resistance, with higher cathodic polarization resistance suggesting cathodic reaction to be the rate-limiting factor. Cathodic R_s was also higher than anodic R_s, attributed to high ohmic resistance caused by waterproof layer. The impedance spectra of anode and cathode of the MFC at the open circuit potential (OCP) revealed that the anodic microbial is preferential towards a neutral pH and microbial activities decreased at higher or lower pH, while the cathodic reaction was improved with rise in pH.

7.5 Electrode Coating

Work by He et al. [35] demonstrated plasma-based ion implementation for modification of carbon paper anode for enhancement of electricity generation in MFC. Pre-acclimated anaerobic sludge was utilized as inoculum for use with acetate-laden synthetic wastewater as substrate, of which application of EIS elucidated as the consistency of ohmic resistance between the experimental and control electrodes but a noticeable reduction of charge-transfer resistance at the electrode/electrolyte interface post N^+ ion implementation. Enhanced anode performance was attributed to changes in the microstructure of the electrode surface following nitridation, thereby resulting in enhanced electron transfer.

In another study by Nandy et al. [22], Fe_2O_3 was introduced as a coating of carbon electrodes in MFCs owing to its easy availability, high surface area, physical and catalytic properties, non-toxicity, low cost, and stability. With activated sludge applied as inoculum, MFC with coated anode and cathode resulted in 78% higher power output in comparison to those with non-coated carbon electrodes. EIS analysis was done on a two-electrode mode and carried out every 15 days, recognized as cycle 1 (immediate post-inoculation), cycle 2 (day 15), and cycle 3 (day 30), respectively. Results indicate the resistance of unmodified electrodes to be higher than those coated with Fe_2O_3 but with significant reductions for all cells for cycle 2 in comparison to cycle 1 as the biofilm develops, resulting in the reduction of R_{ct}. Reduction was more significant in MFC with coated electrodes, attributed to more efficient biofilm formation. However, increased impedance in cycle 3 compared to cycle 2, along with consistent ohmic resistance across all cycles, suggests the presence of thicker and aged biofilm reducing the efficiency of electron transport and depletion of available substrates by electroactive biofilm members in the reactor. COD removal efficiency was recorded at about 51–60% for cells with Fe_2CO_3-coated electrodes and 39% for uncoated carbon electrodes.

Understanding the advantage of utilizing electrodes that promote attachment of electrogenic bacteria on the anode surface for enhancement of electron transfer and reduction of electrode resistance, work by Neethu et al. [36] focused on the feasibility of stainless steel mesh cage filled with sodium alginate beads containing either bacterial inoculum (Ic) from anaerobic sludge of septic tank, activated carbon (AC) or a combination of both (AC-Ic). Anode with AC-Ic beads embedded exhibited a maximum power density of 2.6 W/m^3 and enhanced COD removal efficiency of 91.6 ± 2.1%, owing to enhanced bacterial-electrode interaction. With a three-electrode mode configuration, EIS analysis highlighted the lower solution resistance within MFC with Ic, AC, and AC-Ic added electrodes compared to control bare carbon felt and stainless steel mesh electrodes. This has been attributed to the higher concentration of electrogenic bacteria in an anodic chamber of MFC, thereby resulting in more metabolites and mediators produced in the anolyte. Lowest R_{ct} was exhibited by MFC with stainless steel-based anodes containing AC-Ic beads, presumably due to the increased surface area of anode, resulting in improved microbial attachment.

7.6 Substrate Utilization

In a study by Bhowmick et al. [37], anaerobic mixed bacterial consortia collected from the septic tank were used as inoculum for experiments on the application of fish market wastewater with high ammonium nitrogen content for in-situ suppression of methanogenesis within MFC. In comparison to synthetic wastewater, the application of fish market wastewater led to lower specific methanogenic activity and higher power recovery. Utilization of EIS isolated charge-transfer resistance to be at 4.8 Ω and 102.2 Ω, respectively, for undiluted fish market wastewater-based MFC in comparison to synthetic wastewater-based MFC, suggestive as to the higher electron shuttling present in the former. R_{ct} values of 10.5 Ω and 28.0 Ω for 3 times and 5 times diluted fish market wastewater-based MFC is suggestive of the higher rate of electrochemical response at the interface of anolyte to the anode by lowering the activation barrier to facilitate enhanced growth of electrogenic bacteria as compared with that of synthetic wastewater.

Recognizing the intensive energy necessary to be invested for the treatment of petrochemical wastewater (PCW) to reduce its high chemical oxygen demand (COD), Sarmin et al. [2] looked to achieve simultaneous PCW treatment with electricity generation with anaerobic sludge as biocatalyst in MFC. This was successfully demonstrated with a maximum power density of 0.75 Wm^{-3}, current density of 412 mAm^{-2}, and 0.45 V open circuit voltage. From PCW with initial COD of 45,000 mgL^{-1}, COD removal efficiency was recorded at 40%. EIS analysis later revealed R_{ct} for the electrode/biofilm interface is slightly higher than R_{ct} across biofilm/solution interface on day 3. On day 7, maximum capacitance and constant phase element were achieved, thereby demonstrating its charge-holding capacity. On the other hand, mass transfer resistance was found to be significantly lower on day 7, with good diffusivity attributed to a lower diffusion path, thereby reducing the total resistance of the system.

7.7 *Anode Additives*

In a study by Wang et al. [38], mechanisms and performances of biochar additive in promoting volatile fatty acids (VFA) oxidation via extracellular electron transfer pathway (EET) pathway was looked into. Using food waste sludge as inoculum and typical VFAs of acetate and propionate as substrates, maximum current density was enhanced by 69–200% compared to control groups, with the microbial community on the biofilm of biochar-assisted groups dominated by electroactive *Geobacteraceae*, associated with its enrichment in relative abundance from 4.6% to 31.7% in propionate-added groups. Application of EIS under open circuit potential with a measured frequency range from 1 Hz to 100 kHz and amplitude of 5 mV resulted in an elucidated reduction of R_{ct} in comparison to control groups, indicating the acceleration of electrochemical rate from anolyte to in biofilm-anode interface by biochar assistance, due to higher extracellular electron transfer activity of electroactive microbes on anode surface.

Aside from additives, another study by Bagchi and Behera [39] also looked into bioaugmentation of mixed anaerobic sludge inoculum with the introduction of complementary *Pseudomonas aeruginosa* for pyocyanin production under intermittent aeration within the anode compartment of an MFC. MFC introduced with *P. aeruginosa* under aeration (MFC_{P+A}) was found to produce more electricity than MFC with *P. aeruginosa* without aeration (MFC_P) and without *P. aeruginosa* entirely (MFC_c) by as much as 4% and 31%, respectively. Application of EIS after steady state has been reached showed MFC_c to have the highest solution resistance (7.43 Ω) compared to both augmented MFCs (MFC_P—3.6 Ω, MFC_{P+A}—1.07 Ω). Capacitance was also highest in MFC_{P+A} (5.5 mF) compared to MFC_P (3 mF) and MFC_c (1.5 mF), suggesting better electron holding capacity of biocatalyst in augmented MFCs as opposed to the control.

7.8 *Catholyte Choice Assessment*

Work by Ghadge et al. [40] assessed the choice of hypochlorite as catholyte in comparison to oxygen for waste-activated sludge (WAS) digestion and power production. With WAS as substrate and inoculum, power production achieved (8.7 W/m^3) was two times higher than that of oxygen (4.2 W/m^3) as well as a higher reduction of total chemical oxygen demand, total suspended solids, and volatile suspended solids. Specifically, the application of EIS over a frequency range of 100 kHz to 1 MHz with an AC signal of 10 mV amplitude led to the elucidation of reduced solution resistance (R_s) at high frequency range (from 33 Ω to 20.6 Ω), owing to better electrical conductivity of hypochlorite. Polarization resistance (R_p) was also reduced from 9.6 Ω to 5.9 Ω, primarily due to an increase in the rate of electrochemical reactions as well as lower Warburg diffusion resistance for MFC with hypochlorite as catholyte.

8 Conclusion

The chapter covers the application of sludge as fuel in MFC and has extended several purposes ranging from simple mixed-culture inoculum to studies on the degree of bioremediation and enhancement of MFC components. To each aim, EIS represents a value-added approach for elucidation of the numerous internal resistances at play and the resultant effect each experiment has on the resistance components within. Such an approach offers a promising outlook for further understanding and enhancement of sludge-based MFC in the years to come.

Acknowledgements The authors would like to thank the Universiti Sains Malaysia for the financial support of this study via APEX Era Grant (1001/PTEKIND/881004). The authors have declared no conflict of interest for the manuscript.

References

1. Kim B, Chang IS, Dinsdale RM, Guwy AJ (2021) Accurate measurement of internal resistance in microbial fuel cells by improved scanning electrochemical impedance spectroscopy. Electrochim Acta 366:137388. https://doi.org/10.1016/j.electacta.2020.137388
2. Sarmin S, Ideris AB, Ethiraj B, Amirul Islam M, Yee CS, Rahman Khan MM (2020) Potentiality of petrochemical wastewater as substrate in microbial fuel cell. IOP Conf Ser Mater Sci Eng 736(3):032015. https://doi.org/10.1088/1757-899X/736/3/032015
3. Magar HS, Hassan RYA, Mulchandani A (2021) Electrochemical impedance spectroscopy (EIS): principles, construction, and biosensing applications. Sensors 21(19):6578. https://doi.org/10.3390/s21196578
4. Roy S, Marzorati S, Schievano A, Pant D (2017) Microbial fuel cells, vol 3. Elsevier
5. Logan BE et al (2006) Microbial fuel cells: methodology and technology. Environ Sci Technol 40(17):5181–5192. https://doi.org/10.1021/es0605016
6. Zhao F, Slade RCT, Varcoe JR (2009) Techniques for the study and development of microbial fuel cells: an electrochemical perspective. Chem Soc Rev 38(7):1926–1939. https://doi.org/10.1039/b819866g
7. Kumar SS, Kumar V, Basu S (2019) Electroanalytical techniques for investigating biofilms in microbial fuel cells. In Bioelectrochemical interface engineering, 1st edn. Wiley, Hoboken, NJ, pp 149–163. https://doi.org/10.1002/9781119611103.ch9
8. Du Z, Li H, Gu T (2007) A state of the art review on microbial fuel cells: a promising technology for wastewater treatment and bioenergy. Bioenergy Adv 25(5):464–482. https://doi.org/10.1016/j.biotechadv.2007.05.004

9. He Z, Wagner N, Minteer SD, Angenent LT (2006) An upflow microbial fuel cell with an interior cathode: assessment of the internal resistance by impedance spectroscopy. Environ Sci Technol 40(17):5212–5217. https://doi.org/10.1021/ES060394F
10. Khan MJ, Iqbal MT (2005) Modelling and analysis of electrochemical, thermal, and recetant flow dynamics for a PEM fuel cell system. Fuel Cells 5(4):463–475. https://doi.org/10.1002/fuce.200400072
11. Hoogers G (2003) Fuel cell technology handbook. CRC Press, London
12. Larminie J, Dicks A (eds) (2003). Wiley, Hoboken, NJ
13. Oh S, Min B (2004) Cathode performance as a factor in electricity generation in microbial fuel cells. Environ Sci Technol 38(18):4900–4904. https://doi.org/10.1021/es049422p
14. Liang P, Huang X, Fan MZ, Cao XX, Wang C (2007) Composition and distribution of internal resistance in three types of microbial fuel cells. Appl Microbiol Biotechnol 77(3):551–558. https://doi.org/10.1007/s00253-007-1193-4
15. Varanasi JL, Veerubhotla R, Das D (2018) Diagnostic tools for the assessment of MFC. In: Microbial fuel cell. Springer, Berlin, pp 249–268. https://doi.org/10.1007/978-3-319-66793-5_13
16. Fan Y, Sharbrough E, Liu H (2008) Quantification of the internal resistance distribution of microbial fuel cells. Environ Sci Technol 42(21):8101–8107
17. Cheng S, Liu H, Logan BE (2006) Increased power generation in a continuous flow MFC with advective flow through the porous anode and reduced electrode spacing. Environ Sci Technol 40(7):2426–2432. https://doi.org/10.1021/es051652w
18. Aelterman P, Rabaey K, Pham HT, Boon N, Verstraete W (2006) Continuous electricity generation at high voltages and currents using stacked microbial fuel cells. Environ Sci Technol 40(10):3388–3394. https://doi.org/10.1021/ES0525511
19. Petrone R et al (2013) A review on model-based diagnosis methodologies for PEMFCs. Int J Hydrogen Energy 38(17):7077–7091. https://doi.org/10.1016/j.ijhydene.2013.03.106
20. ter Heijne A, Schaetzle O, Gimenez S, Navarro L, Hamelers B, Fabregat-Santiago F (2015) Analysis of bio-anode performance through electrochemical impedance spectroscopy. Bioelectrochem 106:64–72. https://doi.org/10.1016/j.bioelechem.2015.04.002
21. Fan LP, Li JJ (2016) Overviews on internal resistance and its detection of microbial fuel cells. Int J Circuits Syst Signal Process 10:316–320
22. Nandy A, Sharma M, Venkatesan SV, Taylor N, Gieg L, Thangadurai V (2019) Comparative evaluation of coated and non-coated carbon electrodes in a microbial fuel cell for treatment of municipal sludge. Energies 12(6):1034. https://doi.org/10.3390/en12061034
23. Sekar N, Ramasamy RP (2013) Electrochemical impedance spectroscopy for microbial fuel cell characterization. J Microb Biochem Technol 5(2). https://doi.org/10.4172/1948-5948.s6-004
24. Lasia A (2014) Electrochemical impedance spectroscopy and its applications. Springer, New York
25. Sindhuja M, Kumar NS, Sudha V, Harinipriya S (2016) Equivalent circuit modeling of microbial fuel cells using impedance spectroscopy. J Energy Storage 7:136–146. https://doi.org/10.1016/j.est.2016.06.005
26. Springer TE, Zawodzinski TA, Wilson MS, Gottesfeld S (1996) Characterization of polymer electrolyte fuel cells using AC impedance spectroscopy. J Electrochem Soc 143(2):587–599. https://doi.org/10.1149/1.1836485
27. Page SC, Anbuky AH, Krumdieck SP, Brouwer J (2007) Test method and equivalent circuit modeling of a PEM fuel cell in a passive state. IEEE Trans Energy Convers 22(3):764–773. https://doi.org/10.1109/TEC.2007.895857
28. Cetinkaya AY, Ozdemir OK, Demir A, Ozkaya B (2017) Electricity production and characterization of high-strength industrial wastewaters in microbial fuel cell. Appl Biochem Biotechnol 182(2):468–481. https://doi.org/10.1007/s12010-016-2338-7
29. Baranitharan E et al (2015) Enhanced power generation using controlled inoculum from palm oil mill effluent fed microbial fuel cell. Fuel 143:72–79. https://doi.org/10.1016/j.fuel.2014.11.030

30. Huang G, Zhang Y, Tang J, Du Y (2020) Remediation of Cd contaminated soil in microbial fuel cells: effects of Cd concentration and electrode spacing. J Environ Eng 146(7):04020050. https://doi.org/10.1061/(asce)ee.1943-7870.0001732
31. Meng F et al (2014) Bioelectrochemical desalination and electricity generation in microbial desalination cell with dewatered sludge as fuel. Bioresour Technol 157:120–126. https://doi.org/10.1016/j.biortech.2014.01.056
32. Meng F et al (2017) Bioelectricity generation and dewatered sludge degradation in microbial capacitive desalination cell. Environ Sci Pollut Res 24(6):5159–5167. https://doi.org/10.1007/s11356-016-6853-4
33. Taşkan B, Taşkan E, Hasar H (2020) Electricity generation potential of sewage sludge in sediment microbial fuel cell using Ti–TiO2 electrode. Environ Prog Sustain Energy 39(5). https://doi.org/10.1002/ep.13407
34. He Z, Huang Y, Manohar AK, Mansfeld F (2008) Effect of electrolyte pH on the rate of the anodic and cathodic reactions in an air-cathode microbial fuel cell. Bioelectrochem 74(1):78–82. https://doi.org/10.1016/j.bioelechem.2008.07.007
35. He YR et al (2012) Enhanced electricity production from microbial fuel cells with plasma-modified carbon paper anode. Phys Chem Chem Phys 14(28):9966–9971. https://doi.org/10.1039/c2cp40873b
36. Neethu B, Bhowmick GD, Ghangrekar MM (2020) Improving performance of microbial fuel cell by enhanced bacterial-anode interaction using sludge immobilized beads with activated carbon. Process Saf Environ Prot 143:285–292. https://doi.org/10.1016/j.psep.2020.06.043
37. Bhowmick GD, Neethu B, Ghangrekar MM, Banerjee R (2020) Improved performance of microbial fuel cell by in situ methanogenesis suppression while treating fish market wastewater. Appl Biochem Biotechnol 192(3):1060–1075. https://doi.org/10.1007/s12010-020-03366-y
38. Wang G et al (2021) Poorly conductive biochar boosting extracellular electron transfer for efficient volatile fatty acids oxidation via redox-mediated mechanism. Sci Total Environ 13:151113. https://doi.org/10.1016/j.scitotenv.2021.151113
39. Bagchi S, Behera M (2021) Bioaugmentation using pseudomonas aeruginosa with an approach of intermittent aeration for enhanced power generation in ceramic MFC. Sustain Energy Technol Assess 45:101138. https://doi.org/10.1016/j.seta.2021.101138
40. Ghadge AN, Jadhav DA, Pradhan H, Ghangrekar MM (2015) Enhancing waste activated sludge digestion and power production using hypochlorite as catholyte in clayware microbial fuel cell. Bioresour Technol 182:225–231. https://doi.org/10.1016/j.biortech.2015.02.004

Sewage Sludge Particle Surface Interactions: Technology and Purification Approaches

Amir Muhammad Noh Amin Abdul Rahman, Muhammad Naziff Ahamad Said, and Mohamad Danial Shafiq

Abstract Interparticle surface interactions between the sludge are a dominant factor in the development of purification technology. The interplay between coagulation/flocculation and stabilization is heavily dependent on the colloidal characteristics of the sludge in water, where when the surfaces possess charges, the interactions can be summarized by the classical DLVO theory. The basis of this theory considers many factors including the chemical composition of the sludge particles and water components such as extracellular polymeric substances and heavy metals. Sludge stabilization is an essential procedure in anaerobic digestion, where polymeric coagulants are commonly added to convert the sludge into priceless energy sources. This green energy can also be generated through microbial fuel cell technology by utilizing sludge waste.

Keywords Interparticle surface interaction · Colloidal · Sludge stabilization · Polymeric coagulant · Sewage sludge treatment

1 Introduction to Colloids in Wastewater

Among the characteristics of industrial wastewater are turbidity and high percentage of biochemical oxygen demand (BOD) where the colloidal behavior of the biosolids in water is essential [1]. Figure 1 shows an example of typical wastewater treatment in an industrial plant. The remainder of the treated waste is a dilute suspension of solids that have been recovered during the treatment process, where these wastewater treatment solids are referred to as sewage sludge [2]. At a colloidal scale, the kinetics of sludge particles in wastewater are influenced by the random Brownian motion, which bombards the particles of the dispersed phase with the molecules of the dispersing continuum [3]. These particles are essentially "non-settleable" because of

A. M. N. A. Abdul Rahman (✉) · M. N. Ahamad Said · M. D. Shafiq (✉)
School of Materials and Mineral Resources Engineering, Universiti Sains Malaysia, Engineering Campus, 14300 Gelugor, Penang, Malaysia
e-mail: danialshafiq@usm.my

M. Mohd Zaini Makhtar et al. (eds.), *Microbial Fuel Cell (MFC) Applications for Sludge Valorization*, Green Energy and Technology,
https://doi.org/10.1007/978-981-99-1083-0_9

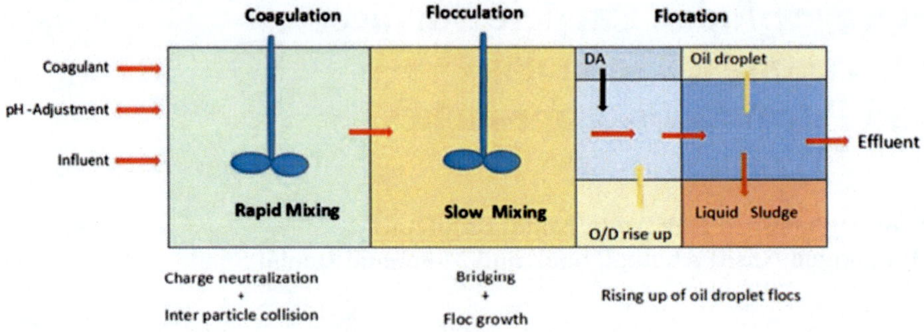

Fig. 1 Example wastewater treatment of industrial mineral oil wastewater treatment DA—dis-dissolved air; O/D—oil droplets. Reprinted with permission from [4]. Copyright 2018 African Journal Online (CCBY)

their significant surface charges, small sizes, and low particle weights. Destabilizing the colloidal particles is critical in removing sludge particles from the wastewater before they enter into water streams, therefore one must understand the interactions between the sludge particles in wastewater at a colloidal scale.

Sewage sludge water treatment is a basic necessity of the present society, and there are various methods of treatment available including photocatalysis, adsorption, membrane technology, ultrasonication, biological techniques, and coagulation/flocculation [5–8]. Coagulation or flocculation is one of the primary recovery and purification processes of wastewater utilized in treatment plants. This is due to its effectiveness, ease of implementation, simple design, and low processing cost [9]. The process employs aggregation of microscopic suspensions when a coagulant of flocculent is added to the water system, where the assembly of aggregates is known as flocs or micro flocs [10, 11].

Coagulation and flocculation occur in successive steps, allowing particle collision and the growth of floc. The process is followed by sedimentation. Several types of coagulants are used in water and wastewater treatment settings such as synthetic and natural coagulants [12]. Aluminum sulfate (alum) is the most common coagulant used for water purification. Other salts, such as ferric sulfate and sodium aluminate are also commonly used in wastewater purification. However, each type of coagulant has its unique properties, where the positive ions will entrap the negatively charged organic matter to destabilize the electrostatic charge to facilitate agglomeration. Table 1 shows the typical coagulant used in wastewater treatment.

Polymeric-based coagulants are used to assist inorganic coagulants, where these species can be anionic, cationic, or approaching the neutral charge composed of extremely high molecular weight structure. Cationic polymers, such as low-to-medium weight polymers, can be used alone or in combination with aluminum ferric coagulants to attract colloid particles via surface charge neutralization. According to Ravina and Moramarco [21], a dual-polymer coagulant system can be used to improve coagulation efficiency [21]. Meanwhile, a high molecular weight anionic polymer may be used to mechanically bridge the micro flocs into large and rapidly

Table 1 Type coagulant and its chemical formula

Type of coagulant	Coagulant	Chemical formula	
Aluminum Series	Aluminum sulfate	$Al_2(SO4)_3 \cdot 18H_2O$	[13]
	Poly aluminum chloride	$[Al_2(OH)_nCl_{6-n}]_m$	[14]
	Polyaluminum Silicate Sulfate	$Al_a(OH)_b(SO_4)_c(SiO_x)_d$	[15]
Iron Series	Ferrous sulfate	$FeSO_4 \cdot 7H_2O$	[16]
	Poly ferric sulfate	$[Fe_2(OH)_n(SO4)_{3-n/2}]_m$	[17]
	Ferric chloride	$FeCl_3 \cdot 6H_2O$	[18]
Organic Series	Polyacrylamide	$(C_3H_5NO)_n$	[19]
	Polyamine	$(NH_2-(CH_2)_n-NH-(CH_2)_n-NH_2)$	[20]

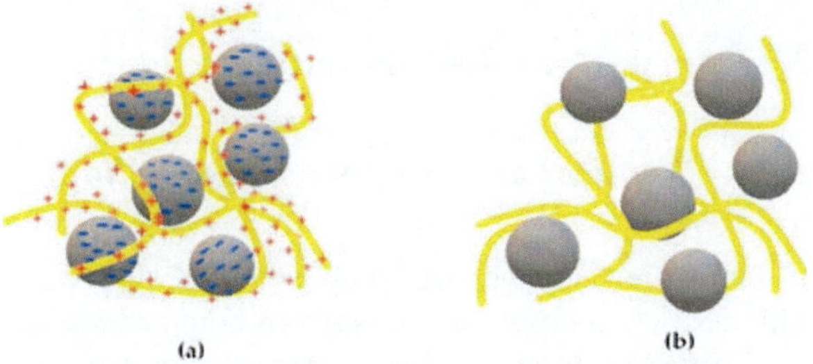

Fig. 2 Scheme of polymeric coagulant (the curved lines represent polymer chains adsorbing to the spherical colloidal particles): **a** charge neutralization and **b** polymer bridging. Reprinted with permission from [23]. Copyright 2020 MDPI (CCBY)

settling flocs. Kamizela and coworkers [22] mentioned that typical polymers used for sludge conditioning are water-soluble, high molecular weight compounds, whose high flocculating efficiency results mainly from their structure (mostly linear), high molecular weights (up to several million), functional groups (e.g., *OH, *COOH, *NH2, *SO3H), and adsorptive properties [22]. Figure 2 illustrates the schematic presentation of polymeric coagulant affecting the interactions of colloidal particles via charge neutralization and bridging mechanisms.

According to Shokri and Fard [24], when metallic coagulants polymerize, they can form strong bonds between colloids through dipole–dipole interactions and hydrogen bonding. As a result, the bridging bonds between colloids become more robust, causing the particles to lose their stability. However, they both also noted that adding polymer and fast mixing materials over the optimal point can lead to the separation of the particle bridges and recovery of colloidal stability [24].

1.1 Theory of Colloids

Generally, the stability of colloidal sludge particles can be explained by the Derjaguin, Landau, Verwey, and Overbeek (DLVO) theory. In DLVO theory, the total energy of adhesion is the result of the van der Waals attractive forces and the electrostatic repulsive interactions due to the interpenetration of the double layer [25–27]. The electrostatic repulsion becomes significant when two colloidal particles approach each other, subsequently interfering with their electrical double layer. Surface charge neutralization is achieved when the bound surface ions (charge) are attracted to the counter ions [28]. This mechanism is based on the addition of coagulant molecules containing ionizable groups to the colloidal particles. The charge of the dissociated molecules must be opposite to the electrostatic signature of the colloid [29]. As the coagulant molecules dissociate in water, the charged molecules can effectively interact with the colloid's Stern layer, thereby replacing the counter ions originally present in this layer [30]. In turn, the depletion of electrostatic charges allowed van Der Waals forces to become dominant, resulting in the coagulation of the colloidal particles [31].

2 Interparticle Interactions

2.1 Chemical and Surface Composition of Sewage Sludge

In the earlier chapter, the composition of sludge has been extensively discussed. This chapter focuses on the surface composition of sludge as a whole, which is contributed by the aforementioned sludge constituents and classes of material. Sludge is a common by-product of water and wastewater treatments, where the production of an abundance of organic and organic wastes is expected. The interactions between sludge particles are often complex as sludge is produced from numerous industrial and residential systems, which consequences in different compositions of sludge particles and their qualities [32]. The exploration of sludge particle composition and interactions is critical in considering the fate of this abundant waste for energy generation, construction, and agricultural uses [33]. The pre-treatment process and parameters before sludge conversion are also influenced by the characteristics of the sludge [34]. The composition of sludge is highly dependent on the surrounding temperature as it is a vital criterion for the living condition of organic matter, and the kinetics of sludge particles in water.

For example, the enzymatic activity of sludge extracts can be enhanced at a moderately elevated temperature, although the temperature range is varied depending on the specific types of enzymes present in the sludge particles. This also correlates with the amount of enzymatic extractable portion in sludge, as heavy metals and other inorganic contaminants are significant too in sludge [35]. Meanwhile, the pH and ionic strength of the water system are equally essential in determining the interactions between sludge particles. The strength of sludge flocs can be altered by the ionic composition in the water system, where a small amount of ions results in weak sludge flocs [36].

2.2 Heavy Metals

Heavy metals are profoundly present in sludge especially those from sewage and industrial wastewater treatment plants, which are predominantly produced from sewers corrosion. These heavy metals are considered environmental threats due to their non-biodegradability and may react with living tissues, eventually causing dire health hazards. The most common toxic heavy metals found in sewage sludge are cadmium (Cd), chromium (Cr), copper (Cu), zinc [37], lead (Pb), nickel (Ni), and mercury (Hg) [38]. These heavy metals are mobile and biologically available to be interacted with the natural and organic matter within the sludge and in water. The degree of toxicity caused by the heavy metals present in water varies depending on the physicochemical properties of the water such as pH and salinity, which affected the geochemical behavior of heavy metals [39]. The geochemical behavior of heavy metals is also influenced by nutrients present in the aquatic ecosystem such as phosphorus (P), nitrogen (N), and organic matter. Importantly, the interaction is driven by the heavy metals' selectivity for various binding and adsorption sites, depending on the ionic characteristics of both metals and other matters such as ionic radii, electronegativity, and hydrolysis constant [39, 40]. The sorption and desorption of heavy metals onto organic matter are also varied based on water electrical conductivity [41]. Metal-binding on activated sludge surfaces also increases with pH from acidic to neutral [42].

Zykova and coworkers reported that the optimum heavy metals adsorption on two microorganisms, Pseudomonas and Micrococcus, are 293 K. They also added four possible mechanisms involved in the resistance of microorganisms to heavy metals. The first one is biosorption; a process caused by metals binding on the cell surface via its functional groups. The second mechanism is bioaccumulation; a process arising from an interaction between metals with the cytoplasm components and enzyme active groups. Then, the redox reaction in the cytoplasm yields a less toxic product, and finally, the accumulation of the metal in the cytoplasm decreases through the efflux system [37].

2.3 Extracellular Polymeric Substances (EPS)

EPS is abundant in activated sludge, which represents up to 80% of the total mass of activated sludge [43]. Initial findings reported that EPS consists of carbohydrates and recent findings reported that EPS is evidently composed of protein and humus-like substances such as uronic acids, and deoxyribonucleic acid (DNA) [44, 45]. EPS is crucial in the formation of activated sludge flocs, where EPS is negatively charged, originated by ionizations of anions from carboxylic and phosphate [46]. The presence of cationic substituents such as calcium and potassium are responsible for the formation of flocs by linking the negatively charged EPS. Many discussions suggested that the formation of flocs is caused by intermolecular interactions between surfaces with net charge with another, forming larger aggregates. The classical DLVO theory is one of the well-acknowledged fundamentals owing to the organization of activated sludge flocs. Although, this theory is only valid for water with low ionic strength in which the flocs stability is jeopardized at ionic concentrated water [47].

However, there are conflicting findings reported on the decline of sludge flocculation with EPS generation. The bioflocculation ability is highly dependent on the total carbohydrates, protein, and uronic acid content. The process is inhibited when the system is acidic, owing to the reduced polymeric interactions and electrostatic repulsion [48]. The authors also added that when the fraction of uronic acid to EPS is high, the microorganisms have negatively charged surfaces, thus inhibiting bioflocculation. Polymeric interactions are enhanced when the protein and carbohydrate concentration in EPS is high, enabling flocculation [49]. Figure 3 compares two types of interactions (a) ionic and (b) hydrogen bonds between EPS in a reactor. Samples in Fig. 3 (a) were added with EDTA to initiate the formation of complex ions. The addition of EDTA did not influence the aggregation behavior of EPS (similar to the control sample). Therefore, this present finding suggests that the ionic interaction is not the dominant driving force of EPS aggregation. Meanwhile, in Fig. 3 (b), urea was added to the sample to evaluate the hydrogen bonding-driven sludge aggregation. The ability of urea to break the hydrogen bonding between EPS is proven when the increasing urea concentration causes a reduction of sludge aggregation over time [48]. The authors also conclude that the bioflocculation of EPS in sludge is mainly controlled by hydrogen bonding between tightly bound-EPS fractions.

Sewage sludge interparticle interactions are vital in many processes including sewage separation and treatment, metabolic activity, and densification and dewatering [50]. These processes are affected by the sewage sludge morphologies and

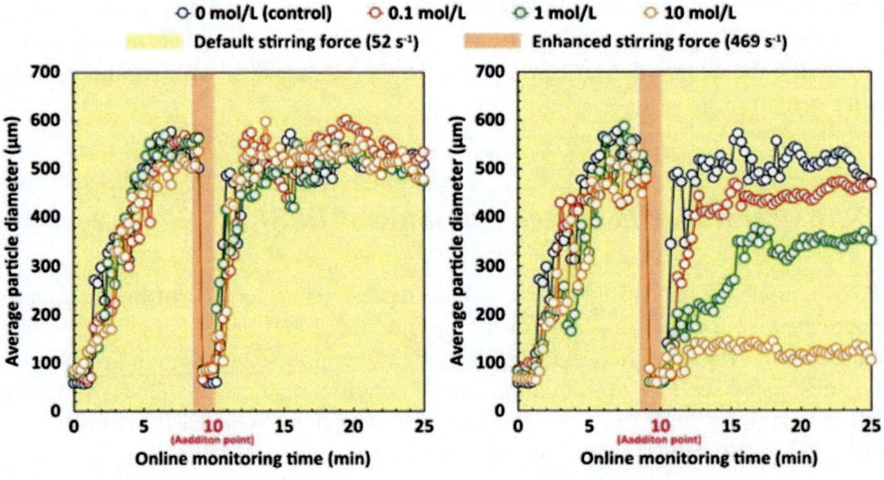

Fig. 3 Average sludge particle diameter when added with **a** EDTA and **b** urea. Reprinted with permission from [48]. Copyright 2016 Nature (CCBY)

particle sizes [50, 51]. Chemical separation is a vital method in treating colloidal wastewater and water systems. The underlying concept behind this method is the way of balancing out the stability of sludge particles in water by adding coagulants, flocculants, and surfactants, converting the fine particles into settable flocs and agglomerates [18]. This is an efficient practice for fine solid waste treatment in water, where this process is employed as a pre-process for water treatment prior to filtration [52]. The underlying origin of coagulant and flocculent effectiveness in treating sludge waste relies on the net surface charge of the sewage sludge particles, where the surface charge is contributed by the presence of the element as described in the earlier subchapter. The coagulant-sludge adsorption characteristics determine the effectiveness of the coagulant and the formation of multicharged complexes [53]. The multicharged complexes embody net interactions of numerous charged species and are controlled by the pH and ionic (co-ions and counter ions) presence in the overall system [54].

The coagulation and flocculation processes would require particle charge neutralization, where the resulting neutralization reaction is simply counterbalanced by the interactions between the charges involved. Typical coagulation and flocculation processes are described in Fig. 4. The optical properties of the sewage sludge-laden water such as turbidity and UV absorption are critical in determining the effectiveness of the coagulant types and dosages. The size of the flocs can be determined using the laser diffraction method, where the intensity of the scattered laser determines the distribution of flocs sizes [50].

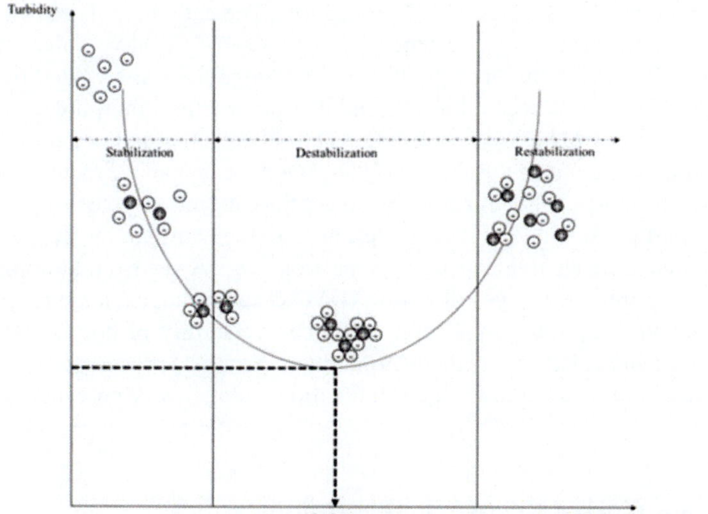

Fig. 4 The coagulation–flocculation processes. Reprinted with permission from [54]. Copyright 2020 MDPI (CCBY)

2.4 Sewage Sludge Stability

Many sewage sludge treatments and managements employ stabilization techniques to convert biomass waste into a useful source of energy, crops, and biobased materials. The selection of an efficient treatment method is highly dependent on the target end product of the sludge conversion. In this subchapter, the technologies discussed are more dedicated to sludge particle stabilization on a colloidal scale. Anaerobic digestion (AD) is the commonly used stabilization technology for the production of biogas by transforming sludge organic solids [55]. AD affects the particle size distribution of sewage sludge particles and dewatering ability by irreversibly transforming the structural matrix of sewage sludge particles and aggregates [56]. The efficiency of AD biological treatment is compromised due to several factors including complex structural components of sewage sludge, presence of extracellular polymeric substances (EPS), and recalcitrance on their cell walls. The sludge AD process typically aims to degrade EPS produced from sewage sludge and eventually modifies the particle size distribution.

Chemical coagulants or flocculants are usually incorporated into the sewage sludge to aid the sedimentation rate prior to the AD procedure. Among the notable coagulants used to improve the process is cationic and anionic polymers. Cationic polymers are a class of charged polymers or polyelectrolytes with ionic charges along their chain. Chu et al. [57] compared the performance of cationic, anionic, and non-ionic polyacrylamide on the effect of AD of wastewater sludge. They revealed that the zeta potential value for all sludge systems was gradually converged to the − 20 mV regime after 10 days of AD procedures. The cationic polymer-flocculated sludge witnessed surface charge tendency towards the negative value, while the anionic counterpart became less negative [57]. Sludge flocs aggregated by cationic polyacrylamide showed larger floc size and denser structural morphology [57].

The usage of polyDADMAC in aiding AD has been recorded as a promising and efficient process. Adequate dosage of polyDADMAC can manage to maintain significantly large flocs size, while further addition of the cationic polyelectrolyte will alter the sludge particle size due to the adsorption of polyelectrolyte on sludge surfaces, resulting in an increase in electrostatic repulsion [58]. Apart from altering the size of the flocs, optimized dosages of polyDADMAC can remove a large proportion of lignosulphonate [59] and polyphenols [60]. The versatility of polyDADMAC has also improved and enhanced methane production in the AD process [61], suggesting its efficiency to be utilized in sludge stabilization and recovery procedures.

3 Sewage Sludge Recovery Technologies

Like any other biomass, sewage sludge is composed of energetic organic and inorganic contents, making it renewable and recoverable. The principal aim of the utilization of renewable sources for energy generation is to mitigate the carbon footprint, and environmental pollution in general, which eventually contributes to the establishment of sustainable living. Sludge is considered a green source of energy as many emerging technologies employ the utilization of various types of sludge of multiple origins with minimal energy leakage and environmental and health hazards. This concern arose when the stabilized sewage sludge for agricultural uses is constrained by stringent legislation, worrying about the trace amounts of toxic content in sludge [62].

Electricity is one type of energy that is extensively produced from sewage sludge. Microbial fuel cells (MFCs) are one developing method to produce electricity from sludges. MFCs convert organic matter in sewage sludge into electrical power, which employs microorganisms in sludge as catalysts to oxidize organic and inorganic substances into electrical energy via the bioelectrochemical catalytic activity of microbes [63, 64]. Muaz and coworkers used dewatered sludge as a substrate for electrical power generation via MFCs and reported that the highest chemical oxygen demand (COD) removal and voltage generated from the biomass were 149.2 mg/L and 852.7 mV [65]. They also revealed that the yield for these two outputs can be optimized to up to 15% by setting the cell electrode distance at 2.93 cm; the moisture content was brought up to 31.99 v/w. %, and operating temperature set at 38.05 °C. The power generated by recent findings has significantly improved from the previous ones, which were evaluated a decade ago. This explains that MFCs technology is persistent and can still be developed to generate electricity using sludge from wastewater without high energy leakage.

The sustainable approach of sewage sludge-recovered energy is essential in determining the sustainability of the process. Apart from generating power, MFCs can be utilized to enhance the sludge recovery process by acting as organic pollutants removal and extracting devices. The biomass energy extraction process hindered the production of sewage sludge waste. Recent technology used an integrated MFCs-membrane bioreactor (MBR) system to reduce the production of sludge [66]. This integration enables synergistic and holistic approaches to utilize energy recovery and membrane fouling issues that are commonly reported in standalone systems. The electric field force generated in the integrated system reduced sludge deposition on the membrane, which can cause fouling as depicted in Fig. 5 [67, 68].

The production of biogas from sludge is also one of the extensively reported yields of sludge treatment and recovery. Sewage sludge-derived biogas (methane) is an energy-rich renewable gas manufactured via anaerobic digestion of sludge in an oxygen-absent condition with the aid of microorganisms [69]. The production of bio-derived methane is considered an economically and environmentally friendly approach of sewage sludge recovery technology. The methanization process is accessible and the end digestion residue is inert, nutrient-rich, and odor-free and can be utilized as agricultural fertilizer [69, 70]. Recent works reported on the enhancement of biogas yield when anaerobic co-digestion (ACD) process is employed. ACD combines two organic substrates: a major substrate such as sludge and manure and an additional substrate [71]. Figure 6 depicts the typical process for biogas production from biomass. The selection of appropriate co-digester is dependent on the optimal carbon-to-nitrogen (C-N) ratio for anaerobic digestion (20–30), in which sewage

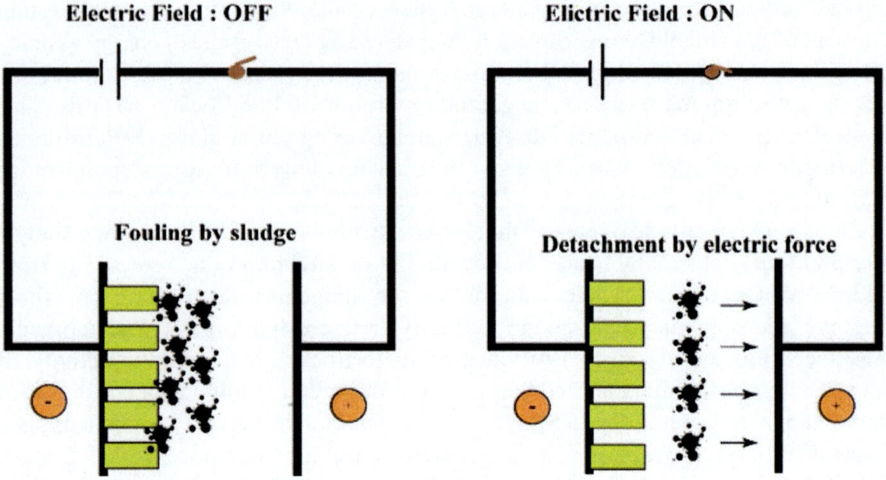

Fig. 5 Membrane sludge fouling hindered by electric force. Reprinted with permission from [67]. Copyright 2019 MDPI (CCBY)

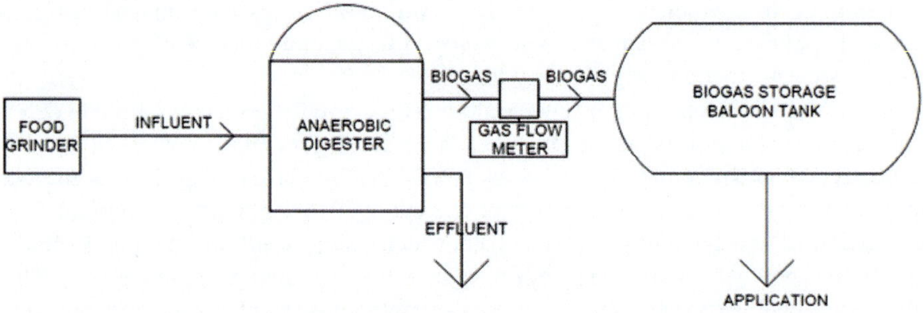

Fig. 6 Typical process flowchart of biogas production from food waste. Reprinted with permission from [73]. Copyright 2019 MDPI (CCBY)

sludge with lower C-N ratio can be co-digested with a substrate with a high C-N ratio [71, 72].

As tabulated in Table 2, the biogas production was enhanced when some fractions of biowastes co-substrate used in the anaerobic digestion of sewage, dewatered, activated, and fecal sludges. The volatile solids removal is also an indication of digester performance, where the volatile solids are degraded and converted into methane [74, 75]. Other than the concentration of co-digester, the key factors that contribute to the digestion performance include mixing parameters, operating temperature, organic loading rate, and hydraulic retention time [71]. Other than MFC, various efficient techniques have been developed to increase power generation using sludge from wastewater. The conversion of sewage sludge into biogas substitutes another route to generate electrical power. Recent findings recapitulated that energy derived from

Table 2 AD performance based on types of substrate used

Main substrate	Co-substrate	Biogas production	Digestion performance	References
Sewage sludge	–5% Food waste 5% Grape residue 5% Crude glycerol	~200 ml/l/d ~600 ml/l/d ~300 ml/l/d ~1800 ml/l/d	46.1% of volatile solids removed 53.5% of volatile solids removed 40.9% of volatile solids removed 44.9% of volatile solids removed	[74]
Dewatered sludge	10% Food waste	Improved 46% from mono-substrate system	Volatile solids removal increased from 38 to 62%	[80]
Waste activated sludge	30% Swine manure	402 mL biogas/ g volatile solid added	–	[81]
Fecal sludge	Up to 50% of Garden waste	50% greater than mono-substrate system	Volatile solids removal increased to up to ~92%	[82]

biogas generates more power compared to landfill gas, predominantly due to the presence of abundant putrescible substances [76, 77]. The utilization of bioenergy has received tremendous feedback due to its advantages towards environmental and economical sustainability [78]. The European Union (EU) has achieved its target of having at least 20% of energy production within the member states be bio-derived [79].

Although biological (anaerobic digestion) and bioelectrochemical (MFC) techniques are the most utilized techniques for sludge recovery into energy sources, thermochemical routes such as combustion, pyrolysis, and gasification are among the preferred methods too. One major drawback of thermochemical conversion is the critical demand for sludge with low moisture content, where sludge drying is also a mandatory process that requires an intense amount of energy [62]. The authors also mentioned that combustion yield of sludge can be used to generate electrical power via heating and aiding waste incineration by reducing the volume. The pyrolysis procedure of sludge is employed for biochar and bio-oil production, predominantly to mitigate environmental pollution and health hazards associated with sludge [83–85].

4 Conclusions

The chapter covers the colloidal behavior of sewage sludge particles, which is a pivot key in the selection and determination of the biomass purification and recovery processes. The underlying origin of sewage sludge colloidal stabilization can be described by the classical DLVO theory, and is largely dependent on the physical properties of the water bodies, key chemical components, and origin of the sludge particles. Charged and uncharged polymeric coagulants or flocculants are commonly employed in sludge-laden wastewater system to efficiently increase the solid content and stabilize the sludge particles. The pre-treatment process is essential in converting the biomass into various sources of energy via microbial fuel cells and anaerobic digestion. Proper understanding of sewage sludge colloidal behavior is vital to ensure the abundant biomass can be fully utilized for a more sustainable and healthy future.

Acknowledgements The authors would like to thank the Universiti Sains Malaysia for the financial support of this study via APEX Era grant (1001/PBAHAN/881008).

References

1. Garg S (2022) Industrial wastewater: characteristics, treatment techniques and reclamation of water. In: Advanced industrial wastewater treatment and reclamation of water. Springer, p 1–23. https://doi.org/10.1007/978-3-030-83811-9_1
2. Negi P, Verma H, Singh SP, Mahapatra BS, Jatav HS (2022) Global scenario of sewage-sludge management. In Sustainable management and utilization of sewage sludge. Springer, pp 383–401. https://doi.org/10.1007/978-3-030-85226-9_18
3. Gong Z, Guo J, Li Q, Xie H (2022) Lipopolysaccharide-stabilized ionic colloids induce biomineralization. Colloids Surf B Biointerfaces 211:112331. https://doi.org/10.1016/j.colsurfb.2022.112331
4. Tetteh EK, Rathilal S (2018) Effects of a polymeric organic coagulant for industrial mineral oil wastewater treatment using response surface methodology (RSM). Water Sa 44(2):155–161. https://doi.org/10.4314/wsa.v44i2.02
5. Lian S, Fan S, Yang Y, Yu B, Dai C, Qu Y (2022) Selenium nanoparticles with photocatalytic properties synthesized by residual activated sludge. Sci Total Environ 809:151163. https://doi.org/10.1016/j.scitotenv.2021.151163Get
6. Mohd Zaini Makhtar M, Tajarudin HA, Samsudin MDM, Vadivelu VM, Shoparwe NF, Zainuddin NI (2021) Membrane-less microbial fuel cell: Monte Carlo simulation and sensitivity analysis for COD removal in dewatered sludge. AIP Adv 11(6):065016. https://doi.org/10.1063/5.0039014
7. Do K-U, Chu X-Q (2022) Performances of membrane bioreactor technology for treating domestic wastewater operated at different sludge retention time. Dev Wastewater Treat Res Process: 107–122. https://doi.org/10.1016/B978-0-323-85583-9.00010-7
8. Ruiz-Hernando M, Vinardell S, Labanda J, Llorens J (2022) Effect of ultrasonication on waste activated sludge rheological properties and process economics. Water Res 208:117855. https://doi.org/10.1016/j.watres.2021.117855
9. Huang Z, Wang Y, Jiang L, Xu B, Wang Y, Zhao H et al (2018) Mechanism and performance of a self-flocculating marine bacterium in saline wastewater treatment. Chem Eng J 334:732–740. https://doi.org/10.1016/j.cej.2017.10.076
10. Jiang J-Q (2015) The role of coagulation in water treatment. Curr Opin Chem Eng 8:36–44. https://doi.org/10.1016/j.coche.2015.01.008
11. Sharp EL, Parson SA, Jefferson B (2006) Coagulation of NOM: linking character to treatment. Water Sci Technol 53(7):67–76. https://doi.org/10.2166/wst.2006.209

12. Alazaiza MYD, Albahnasawi A, Ali GAM, Bashir MJK, Nassani DE, Al Maskari T et al (2022) Application of natural coagulants for pharmaceutical removal from water and wastewater: a review. Water 14(2):140. https://doi.org/10.3390/w14020140
13. Rodriguez AZ, Wang H, Hu L, Zhang Y, Xu P (2020) Treatment of produced water in the Permian basin for hydraulic fracturing: Comparison of different coagulation processes and innovative filter media. Water 12(3):770. https://doi.org/10.3390/w12030770
14. Rydzek G, Ji Q, Li M, Schaaf P, Hill JP, Boulmedais F, Ariga K (2015) Electrochemical nanoarchitectonics and layer-by-layer assembly: from basics to future. Nano Today 10(2):138–167
15. Arnold-Smith AK, Christie RM, Jolicoeur C (1992) Polyaluminium silicate sulphate—a new coagulant for potable and wastewater treatment. In: Chemical water and wastewater treatment II. Springer. pp 203–219. https://doi.org/10.1007/978-3-642-77827-8_14
16. Brandt MJ, Johnson KM, Elphinston AJ, Ratnayaka DD (2017) Storage, clarification and chemical treatment. In: Twort's water supply. Elsevier, pp 323–66. https://doi.org/10.1016/B978-0-7506-6843-9.00015-9
17. Liu C, He Y, Li F, Wang H (2013) Preparation of poly ferric sulfate and the application in micro-polluted raw water treatment. J Chin Adv Mater Soc 1(3):210–218. https://doi.org/10.1080/22243682.2013.835120
18. Barrera-Díaz CE, Balderas-Hernández P, Bilyeu B (2018) Electrocoagulation: fundamentals and prospective. In: Electrochemical water and wastewater treatment. Elsevier, pp 61–76. https://doi.org/10.1016/B978-0-12-813160-2.00003-1
19. Wang H-F, Hu H, Wang H-J, Zeng RJ (2019) Combined use of inorganic coagulants and cationic polyacrylamide for enhancing dewaterability of sewage sludge. J Clean Prod 211:387–395. https://doi.org/10.1016/j.jclepro.2018.11.208
20. Tetteh EK, Rathilal S (2019) Application of organic coagulants in water and wastewater treatment. Org Polym.
21. Ravina L, Moramarco N (1993). Everything you want to know about coagulation & flocculation. Zeta-Meter, Inc., Staunton, pp 1–37
22. Kamizela T, Kowalczyk M (2019) Sludge dewatering: processes for enhanced performance. In: Industrial and municipal sludge. Elsevier, pp 399–423. https://doi.org/10.1016/B978-0-12-815907-1.00018-0
23. Maćczak P, Kaczmarek H, Ziegler-Borowska M (2020) Recent achievements in polymer bio-based flocculants for water treatment. Materials 13(18):3951. https://doi.org/10.3390/ma13183951
24. Shokri A, Fard MS (2022) A critical review in electrocoagulation technology applied for oil removal in industrial wastewater. Chemosphere 288:132355. https://doi.org/10.1016/j.chemosphere.2021.132355
25. Dorobantu LS, Bhattacharjee S, Foght JM, Gray MR (2009) Analysis of force interactions between AFM tips and hydrophobic bacteria using DLVO theory. Langmuir 25(12):6968–6976. https://doi.org/10.1021/la9001237
26. Israelachvili JN (2011) Intermolecular and surface forces. Academic Press, Santa Barbara, CA, USA, p 704
27. Spitzer JJ (2003) A consistent alternative to the DLVO theory. J Adhes 79(10):893–903. https://doi.org/10.1080/714906147
28. Langford A, Bruchsaler M, Gupta M (2022) Suspension properties and characterization of aluminum-adjuvanted vaccines. In: Practical aspects of vaccine development. Elsevier, pp 225–266. https://doi.org/10.1016/B978-0-12-814357-5.00008-8
29. Hussain G, Robinson A, Bartlett P (2013) Charge generation in low-polarity solvents: poly (ionic liquid)-functionalized particles. Langmuir 29(13):4204–4213. https://doi.org/10.1021/la3049086
30. Bajpai M, Katoch SS, Kadier A, Singh A (2022) A review on electrocoagulation process for the removal of emerging contaminants: theory, fundamentals, and applications. Environ Sci Pollut Res 29:1–30. https://doi.org/10.1007/s11356-021-18348-8

31. Mateos H, Palazzo G (2022) Colloidal stability. In: Colloidal foundations of nanoscience. Elsevier, pp 57–83. https://doi.org/10.1016/B978-0-12-822089-4.00001-5
32. Tytła M (2019) Assessment of heavy metal pollution and potential ecological risk in sewage sludge from municipal wastewater treatment plant located in the most industrialized region in Poland—case study. Int J Environ Res Public Health 16(13):2430. https://doi.org/10.3390/ijerph16132430
33. Lee Y-C, Lo S-L, Kuo J, Tsai C-C (2012) Beneficial uses of sludge from water purification plants in concrete mix. Environ Eng Sci 29(4):284–289. https://doi.org/10.1089/ees.2010.0479
34. Dai X, Xu Y, Lu Y, Dong B (2017) Recognition of the key chemical constituents of sewage sludge for biogas production. RSC Adv 7(4):2033–2037. https://doi.org/10.1039/C6RA26180A
35. Liu Z, Smith SR (2019) Enzyme activity of waste activated sludge extracts. Water Sci Technol 80(10):1861–1869. https://doi.org/10.2166/wst.2020.002
36. Ahmadi MM, Soheili M, Esfahani MM (2005) Effect of ionic strength on settling of activated sludge. Iranian J Env Health Sci Eng 2:1–5
37. Zykova I, Maksimuk N, Rebezov M, Kuznetsova E, Derkho M, Sereda T et al (2019) Interaction between heavy metals and microorganisms during wastewater treatment by activated sludge. J Eng Appl Sci 14(11):2139–2145
38. Tytła M, Widziewicz K, Zielewicz E (2016) Heavy metals and its chemical speciation in sewage sludge at different stages of processing. Environ Technol 37(7):899–908. https://doi.org/10.1080/09593330.2015.1090482
39. Miranda LS, Wijesiri B, Ayoko GA, Egodawatta P, Goonetilleke A (2021) Water-sediment interactions and mobility of heavy metals in aquatic environments. Water Res 202:117386. https://doi.org/10.1016/j.watres.2021.117386
40. Kinraide TB, Yermiyahu U (2007) A scale of metal ion binding strengths correlating with ionic charge, Pauling electronegativity, toxicity, and other physiological effects. J Inorg Biochem 101(9):1201–1213. https://doi.org/10.1016/j.jinorgbio.2007.06.003
41. Keshavarzifard M, Moore F, Sharifi R (2019) The influence of physicochemical parameters on bioavailability and bioaccessibility of heavy metals in sediments of the intertidal zone of Asaluyeh region, Persian Gulf, Iran. Geochemistry 79(1):178–187. https://doi.org/10.1016/j.geoch.2018.12.007
42. Wang J, Huang CP (2002) Heavy metal interactions with activated sludge particulates. Proc Water Environ Fed 2002(15):48–60. https://doi.org/10.2175/193864702784247314
43. Tian Y, Zheng L, Sun D-z (2006) Functions and behaviors of activated sludge extracellular polymeric substances (EPS): a promising environmental interest. J Environ Sci 18(3):420–427
44. Higgins MJ, Novak JT (1997) Characterization of exocellular protein and its role in bioflocculation. J Environ Eng 123(5):479–485. https://doi.org/10.1061/(ASCE)0733-9372(1997)123:5(479)
45. Park C, Novak JT (2007) Characterization of activated sludge exocellular polymers using several cation-associated extraction methods. Water Res 41(8):1679–1688. https://doi.org/10.1016/j.watres.2007.01.031
46. Liu X-M, Sheng G-P, Luo H-W, Zhang F, Yuan S-J, Xu J et al (2010) Contribution of extracellular polymeric substances (EPS) to the sludge aggregation. Environ Sci Technol 44(11):4355–4360. https://doi.org/10.1021/es9016766
47. Zita A, Hermansson M (1994) Effects of ionic strength on bacterial adhesion and stability of flocs in a wastewater activated sludge system. Appl Environ Microbiol 60(9):3041–3048. https://doi.org/10.1128/aem.60.9.3041-3048.1994
48. Guo X, Wang X, Liu J (2016) Composition analysis of fractions of extracellular polymeric substances from an activated sludge culture and identification of dominant forces affecting microbial aggregation. Sci Rep 6(1):1–9. https://doi.org/10.1038/srep28391
49. Badireddy AR, Chellam S, Gassman PL, Engelhard MH, Lea AS, Rosso KM (2010) Role of extracellular polymeric substances in bioflocculation of activated sludge microorganisms under glucose-controlled conditions. Water Res 44(15):4505–4516. https://doi.org/10.1016/j.watres.2010.06.024

50. Bieganowski A, Lagod G, Ryzak M, Montusiewicz A, Chomczynska M, Sochan A (2012) Measurement of activated sludge particle diameters using laser diffraction method/Pomiary srednicy czastek osadu czynnego za pomoca metody dyfrakcji laserowej. Ecol Chem Eng 19(4):597. https://doi.org/10.2478/v10216-011-0042-7
51. Wilén B-M, Nielsen JL, Keiding K, Nielsen PH (2000) Influence of microbial activity on the stability of activated sludge flocs. Colloids Surf, B 18(2):145–156. https://doi.org/10.1016/S0927-7765(99)00138-1
52. Yargeau V (2012) Water and wastewater treatment: chemical processes. In: Metropolitan sustainability. Elsevier, pp 390–405. https://doi.org/10.1533/9780857096463.3.390
53. Hutchison PR, Healy TW (1990) Coagulation and flocculation—destabilizing practices? (With particular reference to metal ion coagulants). In: Surface and colloid chemistry in natural waters and water treatment. Springer, pp 119–134. https://doi.org/10.1007/978-1-4899-2510-7_9
54. Kurniawan SB, Abdullah SRS, Imron MF, Said NSM, Ismail NI, Hasan HA et al (2020) Challenges and opportunities of biocoagulant/bioflocculant application for drinking water and wastewater treatment and its potential for sludge recovery. Int J Environ Res Public Health 17(24):9312. https://doi.org/10.3390/ijerph17249312
55. Tyagi VK, Lo SL (2016) Energy and resource recovery from sludge: full-scale experiences. In: Environmental materials and waste. Elsevier, pp 221–244. https://doi.org/10.1016/B978-0-12-803837-6.00010-X
56. Braguglia CM, Mininni G, Rolle E (2006) Influence of anaerobic digestion on particle surface charge and optimal polymer dosage. Water Sci Technol 54(5):43–50. https://doi.org/10.2166/wst.2006.545
57. Chu CP, Lee DJ, Chang B-V, You CH, Liao CS, Tay JH (2003) Anaerobic digestion of polyelectrolyte flocculated waste activated sludge. Chemosphere 53(7):757–764. https://doi.org/10.1016/S0045-6535(03)00553-8
58. Zhang Z, Zhang Q, Chen Y, Li Z (2018) Poly (dimethyldiallylammonium chloride) (poly-DADMAC) assisted cellulase pretreatment for microfibrillated cellulose (MFC) preparation and MFC analysis. Holzforschung 72(7):531–538. https://doi.org/10.1515/hf-2017-0152
59. Feizi ZH, Kazzaz AE, Kong F, Fatehi P (2019) Evolving a flocculation process for isolating lignosulfonate from solution. Sep Purif Technol 222:254–263. https://doi.org/10.1016/j.seppur.2019.04.042
60. Geng C-X, Cao N, Xu W, He C, Yuan Z-W, Liu J-W et al (2018) Molecular characterization of organics removed by a covalently bound inorganic–organic hybrid coagulant for advanced treatment of municipal sewage. Environ Sci Technol 52(21):12642–12648. https://doi.org/10.1021/acs.est.8b03306
61. Zahrim AY, Dexter ZD, Joseph CG, Hilal N (2017) Effective coagulation-flocculation treatment of highly polluted palm oil mill biogas plant wastewater using dual coagulants: decolourisation, kinetics and phytotoxicity studies. J Water Process Eng 16:258–269. https://doi.org/10.1016/j.jwpe.2017.02.005
62. Oladejo J, Shi K, Luo X, Yang G, Wu T (2019) A review of sludge-to-energy recovery methods. Energies 12(1):60. https://doi.org/10.3390/en12010060
63. Gude VG (2016) Microbial fuel cells for wastewater treatment and energy generation. In: Microbial electrochemical and fuel cells. Elsevier, pp 247–285. https://doi.org/10.1016/B978-1-78242-375-1.00008-3
64. Yuan Y, Yuan T, Wang D, Tang J, Zhou S (2013) Sewage sludge biochar as an efficient catalyst for oxygen reduction reaction in an microbial fuel cell. Biores Technol 144:115–120. https://doi.org/10.1016/j.biortech.2013.06.075
65. Muaz MZM, Abdul R, Vadivelu VM (2019) Recovery of energy and simultaneous treatment of dewatered sludge using membrane-less microbial fuel cell. Environ Prog Sustainable Energy 38(1):208–219. https://doi.org/10.1002/ep.12919
66. Karlikanovaite-Balikci A, Özgün ÖK (2020) Generation of electricity and sludge reduction in a microbial fuel cell. J Int Environ Appl Sci 15(3):141–151
67. Wang J, Zhao S, Kakade A, Kulshreshtha S, Liu P, Li X (2019) A review on microbial electrocatalysis systems coupled with membrane bioreactor to improve wastewater treatment. Microorganisms 7(10):372. https://doi.org/10.3390/microorganisms7100372

68. Tian Y, Li H, Li L, Su X, Lu Y, Zuo W et al (2015) In-situ integration of microbial fuel cell with hollow-fiber membrane bioreactor for wastewater treatment and membrane fouling mitigation. Biosens Bioelectron 64:189–195. https://doi.org/10.1016/j.bios.2014.08.070
69. Demirbas A, Taylan O, Kaya D (2016) Biogas production from municipal sewage sludge (MSS). Energ Sources Part A 38(20):3027–3033. https://doi.org/10.1080/15567036.2015.1124944
70. Kalloum S, Bouabdessalem H, Touzi A, Iddou A, Ouali MS (2011) Biogas production from the sludge of the municipal wastewater treatment plant of Adrar city (southwest of Algeria). Biomass Bioenerg 35(7):2554–2560. https://doi.org/10.1016/j.biombioe.2011.02.012
71. Chow WL, Chong S, Lim JW, Chan YJ, Chong MF, Tiong TJ et al (2020) Anaerobic co-digestion of wastewater sludge: a review of potential co-substrates and operating factors for improved methane yield. Processes 8(1):39. https://doi.org/10.3390/pr8010039
72. Mao C, Feng Y, Wang X, Ren G (2015) Review on research achievements of biogas from anaerobic digestion. Renew Sustain Energy Rev 45:540–555. https://doi.org/10.1016/j.rser.2015.02.032
73. Zulkifli AA, Mohd Yusoff MZ, Abd Manaf L, Zakaria MR, Roslan AM, Ariffin H et al (2019) Assessment of municipal solid waste generation in Universiti Putra Malaysia and its potential for green energy production. Sustainability 11(14):3909. https://doi.org/10.3390/su11143909
74. Maragkaki AE, Fountoulakis M, Kyriakou A, Lasaridi K, Manios T (2018) Boosting biogas production from sewage sludge by adding small amount of agro-industrial by-products and food waste residues. Waste Manage 71:605–611. https://doi.org/10.1016/j.wasman.2017.04.024
75. Wang G, Dai X, Zhang D, He Q, Dong B, Li N et al (2018) Two-phase high solid anaerobic digestion with dewatered sludge: improved volatile solid degradation and specific methane generation by temperature and pH regulation. Biores Technol 259:253–258. https://doi.org/10.1016/j.biortech.2018.03.074
76. Ayodele TR, Alao MA, Ogunjuyigbe ASO, Munda JL (2019) Electricity generation prospective of hydrogen derived from biogas using food waste in south-western Nigeria. Biomass Bioenerg 127:105291. https://doi.org/10.1016/j.biombioe.2019.105291
77. Cudjoe D, Han MS, Nandiwardhana AP (2020) Electricity generation using biogas from organic fraction of municipal solid waste generated in provinces of China: techno-economic and environmental impact analysis. Fuel Process Technol 203:106381. https://doi.org/10.1016/j.fuproc.2020.106381
78. Pfau SF, Hagens JE, Dankbaar B (2017) Biogas between renewable energy and bio-economy policies—opportunities and constraints resulting from a dual role. Energy Sustain Soc 7(1):1–15. https://doi.org/10.1186/s13705-017-0120-5
79. European Environment Agency (2021) Share of energy consumption from renewable sources in Europe 2021. https://www.eea.europa.eu/ims/share-of-energy-consumption-from
80. Dai X, Duan N, Dong B, Dai L (2013) High-solids anaerobic co-digestion of sewage sludge and food waste in comparison with mono digestions: stability and performance. Waste Manage 33(2):308–316. https://doi.org/10.1016/j.wasman.2012.10.018
81. Borowski S, Domański J, Weatherley L (2014) Anaerobic co-digestion of swine and poultry manure with municipal sewage sludge. Waste Manage 34(2):513–521. https://doi.org/10.1016/j.wasman.2013.10.022
82. Afifah U, Priadi CR (eds) (2017). Biogas potential from anaerobic co-digestion of faecal sludge with food waste and garden waste. AIP Publishing LLC. https://doi.org/10.1063/1.4979248
83. Kim Y, Parker W (2008) A technical and economic evaluation of the pyrolysis of sewage sludge for the production of bio-oil. Biores Technol 99(5):1409–1416. https://doi.org/10.1016/j.biortech.2007.01.056
84. Agrafioti E, Bouras G, Kalderis D, Diamadopoulos E (2013) Biochar production by sewage sludge pyrolysis. J Anal Appl Pyrol 101:72–78. https://doi.org/10.1016/j.jaap.2013.02.010
85. Fonts I, Gea G, Azuara M, Ábrego J, Arauzo J (2012) Sewage sludge pyrolysis for liquid production: a review. Renew Sustain Energy Rev 16(5):2781–2805. https://doi.org/10.1016/j.rser.2012.02.070

Antimicrobial-Resistant Microorganisms and the Possibility of Using Microbial Fuel Cell Technology to Reduce Their Transmission in the Environment

Ana Masara Ahmad Mokhtar, Chen Sep Ngee, Mohammed Zharif Asyrani Mohammed Alias, Nor Hawani Salikin, Fatin Nur Izzati Mohd Fadzil, and Nur Azzalia Kamaruzaman

Abstract Antimicrobial resistance (AMR) is a severe problem in Malaysia and worldwide; the World Health Organization (WHO) ranks AMR among the top ten global threats to public health and development. Malaysia's Antimicrobial Resistance Action Plan (MyAP-AMR) 2017–2021 describes AMR as a One Health concern requiring multidisciplinary collaboration across all sectors: humans, animals, plants, and the shared environment. Thus, it is vital to understand how it spreads in the environment and, concurrently, to discover feasible measures for reducing its spread. Among the environmental drivers of AMR is wastewater or sewage sludge. It is well established that exposure to sewage can promote the intake of AMR bacteria, resulting in the development of life-threatening infections such as sepsis in humans. As a result, prompt action is required to ensure the sludge is safe for human use. One possibility is to treat sludge with microbial fuel cell (MFC) technology, a bio-electrical device that uses the natural metabolic activity of electrogenic bacteria (EB) to generate electricity. Current research in the laboratory focuses on the use of EB like Bacillus subtilis to catalyse the conversion of carbon sources in sludge to sustainable energy. However, no study has been conducted to determine the benefits of MFC technology in preventing the spread of AMR in the environment. Nonetheless, due to its potential to trigger the production of bacteriocins that can kill or deactivate AMR microorganisms, it is expected that MFC-treated sludge will reduce AMR transmission to the environment and eventually to humans. In general, this chapter

A. M. Ahmad Mokhtar (✉) · C. S. Ngee · M. Z. A. Mohammed Alias · N. H. Salikin · F. N. I. Mohd Fadzil
Bioprocess Technology Division, School of Industrial Technology, Universiti Sains Malaysia, 11800 Penang, Malaysia

C. S. Ngee
e-mail: anamasara@usm.my

A. M. Ahmad Mokhtar
Green Biopolymer Coating and Packaging Centre, School of Industrial Technology, Universiti Sains Malaysia, 11800 Penang, Malaysia

N. A. Kamaruzaman
National Poison Centre, Universiti Sains Malaysia, Penang, Malaysia

M. Mohd Zaini Makhtar et al. (eds.), *Microbial Fuel Cell (MFC) Applications for Sludge Valorization*, Green Energy and Technology,
https://doi.org/10.1007/978-981-99-1083-0_10

will discuss the fundamentals of AMR and MFC and the additional benefits of MFC technology in reducing AMR transmission.

Keywords Antimicrobial resistance · Microbial Fuel cell · Infections

1 Introduction

Rapid urbanisation and industrialisation have contributed significantly to environmental pollution and human dangers due to their hazardous contents, which include pathogens, irritants, carcinogens, flammable, explosive, and oxidising agents [1]. Thus, proper and efficient waste management should be a major priority today, as it has the potential to have a detrimental influence not only on the environment but also on our society [1]. Fortunately, current wastewater treatment appears to be one of the most effective methods for reducing environmental pollution and health hazard. Additionally, the treated sludge has also been regarded as a critical biological resource for managing successful agriculture by improving crop yields and benefiting society and the economy [2]. However, although the sludge has been sufficiently treated before being applied to the soil, it does not completely eliminate the pathogen, heavy metals, organic chemicals, or chemical irritants present in the sewage sludge. For example, while sewage treatment is known to reduce pathogens, some pathogens such as *Clostridium perfringens* and hepatitis A virus (HAV) persist in the treated sewage sludge for an extended length of time and are resistant to existing wastewater treatment methods [3–5].

Additionally, numerous studies have demonstrated the importance of wastewater as an important reservoir of antimicrobial resistance (AMR), as it provides an excellent setting for the survival of AMR bacteria (ARBs) and AMR genes (ARGs). While the treatment process can assist in eliminating or reducing the ARB load, it has a negligible effect on ARGs. ARGs are not biodegradable and can be transferred through horizontal gene transfer to other bacteria, especially the Gram-negative bacteria [6]. Ultimately, it will promote the spreading of pathogenic AMR or ARBs to animals and humans via consumption of infected vegetables or water, inhalation, or direct skin contact, negatively harming their health [7–9]. Thus, it is vital to develop novel strategies for limiting the spread of AMR or ARB to the environment, such as by the employment of viruses, bacteriocins, or predatory bacteria. Interestingly, these alternatives can be applied using microbial fuel cell (MFC) technology, a bio-electrochemical system that employs microorganisms as catalysts to convert chemical energy stored in organic or inorganic substances to electrical energy [10, 11]. Thus, the review will evaluate the adverse effects of ARB on humans and the potential for risk reduction through the application of MFC technology.

2 Antimicrobial Resistance Microorganisms in the Environment and Its Implication to Humans

AMR, often known as drug resistance, is widely recognised as a global threat to human health that demands immediate action in countries worldwide. AMR is a major concern because it could lead to a post-antibiotic era where antibiotics are no longer effective. AMR refers to resistance in a variety of microorganisms to various antimicrobials, including antibacterial, antiviral, antiparasitic, and antifungal medications [12]. It happens when bacteria, viruses, fungi, and parasites acquire resistance to most antimicrobials used to treat infections [13]. Antibiotics and other antimicrobial treatments become ineffective due to the formation and spread of drug-resistant bacteria, resulting in antimicrobial resistance, which continues to undermine the ability to treat illnesses [14].

AMR is a natural occurrence aided by several variables: (a) Misuse and overuse of antimicrobials in clinics and animals' healthcare; (b) poor access to clean water, sanitation, and hygiene for both humans and animals; (c) inadequate infection and disease prevention in healthcare facilities and farms; and (d) lack of awareness and knowledge of antimicrobials usage among the public are all driving factors in the development of antimicrobial-resistant pathogens [15, 16]. Increased antimicrobial resistance can lead to many issues: (a) some severe infections being more difficult to control; (b) remaining inside the body for long periods; and (c) more extended hospital stays, all of which significantly impact patients' quality of life and place a strain on medical care, which is directly linked to high expenditures and high risk of infection-related mortality [17, 18].

3 Mechanism of Resistance

AMR has multiple key mechanisms, including drug uptake restriction, drug target alteration, drug inactivation, and active efflux of a drug [19]. These processes can be classified as innate resistance and acquired resistance. Intrinsic resistance is the innate ability of bacteria to resist the efficacy of a certain antibiotic through inherent structural or functional characteristics, and the mechanisms involved are drug uptake limitation, drug inactivation, and active efflux of a drug [19, 20]. Whereas acquired resistance can be acquired through mutational alterations or horizontal gene transfer, the processes involved are drug target modification, drug inactivation, and active efflux of a drug [19, 21].

3.1 Restricting drug's Entry

Antimicrobial chemicals must gain access to the bacterial cell to disrupt the bacteria's regular functions. Microbes may acquire resistance mechanisms by restricting antimicrobial drug uptake [21] (Fig. 1). This method is frequent in gram-negative bacteria and involves changes in the outer membrane's lipid composition, porin channel selectivity, and porin channel concentrations [19]. Lipopolysaccharide (LPS) is typically found in gram-negative bacteria to act as a barrier to particular compounds [19]. This layer gives inherent resistance to a subset of large antibacterial agents. However, the main pathway by which these antibiotics typically cross the Gram-negative bacteria with large outer membranes is via porin channels [22]. Porin channels allow access to hydrophilic compounds such as β-lactam, tetracyclines, and certain fluoroquinolones in gram-negative bacteria [21]. However, porin alterations can impair drug uptake in two ways: decreased porin protein expression or mutations that alter the porin channel's selectivity. For instance, gram-negative bacteria typically impede the uptake of certain antibiotics, such as aminoglycosides and ß-lactams, by altering porin channels' frequency, size, and selectivity in the cell membrane [19]. Enterobacteriaceae and *Pseudomonas aeruginosa* develop resistance by altering the expression of the porin protein, whereas, in *E. aerogenes*, porin mutations were shown to alter the shape of the porin channel, imparting resistance to imipenem and some cephalosporins [19]. All these changes are necessary to prevent antibiotics from accessing the drug binding sites, such as ribosomes and penicillin-binding proteins (PBPs).

3.2 Modification of Drug Targets

Antimicrobial agents are typically directed against specific targets, and structural alterations might impair drug binding, rendering the drug ineffective. Certain resistant bacteria have the ability to change the antimicrobial drug's target, hence conferring resistance. They may resist antimicrobials by reprogramming or concealing binding target sites to avoid detection. Additionally, bacteria have an evolutionary advantage in that they can gain drug resistance due to spontaneous genes modification, promoting structural changes of these antimicrobial binding sites [22]. For example, genetic mutations will influence the active sites of PBPs, which are transpeptidases involved in the synthesis of peptidoglycan in the cell wall [19] (Fig. 1). As a result, this can impede the binding of β-lactam antibiotics and result in multidrug resistance, which is frequently observed in gram-positive bacteria such as *Streptococcus pneumoniae* [19, 22].

On the contrary, Methicillin-resistant *Staphylococcus aureus* (MRSA) by gaining a new low-affinity PBP rather than structurally changing their existing PBPs [23]. Other examples of this type of resistance mechanism include alteration of peptidoglycan subunit peptide chains, which confers resistance to glycopeptides; prevent drug interaction with the ribosome by affecting ribosomal subunits via ribosomal mutation or ribosomal protection, conferring resistance to macrolides, tetracyclines, and aminoglycosides; alteration in LPS structure of gram-negative bacteria, giving resistance to polymyxins; alterations in RNA polymerase, which provide resistance to fluoroquinolones; and alteration in metabolic enzymes, giving resistance to sulfa drugs, sulfones, and trimethoprim [19, 22]. As a result, the antimicrobial drugs' capacity to bind to their target molecules will be diminished.

3.3 Drug Inactivation

Antimicrobial resistance genes or ARGs may encode enzymes capable of chemically altering or degrading the drug via hydrolysis (Fig. 1). Bacteria inactivate drugs in two ways: either by degrading the drug or by transferring a chemical group to the drug, so inactivating it. A common example is the hydrolytic deactivation of the ß-lactam ring in penicillins and cephalosporins by ß-lactamases, a drug hydrolysing enzyme [22]. When the ß-lactam bond is broken, the antimicrobial drug's antibacterial activity is lost. On the contrary, drug inactivation via enzymatic transfer of a chemical group to the drug can result in drug inactivation by interrupting the drug's interaction with its bacterial target. Acetyl, phosphoryl, and adenyl groups are the most common chemical groups implicated in this resistance mechanism [19, 22], with acetylation is the most diversely employed mechanism, and its efficacy against aminoglycosides, chloramphenicol, streptogramins, and fluoroquinolones have been demonstrated [19].

3.4 Active Efflux of a Drug

Certain bacteria have membrane proteins that operate as an export or efflux pump for certain antimicrobials (Fig. 1), actively transporting the drug out of the cell and preventing the drug from building up in the cells to a level that would be harmful to the bacterium [19, 22]. This mechanism has been observed in *E. coli* and other Enterobacteriaceae against tetracyclines, Enterobacteriaceae against chloramphenicol, Staphylococci against macrolides, and *Staphylococcus aureus* and *Streptococcus pneumoniae* against fluoroquinolones [19, 24]. There are five leading families of efflux pumps, classified according to their structure and energy source in bacteria. These five families are as follows: the ABC family; the multidrug and toxic compound extrusion (MATE) family; the small multidrug resistance (SMR) family; the major facilitator superfamily (MFS) family; and the resistance-nodulation-cell division (RND) family. Most of these efflux pump families are composed of a single component and function by transporting substrates through the cytoplasmic membrane [19].

The transfer of ARGs between bacteria has been described, and this process may contribute to the rapid spread of ARGs between bacteria [25, 26]. The majority

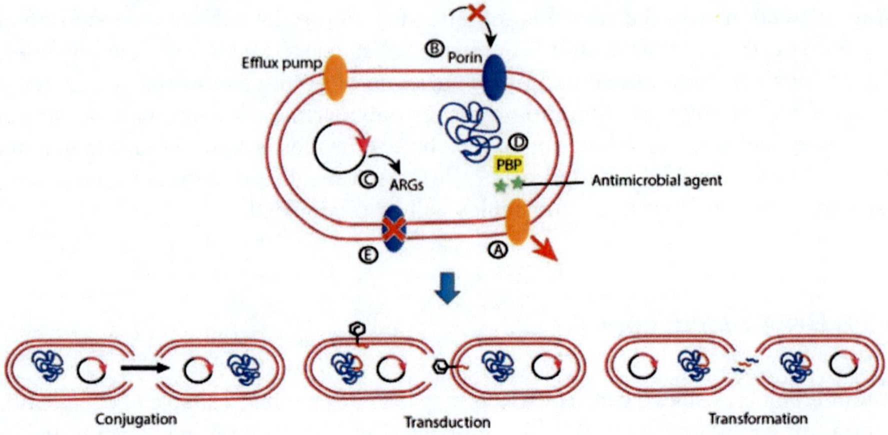

Fig. 1 Possible mechanism of antimicrobial resistance. (A) Altered affinity or increased expression of efflux pump, (B) decreased membrane permeability through modified porins, (C) increased production of antimicrobial resistance genes (ARGs), (D) structural changes or mutations of PBPs, and (E) modification of drug targets may promote ABR among bacteria. Transfer of ARGs between bacteria can occur via conjugation, transduction, and transformation

of ARGs are found on mobile genetic elements like plasmids or transposons, and they are transferred between bacteria via various mechanisms, including transformation, transduction, and conjugation [26–28] (Fig. 1). In transformation, bacteria acquire and incorporate an extracellular DNA segment, previously released into the environment by other organisms [29, 30]. Transduction involves the transmission of a DNA segment or plasmid harbouring ARGs between bacteria by a phage [31]. Finally, conjugation facilitates genetic exchange by elongating a pilus in Gram-negative bacteria or producing sex pheromones in Gram-positive bacteria [32]. Once resistant, a bacterium can rapidly multiply and pass on the resistance determinants to its progeny [33].

4 Antimicrobial Resistance Transmission in the Environment Through Wastewater

AMR has emerged as a significant global concern in recent years, affecting humans, animals, and the environment. The inappropriate and excessive use of antibiotics in a variety of sectors, including agriculture, veterinary medicine, and healthcare, is at the root of the global epidemic that has emerged in AMR transmission in the environment [34], implying that the environment serves as an AMR reservoir and is critical for transmitting AMR microorganisms with ARGs to animals and humans [34, 35]. AMR is facilitated by several factors, including antibiotics, antimicrobial genes, heavy metals, and biocides [34], and it can be transmitted to the environment in a variety of ways, including through wastewater from hospitals, healthcare-related companies, livestock farms, and agriculture [36].

Antibiotics are frequently used to treat human illnesses in the healthcare sector [37]. Antibiotic abuse and overuse in humans, on the other hand, may result in the selection of resistant strains [38]. Additionally, veterinary drugs used in animal husbandry were originally intended to prevent animal diseases but are now being inappropriately used for other purposes, such as feed additives and growth stimulants [39]. As a result, humans and animals expel biologically active antibiotics in their urine and faeces, which are then discharged into wastewater treatment facilities (WWTPs) [40] (Fig. 2). Additionally, the pharmaceutical industry's inappropriate disposal of unused or expired medicines may also contribute to the discharge of antibiotics and AMR microorganisms into the environment. Besides, antibiotics have a half-life of a few hours to hundreds of days, which means they will linger in wastewater and be considered persistent pollutants in the environment [41]. To survive and grow in sewage, bacteria will develop resistance mechanisms to AMR drugs. Susceptible bacteria will be killed, or their activity will be suppressed under the impact of these antibiotics. Meanwhile, bacteria that are intrinsically or acquired resistant to antibiotics have a better chance of survival and expansion [42]. The surviving AMR microorganisms with ARGSs will spread the genes to other microbes, thereby making wastewater a reservoir for AMR microorganisms. Besides, due to the antibiotic's slow decomposition, it may find its way into groundwater or aquatic systems throughout the wastewater treatment process [35], where it can be disseminated to humans, animals, and the environment.

AMR microorganisms can be transmitted to the sewage system through infected human secretion, clinical or industrial settings. The ARBs will then transfer to the wastewater treatment plants (WWTPs) and eventually transmitted to sewage workers and residents who live near the WWTPs. This can be occurred either through drinking infected water, consume infected vegetables, or via recreational use

Additionally, biocides are a factor in AMR. Biocides are antimicrobial chemical compounds frequently used in the home, industry, and healthcare to control infections and microbiological contamination [43]. Ethanol, formaldehyde, chlorhexidine, triclosan, and quaterium ammonium compounds (QACs) are all examples of common biocides [34]. Improper biocide disposal in WWTPs can increase the number of biocides entering the environment, raising the likelihood of causing AMR in microorganisms residing in the WWTPs [40]. For example, chlorine resistance in Salmonella typhi demonstrates how improper biocide use can exert selective pressure on bacteria, which then respond by developing resistance mechanisms [43]. When mixed with other biocides such as QACs and chlorhexidine, triclosan has been proven to cause antibiotic resistance in microbes. Additionally, sub-lethal doses of biocides contribute to selecting mutations conferring antibiotic resistance [34].

Heavy metals are also a vector for AMR transmission in the environment. Heavy metals are non-antibiotic antimicrobial compounds that have been extensively used in agricultural and industrial applications for a variety of purposes [44]. Interestingly, they can also act as a selector for ARGs, possibly by physically associated with plasmids or chromosomes containing ARGs [45]. For example, MRSA

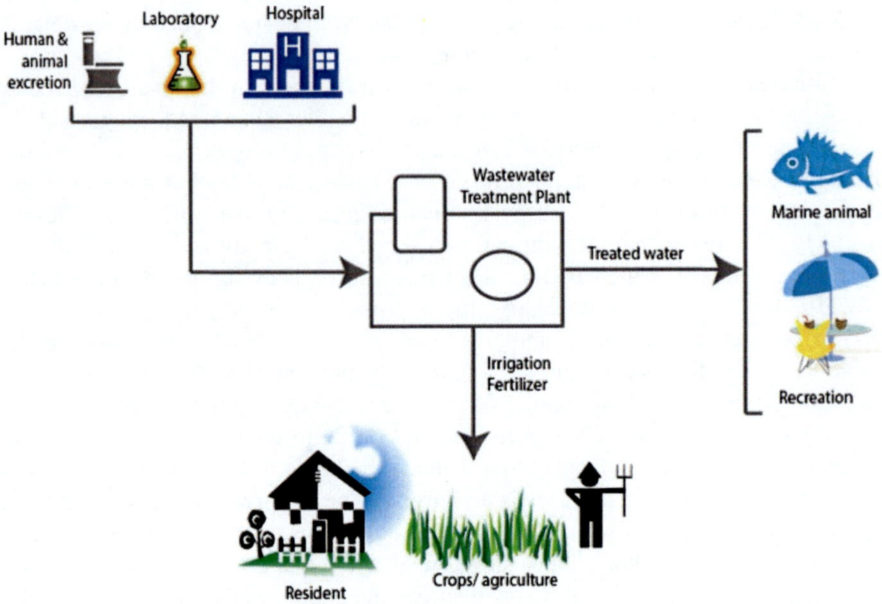

Fig. 2 Potential transmission of antimicrobial resistance microorganisms in the environment.

isolated from livestock possessed plasmids encoding resistance genes to Cu and Cd (*copA*, *cadDX*, and *mco*), as well as numerous antimicrobials such as Macrolides, Lincosamides, Streptogramin B, Tetracyclines, Aminoglycosides, and Trimethoprim (*erm(T)*, *tet(L)*, *aadD*, and *dfrK*) [46, 47]. Thus, incorrect disposal of heavy metals to WWTPs might increase metals entering the environment and the development of AMR microorganisms [40].

The environment is considered a route of transmission for ARBs to humans, facilitated by WWTPs. This finding is consistent with some studies demonstrating the occurrence of antibiotic-resistant bacteria such as *E. coli* against cephalexin, ciprofloxacin, and ampicillin in treated sewage produced by Penang WWTPs [36]. Additionally, humans may be exposed to these bacteria while participating in recreational activities in contaminated surface water, ingesting contaminated drinking water, or consuming fresh fish products [48]. In agriculture, sludge is frequently applied as fertiliser due to its nutritional contents. However, this might also facilitate the spread of AMR microorganisms to animals or humans through the consumption of contaminated crops or direct skin contact [37]. Additionally, the leading cause of AMR emergence worldwide is the improper use of antimicrobial drugs by humans in hospitals to treat infections [13]. Although the primary purpose is to treat the infections caused by pathogenic microorganisms, these microorganisms continually mutate and evolve, enabling them to adapt to their environment and eventually develop resistance to specific antibiotic treatments. As a result, the antimicrobial-resistant bacteria can spread from the infected patients to other people via unclean hands or contaminated objects. Patients who remain untreated for AMR bacteria will subsequently convey these resistant microorganisms to others, increasing the danger of disease spread, severe sickness, and even death.

5 The Effect of Antimicrobial Resistance Microorganisms on Humans

Usually, antibiotics, one of the antimicrobial agents, are prescribed to treat a wide range of illnesses and surgical procedures, including organ transplants, blood infections, pneumonia, and cancer treatment [13]. However, improper use of this drug may cause pathogenic microorganisms to evolve to become resistant to antimicrobial treatment, making infections more difficult to treat and raising the risk of disease spread, severe illness, and death [13]. In general, the more antibiotics used, the more bacteria adapt and develop new ways to survive, leading to antibiotic resistance. As a result, some bacteria live and multiply rather than being eradicated by antibiotics, causing significantly more harm and may impact patients' quality of life, such as increased healthcare expenses, hospitalisation length, and mortality [49].

MRSA, one of the most well-known examples of AMR, has been linked to high mortality rates across the globe each year, rendering treatment ineffective [50]. Additionally, 4.1% of newly diagnosed tuberculosis cases are multidrug resistant. It is anticipated to grow dramatically by 2040, particularly in nations with a high prevalence of tuberculosis, such as India and the Philippines [49, 51]. Compared to non-resistant bacteria, resistant bacteria quadruple the likelihood of getting a severe medical issue and triple the possibility of dying. Naturally, these negative consequences will be magnified as the severity of resistant infections and the host's sensitivity increase [52, 53]. This is consistent with the increased morbidity and fatality rates experienced by infected patients over time. It is estimated that if significant action against AMR is not taken by 2050, roughly ten million people will die [49, 54].

Additionally, AMR may result in inadequate treatment of sepsis, which is among the top cause of death in hospitals, with a mortality rate of 19.7 worldwide [55]. Sepsis is a potentially fatal organ dysfunction caused by an abnormal host response to infection [56]. This is a common complication seen in immunocompromised cancer patients and recipients of hematopoietic stem cell transplants [57]. As a result, the global spread of AMR microorganisms, particularly among immune-deregulated patients, may result in a delay in commencing effective empirical antibiotic therapy, potentially resulting in infection and hence sepsis [57]. Thus, it is vital to decrease AMR transmission in the environment, which can begin with appropriately managing sludge disposal and treatment; and ensuring that it is safe for human application.

6 Microbial Fuel Cell (MFC) technology and Its Application in Minimising Antimicrobial Resistance Transmission in the Environment

MFC technology is a relatively new biotechnology that can remediate wastewater while generating electricity. MFCs enhance electricity generation during wastewater treatment by exploiting residual sludge substrates, such as acetate or glucose, using electrogenic bacteria's (EB's) enzymatic activity [58]. EB is a type of organism capable of transferring electrons extracellularly across the cell membrane to electron acceptors, such as electrodes and oxide minerals, both under anaerobic and microaerobic environments [59, 60]. These bacteria can utilise this extracellular electron transfer (EET) to get energy for development or facilitate cell-to-cell communication [61].

Under aerobic circumstances, microorganisms near the anode decompose organic materials, and the resulting electrons are either transported directly through cell components such as proteases or nanowires on the membrane surface or indirectly via the electron shuttle [62]. The cathode accepts electrons and protons from the anode and initiates a reduction process [63]. The equation below illustrates the reaction of a typical MFC at the anode and cathode when acetate is used [64].

$$\text{Anode reaction}: \ CH_3COO^- + 2H2O \ {-}{-}> \ 2CO_2 + 7H^+ + 8e^-$$

$$\text{Cathode reaction}: \ O_2 + 4e^- + 4H^+ {-}{-}> \ 2H_2O$$

MFCs could generate energy from diverse substrates, from pure chemicals to complex mixtures of organic compounds present in wastewater. While substrates with a high concentration of complex organic matter could stimulate the growth of various active microorganisms, simpler substrates are believed to produce more immediate output. Acetate and glucose are often utilised as substrates in MFCs and power generation. Acetate is commonly used as a substrate for benchmarking novel MFC components, reactor designs, or operating conditions due to its inertness to alternative microbial conversions (fermentation and methanogenesis) at ambient temperature [65]. As for glucose, compared to anaerobic sludge with a limited substrate supply, the introduction of glucose can increase power output by up to 161 mW/m2 [66].

There are three basic reactor configurations: (a) uncoupled bioreactor MFC in a bioreactor followed by a chemical fuel cell, (b) integrated bioreactor MFC, and (c) MFC with bacteria-anode interaction (Fig. 3). In an uncoupled MFC, a biofuel, such as methane gas, is created in a bioreactor before a chemical fuel cell takes place. One of the major disadvantages of this configuration is mainly the low conversion efficiencies of the biological substrate to hydrogen and the requirement for high fuel cell temperatures to achieve sufficient hydrogen oxidation. The second design is identical to the first, except that the fermentation (mainly to hydrogen gas) occurs

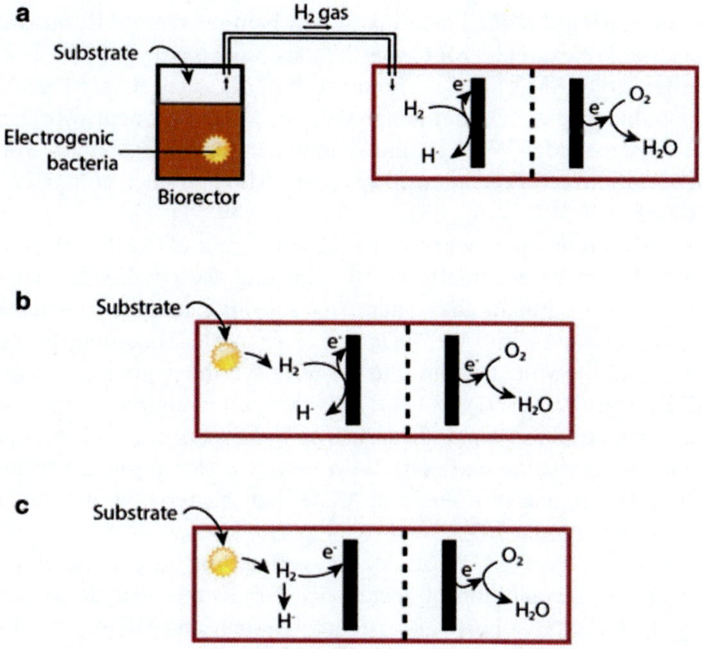

Fig. 3 Three basic reactor configurations: (A) uncoupled bioreactor MFC in a bioreactor followed by a chemical fuel cell, (B) integrated bioreactor MFC, and (C) MFC with bacteria-anode interaction. Adapted with permission from [58](CCBY)

within the fuel cell. This type of MFC frequently employs catalysts to create the best environment for hydrogen gas conversion to electricity. The third configuration, dubbed the genuine MFC, involves direct electron transfer from the bacteria to the anode without an intermediate fermentation product [58].

MFC technology effectively displaces non-renewable fossil fuels such as natural gas and coal and reduces greenhouse gas emissions that contribute to global warming and climate change [67, 68]. However, despite its benefits in green energy production, no study has been undertaken to determine its role in minimising the transmission of AMR bacteria found in sewage sludge to the environment. Yet, it has been demonstrated that several EB employed in MFC, such as *Bacillus subtilis*, release bacteriocin, which may substantially inhibit AMR transmission in the environment.

6.1 *The Application of Microbial Fuel Cells Technology in Reducing Antimicrobial Resistance in the Environment*

One possible alternative is to use MFC. As previously said, MFCs are well renowned for their ability to degrade pollutants while simultaneously generating significant amounts of electric energy. Notably, MFC is also known to break down antibiotics and ARGs, increasing the likelihood that it will help prevent AMR transmission in the environment. For example, MFC was demonstrated to remove 85.1% and 65.5% of sulfamethoxazole (SMX) and norfloxacin (NFLX), respectively [71, 72]. Additionally, the number of ARGs and integrons after MFC treatment was significantly less than that discovered in WWTPs. For example, the relative abundance of the *intI1* is between 63.11 and 652.00 copies/mL(g) in the MFC product, compared to 109 to 1011 copies/mL in WWTPs [62, 73].

There are numerous approaches to increase the rate of AMR bacteria removal by MFC, which can be accomplished by adjusting the conditions that influence antibiotic removal: (a) raising the voltage, (b) selecting suitable substrates, and (c) including some additives into the system. For example, according to Yang et al. (2018), increasing the voltage from 0 to 1.5 V raised the degradation rate of sulfadiazine (SDZ) from 79.3 to 91.9% [74]. This increase in antibiotic removal may be because certain microbial communities, such as *Methylococcus capsulatus, Dechloromonas, Byssovorax cruenta, and Longilinea arvoryzae* are positively influenced by electrical stimulation, and this would promote their bacterial activity to accelerate biological metabolism [62, 75].

Additionally, the type of substrate may affect the rate of antibiotic clearance. For example, the addition of acetate accelerates the breakdown of chloramphenicol by up to 96.53% [76]. Consistently, Zeshan and Ullah (2020) observed that acetate-fed MFCs generated maximum voltage and power densities faster than glucose- or sucrose-fed MFCs, implying that different types of substrates may affect MFCs performance and hence their potential to remove antibiotics [77]. Additionally, certain inorganic compounds, such as nitrite and copper aid in reducing ARGs. For example, it has been demonstrated that the addition of nitrite increases sludge's hydrolysis rate and decreases the requirement for carbon sources, encouraging microorganism growth and metabolism and consequently improving MFC performance [78].

The presence of EB in MFC may potentially affect the antibiotic's clearance rate. *Bacillus* species, for example, are capable of creating a variety of antimicrobial compounds (AMCs) that are useful in the food and pharmaceutical industries [79]. Additionally, strains of the *Bacillus subtilis* group have been recognised for decades to produce a vast array of secondary metabolites capable of mediating antibiosis. At least 4–5% of the genome of any strain of *Bacillus subtilis* is projected to be devoted to synthesising AMCs, which are primarily antimicrobial peptides (AMPs) [79]. AMPs are often cyclic and hydrophobic in structure, with unusual moieties such as D-amino acids (AA) or intramolecular thioether linkages. Along with AMPs, volatile metabolites are a wide class of antimicrobials that play a range of metabolic and functional functions [79]. These peptides are referred to as "bacteriocins," which are low molecular weight molecules that can inhibit the growth of bacteria closely related to the generating strain [80]. Alternatively, bacteriocins may operate as antimicrobial or lethal peptides, directly reducing competing strains or pathogens in the environment, such as *Klebsiella* sp., a capsulated bacterium, resulting in decreased AMR transmission/

7 Malaysia's Initiatives for Preventing the Spread of Antimicrobial Resistance in the Environment

Antibiotic resistance is reaching alarmingly high levels all around the globe. New resistance mechanisms are evolving and spreading over the world, posing a danger to the health management to handle common ailments resulting from the AMR infection [81]. In addition to the escalating prevalence of debilitating diseases, augmenting cases of antibiotic resistance also cause major economic burden due to AMR containment and high cost of disease treatment [82]. Consequently, it is projected that AMR would put a total of 24 million people (particularly from the low-income countries) into extreme poverty by 2030 [83], further corroborating the urgency to combat the dissemination of AMR worldwide comprehensively.

During an ASEAN meeting in 2017, it was recognised that anti-AMR efforts were still insufficient, and that multi-sectoral collaborations from different stakeholders were required. During the summit, ASEAN leaders agree to tackle AMR using the One Health approach which is aimed to enhance the AMR containment activities, actively engage relevant stakeholders, develop defined objectives, and to implement monitoring and evaluation (M&E) systems [84]. In compliance with the declaration, Malaysia has implemented several actions to cut the AMR distribution by implementing different progressive approaches. These include the initiation of One Health concept involving pharmacist, physicians, patients, and other professionals that was established to allow effective communication among the communities thus achieving better public health outcomes for humans, animals, and the environment [85]. Malaysia government with the help from Ministry of Health (MOH) also initiated the "National Surveillance of Antibiotic Resistance (NSAR)" [86] to monitor the occurrence of AMR in Malaysia as well as the protocol on Antimicrobial Stewardship (AMS) Programme in the healthcare facilities to promote an appropriate utilisation of antibiotics in terms of right choice, route of administration, dosage, and duration for antibiotic prescription.

Following the NSAR programme, the "National Surveillance on Antibiotic Utilisation (NSAU)" was also performed to assess the quantity and trajectory of antibiotic use in Malaysian hospitals of various settings and the potential links that contribute to specific antibiotic use. These statistics have aided local health workers in their clinical practice, particularly in the application of antimicrobial stewardship, at the hospital level [87]. In addition to those policies, the Malaysian Action Plan on Antimicrobial Resistance (MyAP-AMR) 2017–2021 has been conducted in collaboration with numbers of constituents and stakeholders including the Ministry of Health (MOH), Ministry of Agriculture and Agro-based Industry (MOA), Ministry of Higher Education (MoHE), Ministry of Defence (MINDEF), hospitals, professional organisations, the animal food industry, private healthcare facilities, community pharmacists, academic institutions, the private sectors, international partners, NGOs, and civil society [88]. The action plan on antimicrobial resistance (MyAP-AMR) is aimed to decelerate the emergence of AMR and prevent its dissemination via four priority areas: (i) public awareness and education, (ii) surveillance and research, (iii) infection prevention and control, and (iv) appropriate use of antimicrobials.

While there is currently scarce or no report on the clinical assessment of the MFC function in reducing the AMR dissemination, its potential on the accomplishment of AMR reduction is still promising. The inclusion of EB which simultaneously functioning as bacteriocin producer may antagonistically inhibit the resistant bacteria in the wastewater substrate hence attenuating the possibility of resistance genotype transmission among the microbial populations. This action will not only benefit the environment in terms of green renewable electricity generation, but it also represents as a contribution from the engineering sector (other than the healthcare organisation) to decelerate the AMR widespread as has been underlined in the MyAP-AMR that promotes the involvement and collaboration from multiple stakeholders in combating this threatening circumstance. Indeed, restricting the AMR distribution in the community and environment via the function of MFC is interesting, however, an in-depth investigation is required to validate its potential in diminishing the AMR incidents.

8 Conclusion

Antimicrobial resistance (AMR) is a significant global public health problem, ranking among the top ten. This study covers the fundamentals of AMR and the possibility for wastewater to act as a carrier of AMR in the environment. Compared to other conventional wastewater treatment procedures, MFCs can help minimise AMR by boosting antibiotic removal rates. These antibiotics, which are frequently generated from hospitals, are greatly concentrated in sludge, favouring the growth of ARGs among the bacteria that live there. Additionally, this chapter discussed strategies to increase the performance of existing MFCs, primarily by increasing electrical stimulation, selecting appropriate substrates, and incorporating some additives, all of which influence the microbial population and its metabolic activities. It is hypothesised that boosting antibiotics' clearance efficiency may help prevent the creation of ARGs. Nonetheless, it is vital to improve the current design of MFCs and comprehend their operation, and by doing so will reduce AMR transmission in the environment.

Acknowledgements This work was supported by the Universiti Sains Malaysia (Apex ERA) (Grant numbers No. 1001.PTEKIND.881006) granted to AMAM.

References

1. Straub TM, Pepper IL, Gerba CP (1993) Hazards from Pathogenic Microorganisms in Land-Disposed Sewage Sludge. In: Reviews of Environmental Contamination and Toxicology. pp 55–91
2. Usman K, Khan S, Ghulam S et al (2012) Sewage Sludge: An Important Biological Resource for Sustainable Agriculture and Its Environmental Implications. Am J Plant Sci 03:1708–1721. https://doi.org/10.4236/ajps.2012.312209
3. Al-Gheethi AA, Efaq AN, Bala JD et al (2018) Removal of pathogenic bacteria from sewage-treated effluent and biosolids for agricultural purposes. Appl Water Sci 8:74. https://doi.org/10.1007/s13201-018-0698-6
4. Alonso E, Santos A, Riesco P (2004) Micro-Organism Re-Growth in Wastewater Disinfected by UV Radiation and Ozone: A Micro-Biological Study. Environ Technol 25:433–441. https://doi.org/10.1080/09593332508618452
5. Ouardani I, Turki S, Aouni M, Romalde JL (2016) Detection and Molecular Characterization of Hepatitis A Virus from Tunisian Wastewater Treatment Plants with Different Secondary Treatments. Appl Environ Microbiol 82:3834–3845. https://doi.org/10.1128/AEM.00619-16
6. Sun D, Jeannot K, Xiao Y, Knapp CW (2019) Editorial: Horizontal Gene Transfer Mediated Bacterial Antibiotic Resistance. Front Microbiol 10:1933
7. Rai P, Tripathi B (2007) Microbial contamination in vegetables due to irrigation with partially treated municipal wastewater in a tropical city. Int J Environ Health Res 17:389–395. https://doi.org/10.1080/09603120701628743
8. Garg NK (2009) Multicriteria Assessment of Alternative Sludge Disposal Methods Neeraj Kumar Garg
9. Lundholm M, Rylander R (1983) Work related symptoms among sewage workers. Br J Ind Med 40:325–329. https://doi.org/10.1136/oem.40.3.325
10. Mohd Zaini Makhtar M, Tajarudin HA (2020) Electricity generation using membrane-less microbial fuel cell powered by sludge supplemented with lignocellulosic waste. International Journal of Energy Research 44:3260–3265. https://doi.org/10.1002/er.5151
11. Shamsudin NA, Sabri MNIM, Tajarudin HA et al (2021) The Future Promising Alternative Renewable Energy from Microbial Fuel Cell. Springer
12. Hocking L, Ali G-C, d'Angelo C, et al (2021) A rapid evidence assessment exploring whether antimicrobial resistance complicates non-infectious health conditions and healthcare services, 2010–20. JAC-Antimicrobial Resistance 3:dlab171. https://doi.org/10.1093/jacamr/dlab171
13. Prestinaci F, Pezzotti P, Pantosti A (2015) Antimicrobial resistance: a global multifaceted phenomenon. Pathogens and Global Health 109:309–318. https://doi.org/10.1179/2047773215Y.0000000030
14. Ayukekbong JA, Ntemgwa M, Atabe AN (2017) The threat of antimicrobial resistance in developing countries: causes and control strategies. Antimicrob Resist Infect Control 6:47. https://doi.org/10.1186/s13756-017-0208-x
15. Larson E (2007) Community Factors in the Development of Antibiotic Resistance. Annu Rev Public Health 28:435–447. https://doi.org/10.1146/annurev.publhealth.28.021406.144020
16. Noyes NR, Slizovskiy IB, Singer RS (2021) Beyond Antimicrobial Use: A Framework for Prioritizing Antimicrobial Resistance Interventions. Annual Review of Animal Biosciences 9:313–332. https://doi.org/10.1146/annurev-animal-072020-080638
17. Llor C, Bjerrum L (2014) Antimicrobial resistance: risk associated with antibiotic overuse and initiatives to reduce the problem. Therapeutic advances in drug safety 5:229–241. https://doi.org/10.1177/2042098614554919
18. Michael CA, Dominey-Howes D, Labbate M (2014) The antimicrobial resistance crisis: causes, consequences, and management. Front Public Health 2:145. https://doi.org/10.3389/fpubh.2014.00145
19. Reygaert WC (2018) An overview of the antimicrobial resistance mechanisms of bacteria. AIMS microbiology 4:482–501. https://doi.org/10.3934/microbiol.2018.3.482

20. Pang Z, Raudonis R, Glick BR, et al (2019) Antibiotic resistance in Pseudomonas aeruginosa: mechanisms and alternative therapeutic strategies. Biotechnology Advances 37:177–192. https://doi.org/10.1016/j.biotechadv.2018.11.013
21. Hoffman SB (2001) Mechanisms of Antibiotic Resistance. Compend Contin Educ Pract Vet 23:464–472
22. Kapoor G, Saigal S, Elongavan A (2017) Action and resistance mechanisms of antibiotics: A guide for clinicians. J Anaesthesiol Clin Pharmacol 33:300–305. https://doi.org/10.4103/joacp.JOACP_349_15
23. Lowy FD (2003) Antimicrobial resistance: the example of Staphylococcus aureus. J Clin Investig 111:1265–1273. https://doi.org/10.1172/JCI18535
24. Munita JM, Arias CA (2016) Mechanisms of Antibiotic Resistance. Microbiology spectrum 4. https://doi.org/10.1128/microbiolspec.VMBF-0016-2015.10.1128/microbiolspec.VMBF-0016-2015
25. Meletis G (2016) Carbapenem resistance: overview of the problem and future perspectives. Therapeutic advances in infectious disease 3:15–21. https://doi.org/10.1177/2049936115621709
26. Poirel L, Pitout JD, Nordmann P (2007) Carbapenemases: molecular diversity and clinical consequences. Future Microbiol 2:501–512. https://doi.org/10.2217/17460913.2.5.501
27. Papp-Wallace KM, Endimiani A, Taracila MA, Bonomo RA (2011) Carbapenems: past, present, and future. Antimicrob Agents Chemother 55:4943–4960. https://doi.org/10.1128/AAC.00296-11
28. Queenan AM, Bush K (2007) Carbapenemases: the versatile beta-lactamases. Clin Microbiol Rev 20:440–458. https://doi.org/10.1128/CMR.00001-07
29. Hannan S, Ready D, Jasni AS et al (2010) Transfer of antibiotic resistance by transformation with eDNA within oral biofilms. FEMS Immunol Med Microbiol 59:345–349. https://doi.org/10.1111/j.1574-695X.2010.00661.x
30. Tenover FC (2006) Mechanisms of antimicrobial resistance in bacteria. American Journal of Infection Control 34:S3–S10. https://doi.org/10.1016/j.ajic.2006.05.219
31. Fusté E, López-Jiménez L, Segura C et al (2013) Carbapenem-resistance mechanisms of multidrug-resistant Pseudomonas aeruginosa. J Med Microbiol 62:1317–1325. https://doi.org/10.1099/jmm.0.058354-0
32. ProfMR S, Jalilzadeh K, Yamakanamardi SM (2009) Horizontal transfer of antibiotic resistance genes among gram negative bacteria in sewage and lake water and influence of some physico-chemical parameters of water on conjugation process. J Environ Biol 30(1):45–49
33. Partridge SR, Ginn AN, Wiklendt AM, et al (2015) Emergence of blaKPC carbapenemase genes in Australia. International Journal of Antimicrobial Agents 45:130–136. https://doi.org/10.1016/j.ijantimicag.2014.10.006
34. Samreen, Ahmad I, Malak HA, Abulreesh HH (2021) Environmental antimicrobial resistance and its drivers: a potential threat to public health. Journal of Global Antimicrobial Resistance 27:101–111. https://doi.org/10.1016/j.jgar.2021.08.001
35. Kraemer SA, Ramachandran A, Perron GG (2019) Antibiotic Pollution in the Environment: From Microbial Ecology to Public Policy. Microorganisms 7:180. https://doi.org/10.3390/microorganisms7060180
36. Ahmad Mokhtar AM bin, Makhtar MMZ, Ahmad Mokhtar AM (2021) Waste and health: Sewage sludge and its hazard to human. Springer
37. Ben Y, Fu C, Hu M, et al (2019) Human health risk assessment of antibiotic resistance associated with antibiotic residues in the environment: A review. Environmental Research 169:483–493. https://doi.org/10.1016/j.envres.2018.11.040
38. Ventola CL (2015) The antibiotic resistance crisis: part 1: causes and threats. P & T : a peer-reviewed journal for formulary management 40:277–283
39. Chattopadhyay MK (2014) Use of antibiotics as feed additives: a burning question. Front Microbiol 5:334
40. Singer AC, Shaw H, Rhodes V, Hart A (2016) Review of Antimicrobial Resistance in the Environment and Its Relevance to Environmental Regulators. Front Microbiol 7:1728. https://doi.org/10.3389/fmicb.2016.01728

41. Sanderson H, Brown RS, Hania P, et al (2019) Chapter 7 - Antimicrobial Resistant Genes and Organisms as Environmental Contaminants of Emerging Concern: Addressing Global Public Health Risks. In: Roig B, Weiss K, Thireau V (eds) Management of Emerging Public Health Issues and Risks. Academic Press, pp 147–187
42. Fair RJ, Tor Y (2014) Antibiotics and bacterial resistance in the 21st century. Perspectives in medicinal chemistry 6:25–64. https://doi.org/10.4137/PMC.S14459
43. Jones IA, Joshi LT (2021) Biocide Use in the Antimicrobial Era: A Review. Molecules (Basel, Switzerland) 26:2276. https://doi.org/10.3390/molecules26082276
44. Cheng G, Ning J, Ahmed S et al (2019) Selection and dissemination of antimicrobial resistance in Agri-food production. Antimicrob Resist Infect Control 8:158. https://doi.org/10.1186/s13756-019-0623-2
45. Yu Z, Gunn L, Wall P, Fanning S (2017) Antimicrobial resistance and its association with tolerance to heavy metals in agriculture production. Food Microbiology 64:23–32. https://doi.org/10.1016/j.fm.2016.12.009
46. Gómez-Sanz E, Kadlec K, Feßler AT et al (2013) Novel erm(T)-carrying multiresistance plasmids from porcine and human isolates of methicillin-resistant Staphylococcus aureus ST398 that also harbor cadmium and copper resistance determinants. Antimicrob Agents Chemother 57:3275–3282. https://doi.org/10.1128/AAC.00171-13
47. Argudín MA, Tenhagen B-A, Fetsch A et al (2011) Virulence and resistance determinants of German Staphylococcus aureus ST398 isolates from nonhuman sources. Appl Environ Microbiol 77:3052–3060. https://doi.org/10.1128/AEM.02260-10
48. Huijbers PMC, Blaak H, de Jong MCM et al (2015) Role of the Environment in the Transmission of Antimicrobial Resistance to Humans: A Review. Environ Sci Technol 49:11993–12004. https://doi.org/10.1021/acs.est.5b02566
49. Dadgostar P (2019) Antimicrobial Resistance: Implications and Costs. Infection and drug resistance 12:3903–3910. https://doi.org/10.2147/IDR.S234610
50. Chatterjee A, Rai S, Guddattu V et al (2018) Is methicillin-resistant Staphylococcus Aureus infection associated with higher mortality and morbidity in hospitalized patients? A cohort study of 551 patients from South Western India. Risk management and healthcare policy 11:243–250. https://doi.org/10.2147/RMHP.S176517
51. Friedrich MJ (2017) Drug-Resistant Tuberculosis Predicted to Increase in High-Burden Countries. JAMA 318:231. https://doi.org/10.1001/jama.2017.9086
52. Cecchini M, Langer J, Slawomirski L (2015) ANTIMICROBIAL RESISTANCE IN G7 COUNTRIES AND BEYOND: Economic Issues, Policies and Options for Action
53. Friedman ND, Temkin E, Carmeli Y (2016) The negative impact of antibiotic resistance. Clinical Microbiology and Infection 22:416–422. https://doi.org/10.1016/j.cmi.2015.12.002
54. Laxminarayan R, Duse A, Wattal C, et al (2013) Antibiotic resistance—the need for global solutions. The Lancet Infectious Diseases 13:1057–1098. https://doi.org/10.1016/S1473-3099(13)70318-9
55. Rudd KE, Johnson SC, Agesa KM et al (2020) Global, regional, and national sepsis incidence and mortality, 1990–2017: analysis for the Global Burden of Disease Study. The Lancet 395:200–211. https://doi.org/10.1016/S0140-6736(19)32989-7
56. Labib A (2019) Sepsis Care Pathway 2019. Qatar medical journal 2019:4. https://doi.org/10.5339/qmj.2019.qccc.4
57. Gudiol C, Albasanz-Puig A, Cuervo G, Carratalà J (2021) Understanding and Managing Sepsis in Patients With Cancer in the Era of Antimicrobial Resistance. Front Med 8:636547. https://doi.org/10.3389/fmed.2021.636547
58. Rabaey K, Lissenns G, Verstraete W (2011) Microbial fuel cells: Performances and perspectives. Biofuels for Fuel Cells: Biomass Fermentation Towards Usage in Fuel Cells
59. Sacco NJ, Bonetto MC, Cortón E (2017) Isolation and Characterization of a Novel Electrogenic Bacterium, Dietzia sp. RNV-4. PLOS ONE 12:e0169955-
60. Tahernia M, Mohammadifar M, Gao Y et al (2020) A 96-well high-throughput, rapid-screening platform of extracellular electron transfer in microbial fuel cells. Biosens Bioelectron 162:112259. https://doi.org/10.1016/j.bios.2020.112259

61. Torres CI, Marcus AK, Lee H-S et al (2010) A kinetic perspective on extracellular electron transfer by anode-respiring bacteria. FEMS Microbiol Rev 34:3–17. https://doi.org/10.1111/j.1574-6976.2009.00191.x
62. Chen P, Guo X, Li S, Li F (2021) A review of the bioelectrochemical system as an emerging versatile technology for reduction of antibiotic resistance genes. Environment International 156:106689. https://doi.org/10.1016/j.envint.2021.106689
63. Mandal S, Das N (2018) Application of microbial fuel cells for bioremediation of environmental pollutants: An overview. Journal of microbiology, biotechnology and food sciences 7:. https://doi.org/10.15414/jmbfs.2018.7.4.437-444
64. Mohd Zaini Makhtar M, Don M, Tajarudin H (2018) Microbial Fuel Cell (MFC) Development from Anaerobic Digestion System. In: Green Energy and Technology. pp 9–31
65. Aelterman P, Rabaey K, Pham HT et al (2006) Continuous Electricity Generation at High Voltages and Currents Using Stacked Microbial Fuel Cells. Environ Sci Technol 40:3388–3394. https://doi.org/10.1021/es0525511
66. Hu Z (2008) Electricity generation by a baffle-chamber membraneless microbial fuel cell. Journal of Power Sources - J POWER SOURCES 179:27–33. https://doi.org/10.1016/j.jpowsour.2007.12.094
67. Mohd Zaini Makhtar M, Vadivelu VM (2019) Membraneless Microbial Fuel Cell: Characterization of Electrogenic Bacteria and Kinetic Growth Model. J Environ Eng 145:04019015. https://doi.org/10.1061/(asce)ee.1943-7870.0001522
68. Anderson TR, Hawkins E, Jones PD (2016) CO2, the greenhouse effect and global warming: from the pioneering work of Arrhenius and Callendar to today's Earth System Models. Endeavour 40:178–187. https://doi.org/10.1016/j.endeavour.2016.07.002
69. Thi T, do, López E, Rodríguez-Rojas A, et al (2011) Effect of recA inactivation on mutagenesis of Escherichia coli exposed to sublethal concentrations of antimicrobials. J Antimicrob Chemother 66:531–538. https://doi.org/10.1093/jac/dkq496
70. Andersson DI, Hughes D (2014) Microbiological effects of sublethal levels of antibiotics. Nat Rev Microbiol 12:465–478. https://doi.org/10.1038/nrmicro3270
71. Xue W, Li F, Zhou Q (2019) Degradation mechanisms of sulfamethoxazole and its induction of bacterial community changes and antibiotic resistance genes in a microbial fuel cell. Bioresource Technology 289:121632. https://doi.org/10.1016/j.biortech.2019.121632
72. Ondon BS, Li S, Zhou Q, Li F (2020) Simultaneous removal and high tolerance of norfloxacin with electricity generation in microbial fuel cell and its antibiotic resistance genes quantification. Bioresource Technology 304:122984. https://doi.org/10.1016/j.biortech.2020.122984
73. Yan W, Guo Y, Xiao Y, et al (2018) The changes of bacterial communities and antibiotic resistance genes in microbial fuel cells during long-term oxytetracycline processing. Water Research 142:105–114. https://doi.org/10.1016/j.watres.2018.05.047
74. Yang X-L, Zhang S, Li H, et al (2018) Effects of voltage on sulfadiazine degradation and the response of sul genes and microbial communities in biofilm-electrode reactors. Ecotoxicology and Environmental Safety 151:272–278. https://doi.org/10.1016/j.ecoenv.2018.01.016
75. Liu S, Song H, Wei S, et al (2015) Effect of direct electrical stimulation on decolorization and degradation of azo dye reactive brilliant red X-3B in biofilm-electrode reactors. Biochemical Engineering Journal 93:294–302. https://doi.org/10.1016/j.bej.2014.11.002
76. Zhang Q, Zhang Y, Li D (2017) Cometabolic degradation of chloramphenicol via a meta-cleavage pathway in a microbial fuel cell and its microbial community. Bioresource Technology 229:104–110. https://doi.org/10.1016/j.biortech.2017.01.026
77. Ullah Z, Zeshan S (2019) Effect of substrate type and concentration on the performance of a double chamber microbial fuel cell. Water Sci Technol 81:1336–1344. https://doi.org/10.2166/wst.2019.387
78. Ruslan AR, Vadivelu VM (2019) Nitrite pre-treatment of dewatered sludge for microbial fuel cell application. Journal of Environmental Sciences 77:148–155. https://doi.org/10.1016/j.jes.2018.06.023

79. Caulier S, Nannan C, Gillis A et al (2019) Overview of the Antimicrobial Compounds Produced by Members of the Bacillus subtilis Group. Front Microbiol 10:302
80. Benítez-Chao DF, León-Buitimea A, Lerma-Escalera JA, Morones-Ramírez JR (2021) Bacteriocins: An Overview of Antimicrobial, Toxicity, and Biosafety Assessment by in vivo Models. Front Microbiol 12:630695. https://doi.org/10.3389/fmicb.2021.630695
81. Organization WH (2020) Antibiotic resistance. https://www.who.int/news-room/fact-sheets/detail/antibiotic-resistance. Accessed 11 Jan 2022
82. Zhen X, Lundborg CS, Sun X et al (2019) Economic burden of antibiotic resistance in ESKAPE organisms: a systematic review. Antimicrob Resist Infect Control 8:1–23
83. Adeyi OO, Baris E, Jonas OB et al (2017) Drug-resistant infections: a threat to our economic future. World Bank Group, Washington, DC
84. ASEAN H of state of MS of ASEAN Leaders' Declaration on Antimicrobial Resistance (AMR): Combating AMR through One Health Approach. https://asean.org/wp-content/uploads/2017/11/3.-ADOPTION_2017_ALD-on-AMR_Endorsed-13th-AHMM.pdf. Accessed 12 Jan 2022
85. Mariappan V, Vellasamy KM, Mohamad NA et al (2021) OneHealth Approaches Contribute Towards Antimicrobial Resistance: Malaysian Perspective. Front Microbiol 12:718774. https://doi.org/10.3389/fmicb.2021.718774
86. (MOH) M of HM (2016) National Antibiotic Surveillance Report 2016. https://www.imr.gov.my/en/component/content/article/75-english-content/national-collaboration/1294-nsar. Accessed 12 Jan 2022
87. (MOH) M of HM (2018) The Report on Antibiotic Utilisation in Malaysian Hospitals 2008–2017
88. Chua AQ, Verma M, Hsu LY, Legido-Quigley H (2021) An analysis of national action plans on antimicrobial resistance in Southeast Asia using a governance framework approach. The Lancet Regional Health-Western Pacific 7:100084

Microbial Fuel Cell (MFC) Innovation in Wastewater Treatment Plant: From Economic Perspective

Soliha Sanusi, Muaz Mohd Zaini Makhtar, and Aziatul Waznah Ghazali

Abstract Non-renewable resources such as petroleum and fossil had reached their limit as the people's population increased yearly. The scarce and limited resources will make these non-renewable assets more expensive in the future, and it is also limited to certain people. Hence, government and all related entities need to find an alternative solution to solve the problem. It is to ensure all people receive the necessity equally and fairly. The microbial fuel cell (MFC) is one of the alternatives in the market that is available to be explored in depth to replace the scarcity of resources. MFC is an innovative electrical power that treats wastewater and other applications through the organic substrates of bacteria-driven oxidation. MFCs are developed by using various materials and a diversity of configurations. It is a capable technology that could be implemented in many applications with broader market expansion. However, developing MFCs, especially the capital cost, is not as cheap as the conventional power supply. Hence, the MFCs manufacturers, producers, and researchers need to find ways to reduce the cost and increase their efficiency to sustain the market for a more extended period. This situation has been explored in the chapter.

Keywords Microbial Fuel Cell (MFC) · Wastewater treatment · Non-renewable resources · Capital cost

---

S. Sanusi · A. W. Ghazali (✉)
Faculty of Economics and Management, Universiti Kebangsaan Malaysia, Bangi, Malaysia

S. Sanusi
e-mail: solihasanusi@ukm.edu.my

Accounting Research Institute, Universiti Teknologi MARA, Shah Alam, Malaysia

M. Mohd Zaini Makhtar
School of Industrial Technology, Bioprocess Technology Division, Universiti Sains Malaysia, Penang, Malaysia

Fellow of Centre of Global Sustainability Studies (CGSS), Universiti Sains Malaysia, Penang, Malaysia

M. Mohd Zaini Makhtar et al. (eds.), *Microbial Fuel Cell (MFC) Applications for Sludge Valorization*, Green Energy and Technology,
https://doi.org/10.1007/978-981-99-1083-0_11

1 Introduction to Microbial Fuel Cell

Microbial Fuel Cells (MFCs) have been introduced in certain countries when the relevant bodies ignore so much sludge. Sludge is created from various industrial processes, on-site sanitation systems, or wastewater treatment. Industrial wastewater treatment plants develop biosolids, namely sludge, as leftovers. The first MFCs were established in the early twentieth century by using an intermediary, a chemical that transmits electrons from the microbes such as pioneer Shewanella and Geobacter in the cell to an anode. MFCs are advanced bioelectrochemical cells that generate electric power by redirecting electrons developed by the microbial oxidation of lowered composites on the anode to oxidized compounds on the cathode through an outer electrical circuit.

It will convert leftover reserves into energy based on inoculated bacteria's metabolic activities, called biocatalysts. The anode electrode is the most important because it generates energy and removes pollutants [1]. Anode will be responsible for the electron transfer rate from the electroactive bacteria to the anode's surface. Nowadays, the material and cost of the anode electrode are pretty expensive in the MFCs field, and it may hamper efforts to expand the commercial use of MFC in the future [2]. In the twenty-first century, MFCs were employed commercially in wastewater treatment and have continued until now.

Sewer water treatment required a superior expense and energy to manage due to contaminants generated by human activities in many commercial and residents' usages. Due to that, biological wastewater treatments replace the current technologies to improve treatment efficiency. Concurrently, MFCs can also be used as a feasible resolution for chemical oxygen demand (COD) elimination and electric power invention [3]. For example, the MFCs can be used as a power supply for many activities, such as charging small electric equipment. Figure 1 shows a sample of schematic diagram for MFC experimental setup [4].

An increasing number of people worldwide has raised environmental problems such as air, water, and land pollution. The elevated power utilization required to operate the human activities and the operational cost to run the MFCs are significant challenges for expanding wastewater management technologies. It has been forecast that treatment expense account for around 3% of the world's electricity spent, and sludge clearance costs are about 50% of sewer water care [5].

The other issue related to wastewater treatment is the unproductive care of the generated greenhouse gases (GHGs) and evaporated ingredients such as ammonia and phosphates. Additional technical problems need to be settled by conventional wastewater treatment [6]. Therefore, the improvement of high-value goods has been researched in MFCs due to the development of a few chemical contents such as hydrogen, hydrogen peroxide, and heavy metals [7]. Hence, MFCs have become the most promising solution in generating bio-electricity with various advantages such as effectiveness, cleanliness, less toxicity, and recyclability [8].

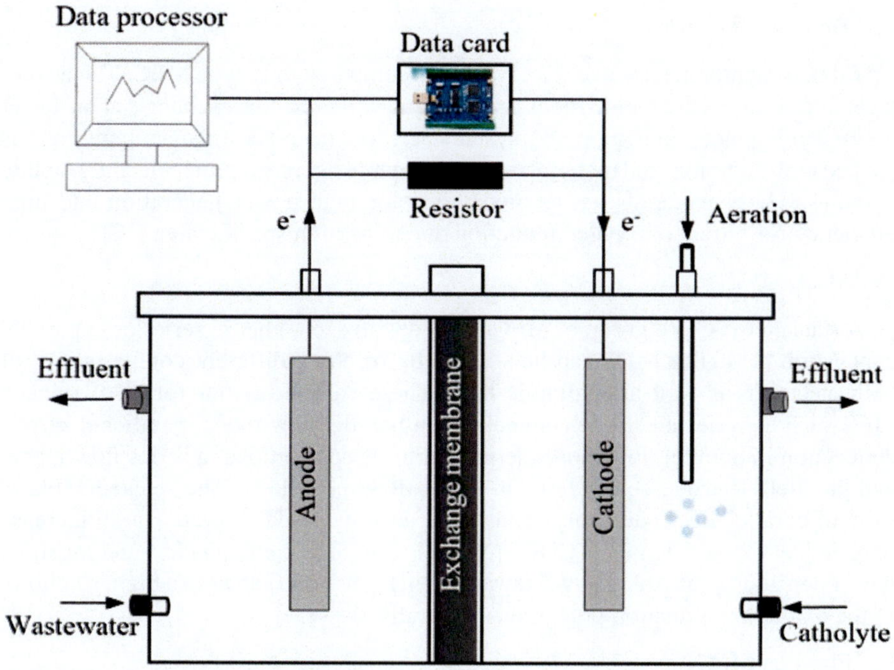

Fig. 1 Scheme of the MFC experimental setup sample. Reprinted with permission from [19]. Copyright 2020 MDPI

2 The Type of Microbial Fuel Cell

A. Single Chamber

The design of single chamber MFC involves only the anodic chamber coupled with an air cathode to which the electrons and protons are [9]. In this design, the anode and cathode were placed in the same chamber and with no membrane installed. Single chamber MFC is the most preferable configuration to be scaled up for wastewater treatment due to its ability to produce high power output, easy structure, and cost effective [10].

B. Double Chamber

Double Chamber MFC is the classical design of MFC that consists of two chambers separated by an ion exchange membrane. The anode and cathode were placed in different chambers. This design has an ability to run in batch modes and continuous modes and currently used for laboratory scale [11]. It shows that double chamber MFC has the extra point compared to single chamber MFC due to its potential to run in batch and continuous modes.

C. Tubular Membrane

Tubular membrane has a lot of advantages compared to H-type MFC. The advantages are it can reduce electrodes spacing and large proton exchange area, lower internal resistance, and improved system performance. The tubular membrane is packed with carbon-based materials such as graphite fibers, graphite felt, and graphite granules. These materials act as biocathode for high power generation and high efficiency for nitrate or oxygen reduction due to its high specific area [12]

D. Stack MFC

A stacked MFC is a group of MFCs linked with each other in series or in parallel connection [13]. The MFC can be stacked by reaching different configurations of both electrode as well as hydraulic flow. These can be of four forms of stacked MFC such as series electrode connections in parallel flow mode, parallel electrode connection in parallel flow mode, series electrode connections in series flow mode, and parallel electrode connections in series flow mode [11]. The stacked MFC in form of parallel electrode connections in series flow mode showed overall increase in chemical oxygen demand (COD) removal, Coulombic efficiencies, and maximal power densities while treating wastewater which was contributed to a higher stability of the oxidation–reduction potentials in overall cells.

E. Flat-Plat MFC

Flat-Plat MFC is designed with single electrode. The cathode electrode is made of a carbon cloth and platinum that act as catalyst which already hot pressed to a Nafion membrane. The cathode electrode is placed in contact with a single sheet of carbon paper that becomes the anode. Oxygen that acts as the electron acceptor is supplied from dry air that passed through the cathode chamber. This type of MFC is made to generate electricity continuously by using the domestic wastewater. The source of microbes to inoculate the MFC are come from the wastewater [14].

3 Cost and Benefits of Current MFCs

Wastewater management and energy produced from MFCs is a technology that proposes numerous advantages to people, society, and industry. It is an environmentally friendly process. During the MFCs activity, the electrical energy generated can be circulated in return to the process, and hence it may reduce the costs related to energy consumption. It also helps us avoid using fossil gases for heating activities.

MFC output has been recognized as small power devices such as sensors, low voltage capacitances, and pocket-sized DC motors. Studies about the benefits of the MFC's application in terms of power utilization and operating expenses have been conducted in numerous past research papers [2]. It helps the users on a small scale in the area generate power for further activities such as charging handphones and lighting purposes. It is beneficial, especially if the area is short of power or isolated from the main areas, such as a small island, remote village, or near a forest. Hence, it can be claimed that MFC is sustainable energy production. Among the critical drivers of MFCs worldwide are increasing initiatives for environmental sustainability, the increasing demand for potable water, and the growing focus on wastewater treatment.

The future application of the MFCs has been mentioned by [15] in their research study. Among the benefits of MFCs are wastewater treatment, renewable power creation from biomass, biosensors or environmental sensors, and product renewal. MFCs can fuel an apparatus that gathers numbers in the ecological atmosphere, especially in waterways or seas. The device requires sufficient energy for day-to-day operations, and consuming a battery by replacing it periodically might be troublesome. Hence, researchers may develop residue fuel cells that can substitute the battery in the sea/river for an extended period. The power of the MFCs was natural and confined in the residue. However, the energy produced is at < 30 mW/m2 due to the minimal biological intensity and elevated inner endurance [15]. Yet, an adjustment to the residue fuel cell circuit will allow the power to be stored and the minimal energy intensity to be compensated for and managed to release data to central sensors [16].

For product recovery, it had been testified that nearly all manure sludges stayed contaminated even after the recovery procedures[17]. Hence, several nations prohibit its application as compost in the agriculture industry. A vital nutrient element such as phosphorus is stuck within the mud. MFC may help to assist the challenge by applying two-chamber MFC with the *Escherichia coli* K12 as the biocatalyst. The electron is given through methylene blue on the anode side. The mud was warmed up and pounded at the cathode chamber, and at the same time, the electron and proton issued by the cathode will be utilized to release phosphate in the type of phosphoric acid from the ferric phosphate hydrate. Twenty-one days operation of this process may generate three percent of ferric phosphate by mass in the sludge and recover about 80% produces with 30 mW/m2 power. This process results in the phosphate that occurred from the cure of the improved phosphoric acid with magnesium chloride.

Still, the capital cost of producing MFC is not as cheap as conventional wastewater treatment due to the cost of electrodes, membranes, and catalysts. The price can be about 30 times greater than the traditional wastewater treatment if the carbon-based substances are treated as conductors [6]. Nonetheless, it depends on the volumes and configurations of the materials being used. Some essential components are the minimum requirements for MFC to work.

The MFC provider may decide on an exceptionally minimal budget for the material they want to use. For example, at the moment, the cost of the electrode assemblies such as stainless steel was calculated based on the manufacturer's decision instead of by looking into the commodity market counter. However, the manufacturer will base on the market price when preparing a customer quotation. Therefore, the MFC producer should alert the commodity market and lock a transaction when the price decreases. As the price fluctuates daily, they may choose the best option at the specific time they need to make an order from the manufacturer.

On the other hand, since the cathodic chamber is optional in MFC, they may withdraw it from the process as anodic alone as a single electrode assembly is enough to ensure the MFC runs. This withdrawal may reduce the cost of MFC, and it can be diverted to other expenses.

There were some studies regarding the return of having MFC in the market and purposely for industrial commercialization [4] [6]. The previous researcher tested the economic viability of MFC capital budgeting through investment analysis methods, which are Net Present Value (NPV) and Internal Rate of Return (IRR)[4]. NPV is the most theoretically reliable analysis [18] as it values the difference between the present value of all cash inflows and all cash outflows. However, two assumptions need to be made if we are using NPV as an analysis tool. The assumption that we need to follow are:

1. All cash flows occur at the end of periods except the initial cash outflow.
2. All cash received will be immediately reinvested at a rate of return equal to the discount rate.

An indicator of whether the MFC is good enough for commercialization is a positive NPV that shows that the project's return exceeds the discount rate. If yes, we are encouraged to continue the MFC. If NPV shows a negative value, we can say that the project's return is lower than the discount rate. Hence, we may stop the project as the cost is beyond their profit. This discount rate is basically equal to the cost of capital, and it is considered the minimum required rate of return. The cost of capital is the average return that a company must pay its long-term creditors and shareholders if the MFC producer wants to do it in a considerable capacity.

For IRR analysis to work for MFC, we need to compute the discount rate of all the investments that will cause the net present value of an investment to be zero[19]. The principle is, if IRR is equal to or greater than the minimum required rate of return, we may continue the project. If the rate is lower than the minimum required rate of return, we may discontinue or reject the project. This IRR works well if a project's cash flows are consistent year-round. It will be difficult to calculate if the cash flow figures are not consistent yearly. A trial and error process must be used to find the internal rate of return. However, the rationale and choice of the discount rate is the most challenging part [20].

Therefore, a few other simple investment analysis tools can be used to measure the return of MFC, such as payback period and simple rate of return. The payback period analysis focuses on the payback period, which is the length of time it takes for a project to recoup its initial cost out of the cash receipts (inflows) that it generates. The formula is by dividing the required investment with the annual net cash inflow. However, the payback period does not consider the time value of money. On the other hand, a simple rate of return focuses on net operating income generated by the MFC. The formula is by dividing annual incremental net operating income by the initial investment.

The above analysis technique is the theoretical part to ensure that the MFCs manufacturers or producers have a profit from their huge investment. But research and development play a main role in reducing MFCs. As noted in the previous study, MFC have an extraordinarily high capital cost, the limited power output of MFCs forestalls the scope for rapid market expansion. Accelerated research funding to enhance the efficiency of MFCs is expected to reduce the required cost. The market for MFCs can be considered at the infancy stage, especially for commercial utilization. It remains to take off as the renewables sector has many superior alternatives.

In terms of cost for each material being used in MFCs, estimation by previous researchers indicates that the cheapest for HF-MFC would be at $1,543 or $4.55 in capital cost per mW generated [21]. In contrast, GB and CF electrode capital costs for assemblies would be 2.3 and 9.1 times higher per mW, respectively. Hence, many producers try to reduce the cathode costs. However, some researchers overlook the low cost of anode material. Anode costs are more than half of the total cost for MFCs to work[21]. A reduction in anode cost will significantly reduce the capital cost for MFCs to operate. Material proposed for the anode to be cheaper, such as copper, felt, stainless steel, and graphene [22]. A reduction in cost by using less expensive material for the anode does not necessarily reduce the power being produced. It was proven but still needs to be checked for further assessment [15].

In this case, they are looking for the minimal or reducing their budget for MFC as the primary purpose of MFC is to replace other power sources but, if possible, with less cost. The different materials may replace other less-cost materials such as using recycled/used food containers.

In terms of the operational cost, it is lower than the capital cost. The capital cost for MFCS is 30 times higher than the method that uses activated mud. The operating expense can be reduced by recovering great-value substances such as Cr or other metallic elements. The cost varies according to the design and the type of method of the MFCs. Henceforth, the important thing for workable wastewater care with electricity manufacture requires capital cost and operational cost to be in balance all the time.

The other way to reduce the operational cost is by using low sludge generation for MFCs instead of activated or anaerobic bacteria. The existence of more contaminants in the wastewater can be tackled by applying preliminary remedies by the researchers. Further technical challenges, such as electrode spacing out can be resolved by using MFCs connected in sequence. It will reduce the electrode distance and maintain the working volume. Another problem faced by the MFC is that the power masses generated cannot meet the needs of the needy to treat the sewer water treatment because the output energy is less than the input power.

Anaerobic bacterial digestion in MFCs is essential for renewable power production and waste remediation. It is clean power and may reduce the compensation expense of electricity due to no drying and recirculation process. It also cuts the sludge treatment cost because it reduces sludge formation. The aerobic activity creates a large amount of waste mud. Using anaerobic bacteria will lower the investment cost than using aerobic treatment since double systems are not needed in this process, namely water handling system and an extra system to manage and pull out the biosolid process. The last advantage is that anaerobic ingestion suggests a possibility for good nutrients [23].

For MFCs to provide a competitive offer in the market and give the best performance to users, researchers need to discover the finest substances and system design of MFCs. It can be achieved by improving the voltage output, system efficiency, steadiness, and durability [23, 24]. The practicality of the MFCs in real-life applications is also vital, as the MFC can work individually in a simple workstation near the office or at home for a single user.

4 MFCs from an Economic Viewpoint

The latest report by GlobeNewsWire 2021 reveals that the wastewater treatment industry will remain the critical necessity generator for the next few decades. Even though commercial fuel has been widely accepted for the transportation industry, research and development regarding MFCs continues due to the interest of companies in the fuel cell area. There are many opportunities to explore to make the MFCs more beneficial, longer sustainable, and cheaper in the future. There are growing efforts towards exploring the possibilities of many applications and materials to maximize fuel cell efficiency, uplifting the prospects of other fuel cell categories like MFCs. It is projected that the MFCs market is set to exceed USD15 Million by the end of 2025. It is also quoted that the wastewater treatment worldwide costs about USD28-30 billion yearly, and it has become a massive opportunity for MFCs manufacturers to work. On the other hand, global MFCs market revenue for 2020 was only around USD9.76 million.

North America and Europe are among the continents that reflect the good growth potential of MFCs, with mediator MFCs ruled supreme over other fuel cell segments in 2020. European countries had started and progressively researched the development of MFCs for industry utilization. The Asia Pacific itself recorded the maximum revenue share in the global MFCs market for 2020. China and India are expected to maintain the top countries in the Asia Pacific for MFCs landscape. Leading industry players include Protonex Technology Corporation, Emefcy, ElectroChem Inc, Cambrian Innovation In., and Triqua International. Global MFCs market from 2018 until 2023 with a five-year annual trend analysis highlights the market size and share for North America, Asia Pacific, Europe, and the rest of the world. It shows that the market for MFCs is expected to increase from year to year at ~ 9.00% compound annual growth rate.

It is also due to the Covid-19 pandemic, whereby all the MFCs research activities, pilot testing, and other experiments were tremendously active during the initial months of 2020. Covid-19, which started in Wuhan, China, had chaos, and many people needed to work from home. Workers need to change their daily lives and working environments immediately, by force, or support. As MFCs can be tested at home as well, the workers may start the MFCs process at the back of their house or in some specific area nearby.

An enhancement of the MFCs market's strong development is also due to strong support from the government, industry, and all related entities to have green fuel aids. In addition, to have a sustainable fuel to substitute non-renewable fuels such as petroleum and gasoline. All the non-renewable and limited resources available will create a global energy crisis if there is no action to find energy substitutes for daily and industry usage. Even now, the non-renewables resources are dwindling at an alarming pace, which makes people worried about the future.

MFCs consumption for wastewater treatment usage also helps in regaining microbes and bacteria appropriate for use as a feedstock for electricity power. Market research future had indicated a summary of MFCs market applications by type, end user, and region. It shows that MFCs can be used as power generation, wastewater treatment, and biosensor applications in the agriculture and food and beverage industries, as well as government and municipal departments. It is famous on five continents worldwide, especially in North America, Europe, Asia Pacific, the Middle East, and Africa, and the last one in South America.

In the local context, Malaysian researchers started MFCs a few years back, with any research done in a laboratory and on a small scale. If we look at the Environmental Performance Index, it seems that Malaysia was ranked number 55 for water and sanitation. Overall, Malaysia was placed at 77 out of 180 countries in the Environmental Performance Index for Malaysia 2018 report, and it had improved into 59th place (improved by 18 rank). It shows that the country can be improved of its aspects to have a good rank in the future. One of them is improving and implementation of MFCs technology for wastewater treatment and electric power supply.

5 Conclusion

The chapter proves that MFCs may become an alternative solution to conventional electrical power supplies for many applications. In addition to their sustainable and green nature and a reduced carbon footprint, their integration into a better economical material or hybrid process may reduce the capital and operational costs as they reduce electricity consumption. An efficient and practical application of MFCs will improve the sustainability of MFCs application in the wastewater treatment process and biosensor. MCFs may integrate with other technologies of water treatment that can be explored in the future. The most important thing is that MFCs are practical to be used despite uncertainty in the future. From the economic perspective, it has been noted that MFCs are accepted for its benefits to replace the conventional electrical supply. Still, the capital and operational cost is haunting it to spread the application in many industries. However, the market of MFCs as large as billions of monies was spent on catering to the water sludge. MFCs are widely accepted in many countries that plan to replace their scarce natural energy supply with other substitutions. Perhaps, in the future, MFC will be one of the leading energy supplies to many people worldwide, especially in countries with limited non-renewable resources. Next, research on the development of cheaper MFC with more efficient ways of storing energy in capacitors can be suggested to boost the power accessibility for the wastewater management process. It should have a superior revitalization value due to the superior capital cost. A more extensive function of MFCs, especially in the industry, is another area to explore with innovative resources and methods that improve MFCs performance.

Acknowledgements The authors would like to thank the Universiti Sains Malaysia for the financial support of this study via APEX Era grant (1001/PGURU/881004)

References

1. Obileke KC, Onyeaka H, Meyer EL, Nwokolo N (2021) Microbial fuel cells, a renewable energy technology for bio-electricity generation: A mini-review. Electrochem Commun 125:107003. https://doi.org/10.1016/j.elecom.2021.107003
2. Yaqoob AA, Ibrahim MNM, Rodríguez-Couto S (2022) Development and modification of materials to build cost-effective anodes for microbial fuel cells (MFCs): An overview. *Biochem Eng J* 164 (September):107779. https://doi.org/10.1016/j.bej.2020.107779.
3. Munoz-Cupa C, Hu Y, Xu C, Bassi A (Feb. 2021) An overview of microbial fuel cell usage in wastewater treatment, resource recovery and energy production. Sci Total Environ 754:142429. https://doi.org/10.1016/J.SCITOTENV.2020.142429
4. Trapero JR, Horcajada L, Linares JJ, Lobato J (2017) Is microbial fuel cell technology ready? An economic answer towards industrial commercialization. Appl Energy 185:698–707. https://doi.org/10.1016/j.apenergy.2016.10.109
5. Saba B, Khan M, Christy AD, Kjellerup BV (2019) Microbial phyto-power systems—A sustainable integration of phytoremediation and microbial fuel cells. Bioelectrochemistry 127:1–11. https://doi.org/10.1016/j.bioelechem.2018.12.005
6. Li WW, Yu HQ, He Z (2014) Towards sustainable wastewater treatment by using microbial fuel cells-centered technologies. Energy Environ Sci 7(3):911–924. https://doi.org/10.1039/c3ee43106a
7. Pandit, S., Chandrasekhar, K., Jadha, D. A., Ghangrekar, M. M., & Das (2019) *Contaminant removal and energy recovery in microbial fuel cells. Microbial Biodegradation of Xenobiotic Compounds,*. 2017.
8. Mansoorian HJ, Mahvi AH, Jafari AJ, Amin MM, Rajabizadeh A, Khanjani N (2013) Bioelectricity generation using two chamber microbial fuel cell treating wastewater from food processing. Enzyme Microb Technol 52(6–7):352–357. https://doi.org/10.1016/j.enzmictec.2013.03.004
9. Sirinutsomboon B (2014) Modeling of a membraneless single-chamber microbial fuel cell with molasses as an energy source. Int J Energy Environ Eng 5(2–3):1–9. https://doi.org/10.1007/s40095-014-0093-5
10. Mohd Zaini Makhtar M, Tajarudin HA (2020) Electricity generation using membrane-less microbial fuel cell powered by sludge supplemented with lignocellulosic waste," *Int J Energy Res* 44(4):3260–3265. https://doi.org/10.1002/er.5151.
11. Zhuang L, Zheng Y, Zhou S, Yuan Y, Yuan H, Chen Y (2012) Scalable microbial fuel cell (MFC) stack for continuous real wastewater treatment. Bioresour Technol 106:82–88. https://doi.org/10.1016/j.biortech.2011.11.019
12. Logan BE, Regan JM (2006) Electricity-producing bacterial communities in microbial fuel cells. Trends Microbiol 14(12):512–518. https://doi.org/10.1016/j.tim.2006.10.003
13. Li Z, Yao L, Kong L, Liu H (2008) Electricity generation using a baffled microbial fuel cell convenient for stacking. Bioresour Technol 99(6):1650–1655. https://doi.org/10.1016/j.biortech.2007.04.003
14. Khalid. Kazem S, Madjid. Fatih M (2014) Passive air breathing flat-plate microbial fuel cell operation. *Journal of Chemical Technology & Biotechnology*. https://doi.org/10.1002/jctb.4325.
15. Makhtar MMZ, Don MM, Tajarudin HA (2018) Microbial fuel cell (MFC) development from anaerobic digestion system. *Green Energy and Technology* 0(9789811081286):9–31. Springer Verlag https://doi.org/10.1007/978-981-10-8129-3_2

16. Aftab S, Shah A, Nisar J, Ashiq MN, Akhter MS, Shah AH (2020) Marketability prospects of microbial fuel cells for sustainable energy generation. Energy Fuels 34(8):9108–9136. https://doi.org/10.1021/acs.energyfuels.0c01766
17. Fischer F, Bastian C, Happe M, Mabillard E, Schmidt N (2011) Microbial fuel cell enables phosphate recovery from digested sewage sludge as struvite. Bioresour Technol 102(10):5824–5830. https://doi.org/10.1016/j.biortech.2011.02.089
18. Marchioni A, Magni CA (2018) Investment decisions and sensitivity analysis: NPV-consistency of rates of return. Eur J Oper Res 268(1):361–372. https://doi.org/10.1016/j.ejor.2018.01.007
19. Magni CA (2010) Average internal rate of return and investment decisions: A new perspective. Engineering Economist 55(2):150–180. https://doi.org/10.1080/00137911003791856
20. Kozlovski A, Daria B, Kozlovskyi S, Lavrov R, Skydan O, Ivanyuta N (2020) Determination of the risk free rate of return on an investment efficiency based on the fractal market hypothesis. Forum Scientiae Oeconomia 8(3):61–72
21. Stoll ZA, Ma Z, Trivedi CB, Spear JR, Xu P (2016) Sacrificing power for more cost-effective treatment: A techno-economic approach for engineering microbial fuel cells. Chemosphere 161:10–18. https://doi.org/10.1016/j.chemosphere.2016.06.072
22. Baudler A, Schmidt I, Langner M, Greiner A, Schröder U (2015) Does it have to be carbon? Metal anodes in microbial fuel cells and related bioelectrochemical systems. Energy Environ Sci 8(7):2048–2055. https://doi.org/10.1039/c5ee00866b
23. Muaz MZM, Abdul R, Vadivelu VM (2019) Recovery of energy and simultaneous treatment of dewatered sludge using membrane-less microbial fuel cell. Environ Prog Sustain Energy 38(1):208–219. https://doi.org/10.1002/ep.12919
24. Tommasi T, Lombardelli G (2017) Energy sustainability of Microbial Fuel Cell (MFC): A case study. J Power Sources 356:438–447. https://doi.org/10.1016/j.jpowsour.2017.03.122

The Environmental Education Element of Secondary School Science Curriculum in Malaysia: Enhancing Students Awareness in Microbial Fuel Cell (MFC) as Green Technology in Treating Dewatered Sludge

Nor Asniza Ishak, Muaz Mohd Zaini Makhtar, and Siti Rohana Man

Abstract The Vision for Shared Prosperity 2030 and the Sustainable Development Goal in SDG 4 is to make Malaysia a country that continues to develop with a fair, equitable and inclusive economic distribution at all levels of income groups, ethnicities, regions and supply chains. Malaysia National Blueprint (2013–2025) emphasizes Environmental Education (EE) among pupils in preparation to provide better job opportunities as well as to enrich the economic prosperity in ensuring future social and economic well-being. The chapter covers the topic of integrating the Microbial Fuel Cell (MFC) as a potential green technology in enhancing the energy recovery in EE is to enhance secondary schools pupils' awareness, as well as to make EE well accepted among young generations as the informal learning and close to the nature and including helping change attitudes and mindsets of human resources, particularly among young generations.

N. A. Ishak (✉) · S. R. Man
School of Educational Studies, Universiti Sains Malaysia,
11800 USM, Gelugor, Penang, Malaysia
e-mail: asnizaishak@usm.my

N. A. Ishak
National Higher Education Research Institute (IPPTN),
11900 Bayan Lepas, Penang, Malaysia

M. Mohd Zaini Makhtar
School of Industrial Technology, Bioprocess Technology Division,
Universiti Sains Malaysia, 11800, Gelugor, Penang, Malaysia

Centre for Innovation and Consultation (CIC), Universiti Sains Malaysia,
11800, Gelugor, Penang, Malaysia

M. Mohd Zaini Makhtar et al. (eds.), *Microbial Fuel Cell (MFC) Applications for Sludge Valorization*, Green Energy and Technology,
https://doi.org/10.1007/978-981-99-1083-0_12

1 Introduction

Since the last century, it is becoming increasingly apparent that human activities have made a significant impact on all the ecosystems worldwide. Today, humans have to deal with a huge range of economic, social and environmental problems that are the consequences of their interactions with the natural world [1]. Some of these problems, such as climate change, biodiversity loss, ozone depletion and food scarcity, are closely linked to the exponential increase in human population, rapid economic development, industrialization, urbanization and various kinds of pollution [2].

The same phenomena is also happening in Malaysia. As Malaysia strives to become a developed nation by 2020, the increase of industries and rapid development has given rise to various environmental problems such as atmospheric pollution, water pollution, coastal pollution and loss of biodiversity [3]. Indeed, the current generation is witnessing an unprecedented economic growth and technological advancement which may have benefited a lot of people, but have also caused detrimental social and environmental consequences [4]. People have started to neglect the importance of maintaining the balance of the natural world, as technology slowly plays more and more significant roles in every aspect of their lives [5]. One of the ways suggested by the renowned independent global non-profit conservation organization, the World Wildlife Fund (WWF) [6], is to create awareness via environmental education (EE).

2 Sustainable Development, Education for Sustainable Development (ESD) and Environmental Education (EE)

To address the environmental problems, humans need to adopt sustainable ways while continuing to develop as a race. In other words, humans need to develop sustainably. The term "sustainable development" was first used in the 1960s and its concepts and guidelines gradually developed and brought to prominence over time during many United Nations conferences and summits from the 1970s to the 2000s. Ultimately, the definition for this term that is widely accepted today is "development that meets the requirements of modern day without jeopardizing the capability of future generations to meet their own requirements" [7]. The understanding of sustainable development must also take green (economic) growth and human development into consideration. Green growth has been identified as an effective way towards sustainable development by fostering economic growth and addressing environmental problems. However, human development is also important and should not be overlooked, because sustainable development is also an issue of social justice, whereby the poor and marginalized often feel the impact of environmental degradation and climate change. In other words, the key issue in Sustainable Development is the need to look into social–ecological relations as well. Eventually, the birth of the 2030 Agenda and the SDGs becomes crucial steps towards human development for everyone [7].

With the aim to evaluate sustainable development around the world, a set of global Sustainable Development Goals (SDGs) that consists of 17 goals and 169 targets was proposed by the Open Working Group created by the United Nations General Assembly [8]. One fundamental part of the 2030 Agenda and SDGs is the recognition of the crucial role that education and ESD play, especially in providing the required platform and mechanisms for all humans to be active agents of change [7].

In the effort to promote sustainable development through education, the term Education for Sustainable Development (ESD) was defined, described and explained in the 40-chapter Agenda 21 during the United Nations Conference on Environment and Development, also known as the Earth Summit, in Rio de Janeiro in 1992. Here, education was agreed to be crucial in achieving sustainable development via all the action areas described in Chapter 36 which call for efforts to enhance learning opportunities by dealing with unsustainable practices and encouraging the quality of life around the world [7].

Subsequently in 2015, the United Nations Educational, Scientific and Cultural Organization (UNESCO) put ESD to the fore at the 2030 Agenda for Sustainable Development and its seventeen SDGs. Among the many targets set by UNESCO, ESD was emphasized in Target 4.7 with the statement, "By 2030, all learners are ensured to obtain the knowledge and skills required to encourage sustainable development through ESD, human rights, gender equality, culture of peace and others" [9].

From the perspective of sustainable development, ESD is based on the values and principles that scaffold sustainable development by preserving the five pillars of sustainability, namely environment, economy, society, peace and partnership. ESD also needs to integrate crucial sustainable development issues and concepts, such as climate change, biodiversity loss, food security, disaster management and resource efficiency, into every form and level of teaching and learning [9].

From the view of education, ESD is a dynamic concept that includes a new vision for a type of education that aspires to empower people from all walks of life to uphold the responsibility of making a sustainable future. Therefore, ESD enables everyone to obtain the knowledge, skills, attitudes and values required to create a sustainable future. ESD also empowers students to make wise decisions and act responsibly for the environment for present and future generations. ESD also equip learners to seek solutions to problems that endanger the sustainability of our planet [7].

However, frustration grew when it was noticed that ESD did not advance quickly enough, partly because nobody had taken actual responsibility for ESD nor raised ample funds to promote it, and partly because nobody knew the way to commence ESD. Subsequently, UNESCO was assigned as the Task Manager for Chapter 36 of Agenda 21, which places education as the focus. Soon, UNESCO started to merge various forms of education, which includes environment, population and development, into one unified concept of ESD. Thereafter, UNESCO defined EE as "an interdisciplinary educational approach with the aim of developing knowledge, skills, values and ethical behaviour favouring an improved quality of life and of the environment as a whole" [7].

Therefore, it is evident that the roots of ESD are firmly planted in EE. In its short 25-year history, EE has gradually but steadily taken its steps towards achieving goals and outcomes similar to those integral to sustainable development [7]. EE has gained international recognition since the 1970s. From the 1977 Tbilisi Declaration, EE was defined as a learning process that improves people's knowledge and awareness for the environment and related challenges, develops the required skills and expertise to overcome the challenges and nurture attitudes, motivations and commitments to make wise decisions and act responsibly [10].

After the 1977 Tbilisi Declaration and the following UNESCO conferences, many countries around the world, such as Egypt [10], Taiwan [11], Greece [12], Serbia [13] and Malaysia as well [14], used the guidelines in the Declaration to develop EE in their countries [12]. Some countries carried out EE programs while others implement EE into the national curriculum, either as a separate subject or cross-curricular.

Despite all these countries reporting the implementation of EE in their countries, the effectiveness of their EE measures or programs has often been overlooked. Hence, there is a dire need to evaluate the effectiveness of EE in all these countries; the same goes to Malaysia. In order to properly gauge the effectiveness of EE in Malaysia, we must first understand the level of EL among Malaysian students [15].

3 Environmental Education (EE) Around the World

Students of today will bring an enormous impact on the environment of tomorrow. Hence, the integration and institutionalization of matters pertaining to sustainable development into EE is more relevant than ever. Thus, it is imperative to measure the EL of students and seek ways to instil pro-environmental behaviour in them via EE [16]. In Turkey, [17] did a review on the EE research in Turkish primary schools. By referring to prior literature, the authors stated that the Turkish government started to pay heed to environmental issues in the 1960s and approved the first environment protection policy in the 1970s. Subsequently, the Ministry of Environment was formed in 1991. These developments influenced the national curricula by forming a political climate in which EE could develop and acquire support from the government. Following the environmental laws that developed from the 1982 Constitution that focused on environmental issues, courses related to the environment began to emerge in the national school curricula. On top of that, a number of national "education for the environment" projects were started in primary schools in the 1990s. The importance of EE was then highlighted in the 1994 Seventh Five Year Development Plan Environment Commission's report. Soon, dissertation studies on EE in the middle schools also began. In 1993, the Turkish Environmental Education Foundation (TURCEV) was formed, and it launched even more EE projects. Efforts to join the European Union (EU) since the 1990s have also helped to increase EE activities and projects in schools nationwide. For the national curricula, the Turkish primary school science curricula developed in 1992, 2000 and 2004 has seen an increase in emphasis on environmental topics by including more and more related topics as the curricula underwent reforms after reforms.

In Greece, the Ministry of Education has introduced EE programs as part of a variety of optional educational programs in secondary schools that are carried out outside the school timetable [18]. The ministry has also established Environmental Education Centres (EECs) in which schoolteachers are recruited to develop, organize and carry out Environmental Education Programs (EEPs), seminars, materials and experiences for school students as well as teaching staff with the aim of raising their environmental awareness. These programs are carried out at the EECs themselves and/or outdoors at specially chosen places such as nature reserves, botanical gardens and agricultural areas [12].

In the United States, the current EE was preceded by three movements, namely the Nature Study movement in the 1890s, the Outdoor Education movement in the 1920s and Conservation Education movement in the 1930s. Philosophies from these movements are still important aspects of EE today. Decades later, events in the 1960s and 1970s such as environmental catastrophes like oil spills and the passing of legislation notably the National Environmental Policy Act of 1969 influenced contemporary EE principles. Then in 1970, President Nixon launched the Environmental Protection Agency (EPA) and passed the 1970 National Environmental Education Act, which led to the formation of the Office of Environmental Education and the first government-sponsored definition of EE. Then in 1971, EE was reinforced with the formation of the National Association for Environmental Education (NAAEE). However, the EE movement encountered major setbacks caused by political agendas in the 1980s but regained momentum in the 1990s when the 1990 National Environment Education Act was passed. Then, EE had its ups and downs in the political arena since 2000 to the current day, from the reintroduction of the 2001 No Child Left Behind Act that caused teachers to focus less on EE to the 2008 No Child Left Inside Act that ensured students achieve basic EL. Eventually, when President Obama signed the 2015 Every Student Succeeds Act, EE programs finally received federal funding for the first time. In short, EE in the United States has been very inconsistent due to politics and finally regained attention in recent years [1].

According to [19], Taiwan is the 6th country in the world that enacted a law for EE in June 2011 and the Environmental Protection Administration (EPA) is assigned to be the statutory authority of the Environmental Education Act (EEA). This act developed a regulation that requires every student and teachers in primary and secondary schools to take part in any type of EE programs every year. Therefore, schools must include EE programs into the school calendar according to the guidelines by their Ministry of Education.

Other than Taiwan, Japan and Korea are another two Asian countries (out of the eight worldwide) that have enacted an Environment Education Act (EEA), as of 2019. According to [20], Japan enacted it in 2003 and revised it in 2011 while Korea enacted it in 2008 and revised it in 2015. These 3 countries have pioneered EE efforts around the world. In Japan, lawmaker-initiated legislation enacted their EEA titled "Act for Enhancing Motivation of Environmental Conservation and Promoting Environmental Education" in July 2003 under the jurisdiction of five ministries. A revision was made in June 2011 that included growing the needs of EE in schools and the title was changed to "Act on the Promotion of Environmental Conservation

Activities through Environmental Education" or APEE. In Korea, the "Environmental Education Promotion Act of Korea" or KEEPA was enacted in 2008. KEEPA has since served as a legal framework for EE policies in Korea and has also provided legislative instruments for EE there. Under the collaboration of the Ministry of Environment and the Ministry of Education, KEEPA aims to promote EE and sustainable development via formal EE. Similar to Greece, KEEPA has supported informal EE by establishing EE centres at the national and regional level. In short, KEEPA covers both formal and informal EE while the national curricula focus on formal EE.

In Serbia, [13] found that Serbian students lack the knowledge to contribute to the development of healthy lifestyles and environmental awareness despite Human Health and Preservation of the Environment being two integral topics in the Serbian primary school curriculum. The authors studied the challenges faced while implementing the intended curriculum of Environmental Education in Serbian primary schools. One of the biggest challenges is that the core content being taught, mostly via science subjects, was mostly geared towards the cognitive dimensions, resulting in less focus on the students' pro-environmental attitudes and behaviour. However, the latest changes in the primary school policy and curriculum that acknowledges the significance of Environmental Education can be deemed as progress while awaiting changes in school practise.

In Israel, according to [21], environmental issues have already been embedded into the national curriculum in Israel since the 1970s, via both infusional and impositional methods. The infusional method refers to the combination of environmental topics and units within other subjects of the national curriculum while the impositional method refers to the study of environmental topics as an elective subject going into the matriculation exams. In recent years, the increase in public environmental awareness led to the increase in governmental commitment to improve EE in the public educational system. This is reflected by the increase in schools that embed EE into their curriculum. Looking back, in 1994, the Ministry of Education (MOE) stated that every school was required to put environmental topics in school activities. In 2003, the MOE set out guidelines to integrate sustainable development principles into the EE in the educational system. Then in 2007, the MOE asked primary schools to dedicate at least 30 hours of EE in their school year annually.

Karimzadegan and Meiboudia [22] did a review on the EE in Iran. According to the researchers, Iran's record of formal environmental training can be found in the initial educational programs by their Ministry of Education (MOE). For instance, since the 1980s environmental subjects have begun to be included in Iran's educational programs while environmental topics are found in subjects such as science, geography and biology in the national curriculum. The educational programs are backed up by textbooks approved and published yearly by the MOE which are used in every school in Iran as their formal books.

This brief review of the implementation of EE in various countries shows that EE is carried out differently in different countries. Some countries integrate EE in their national curriculum while others carry out EE as programs, school activities and elective subjects. Nevertheless, each country has shown improvement in the intention and implementation of EE to raise EL as the way to solve environmental problems.

4 Benefits/Effects of Environmental Education (EE) Programs

In Taiwan, [11] studied the effects of an EE program on the Responsible Environmental Behaviour and its related Environmental Literacy variables of Taiwanese college students. The results of the study revealed that there is a significant improvement in the students' Responsible Environmental Behaviour, Intention to Act, Perceived Knowledge of Environmental Issues, Skills in Using Environmental Action Strategies and so on after the Environmental Education Program. In this study, a quasi-experimental design was used, whereby 121 voluntary students were separated into an experimental class and a control class. Both classes went through 16 weeks of Environmental Education instruction, delivered by the same instructor. The purpose of the study was not revealed to the students so that the Hawthorne effect can be decreased to the minimum. A pre-test was conducted for both classes and it showed no significant differences between them. At the end of the instruction, a post-test was carried out to test the effects of EE on the students' EL. Two months after the end of the course, a follow-up test was carried out to gauge the continuous effect of the course. An analysis of covariance (ANCOVA) was carried out to analyze the immediate effects of educational intervention to promote EL in students by taking the pre-test scores as the covariates and post-test scores as the dependent variables. A paired t-test was then done to analyze the continuous effects on the students' EL variables in the experimental group. The results show that the EE program has successfully increased the students' Responsible Environmental Behaviour (REB), Environmental Responsibility, Intention To Act, Perceived Knowledge of Environmental Issues and Skills in Using Environmental Action Strategies. These effects were still retained 2 months after the program has ended.

In Egypt, [10] studied the effects of an EE program on the knowledge and attitudes of preparatory school students in Alexandria, Egypt. The results of the study showed that after six sessions of EE sessions, students had a very significant improvement in their knowledge and attitudes towards the environment. The authors also emphasized on the need to develop and implement EE programs as part of the regular school curriculum. This study was carried out in three phases. In phase one or the pre-intervention phase, 543 preparatory school students chosen by stratified random sampling took a questionnaire that included socio-economic level, environmental knowledge and attitude towards environmental concepts. In phase two or intervention phase, 150 out of the 543 students underwent the EE program that lasted 6 sessions of 30 min lecture and 15 min discussion, twice a week. In phase three or post-intervention phase, the students were asked to answer the questionnaire again. The results showed that initially students had low environmental knowledge and negative attitude. These were positively correlated to the socio-economic level of the students. However, these significantly improved after the intervention of the EE program, therefore highlighting the need for such program as part of the school curriculum.

In Portugal, [23] carried out a study on the influence of both an EE project and commitment interventions on the conservation behaviours in teenagers. The EE project named "SOS Climate Change and Biodiversity" took place in the Botanic Garden of Lisbon University because real settings let participants relate environmental problems to their everyday lives. In this study, a total of 418 students aged from 11 to 15 years (6th–9th grade in Portugal) from 21 classes from 4 different Lisbon schools were invited to take part. The students were split into 2 groups to test the effectiveness of the EE project. Group 1 consisted of 12 classes (248 students) who took part in the EE project while Group 2 comprised 9 classes (170 students) who served as the control group. To test the influence of commitment on behaviour, 136 students (80 from group 1 and 56 from group 2) signed a public commitment to save water and energy at home for a month. Another 135 students (80 from group 1 and 55 from group 2) signed a private commitment while 147 students (88 from group 1 and 59 from group 2) did not sign any commitment. MANOVA (Multivariate Analysis of Variance) tests were used to compare students with EE and without EE against 3 types of commitments. The results showed that students with EE saved more energy than those who did not while students who signed the public commitment saved more water than those who did not. Therefore, these results suggest that combining EE and signing public commitment can be used as an effective tool to encourage pro-conservation behaviours among teenagers.

Back in Malaysia, [24] studied the perceptions of students of using an EE Kit developed by the World Wildlife Fund (WWF) Malaysia in their schools. Their study involved 14- to 16-year-old students from 3 schools in the states of Pahang, Kuala Lumpur and Kedah; one urban and two rural. In their qualitative analysis, it was found that the majority of the students interviewed had a positive perception, which also means that they welcomed the usage of the EE kit in their schools as a supplement to the EE component already in their national curriculum to enhance their understanding of environmental issues. However, most of them opined that the content in the EE kit is not sufficient and they hoped WWF-Malaysia would improve the EE kit. Some others also stated that even the EE component in their national curricula is not enough to raise their EL. Therefore, from this study the researchers concluded that an enhanced version of the EE kit would serve as an important tool to supplement the already insufficient EE in the national curriculum in order to improve the EL of students in Malaysia.

5 The Role of Environmental Education in Malaysia

Education plays a key role in developing public knowledge and awareness of all kinds of issues; the same goes for environmental issues [25]. Through EE, we can develop a more environmentally literate citizenry who will be able to address various environmental challenges and problems [26]. Therefore, it is undeniable that schools around the world play an important role to deliver EE to students with the aim of nurturing them into citizens who are sensitive towards issues pertaining to the environment [15]. Therefore, the goal of EE is to improve students' environmental literacy by increasing their knowledge, cognitive skills, attitudes and behaviours. Each of these components needs to be developed so that a citizenry, who can address the recent and future global environmental issues, can be developed [1]. In addition, the aims of EE include fostering students' environmental awareness and ability to find solutions for various environmental problems by imparting knowledge to change their beliefs, attitudes and behaviours towards the environment [27]. On the other hand, in the framework of international agendas, the 17 Sustainable Development Goals (SDGs), as a current international political framework, are proposed as a collection of global aspirations aimed at attaining a sustainable future. In this international agenda it is recognized that education plays an extremely relevant role. Nevertheless, it is convenient to remember that although this agenda is pertinent to prepare our future through the implementation of urgent educational interventions, we cannot ignore the fact that the problems we are facing as humankind are the direct result of the dominant economic system and its social and environmental unsustainability.

Despite undergoing rapid industrialization towards becoming a developed nation, Malaysia has always tried to balance its economic and social development with environmental preservation. The country believes, other than carrying out environmental protection measures, in improving our national education system as part of our national development is also a very important step because education is a powerful tool that can shape future citizens to support environmental conservation [4].

Although EE has long been introduced globally since the eighteenth century, it is considerably new in Malaysia, which started in the year 1979, whereby a variety of plans and programs have been carried out to produce an environmentally literate society in Malaysia, involving schools, local and international community [28]. Globally, children acquire their environmental knowledge and attitudes mostly via the public-school curricula and not via specialized Environmental Education programs [29]. The same applies to Malaysia.

The history of EE actually began with the formation of a cabinet committee in 1974 to review the National Education Policy of Malaysia. Subsequently, the results of the committee report were published in 1979 to determine the implementation of EE in the structure of new curricula, KBSR for primary schools, and KBSM for secondary schools, which began in 1982. Then, from 1982 to 1994, the subject "Man and Nature" was implemented as a subject that emphasize on environment in primary schools, but it included elements of history, geography, science, health and civic education. It was then replaced with science and local studies from 1995 onwards. In 1998, the curriculum development centre, the Ministry of Education Malaysia established a set of guidelines for teachers to teach EE by publishing a book titled "KBSR Teacher's Handbook: Environmental Education across the Curriculum". EE teaching strategies in Malaysia are executed through teaching and learning in the classroom and also outside the classroom. In the classroom, the element of EE is integrated in every subject taught in the school, such as Science, Mathematics, Music, English and Islamic Education, thereby imparting environmental knowledge, proficiency, skills and inculcating environmental values and attitudes in students [28].

Besides formal education taught in the classroom, EE was also integrated in co-curricular activities, such as activities of societies and clubs, sports and games and uniformed units activities. The teachers-advisor will integrate pro-environmental values and attitudes in the activities carried out. However, although more than 30 years have passed, the environmental awareness and attitudes towards the environmental sustainability among Malaysians are still very low, compared to other developed countries such as Japan, South Korea and Australia, despite Malaysia has been striving to become a developed country by the year 2020 [28].

According to [5], EE had been carried out formally by governmental organizations in Malaysia, such as the Ministry of Education (MOE) in collaboration with other government agencies such as the Department of Environment (DOE) and the Ministry of Higher Education (MOHE). Non-formal ways have also been carried out by non-governmental organization (NGOs) such as Environmental Protection Society of Malaysia (EPSM), the Environmental Management and Research Association of Malaysia (ENSEARCH), the Malaysian Nature Society (MNS) and World Wildlife Fund (WWF) Malaysia by raising environmental awareness through campaigns and publishing and distributing handout materials to citizens. Apart from disseminating information about the environment, these NGOs tend to be more action-oriented compared to the more formal ways by the governmental agencies. [28] also suggested outdoor education as one of the best ways to teach people especially young children about the environment as it is a hands-on education and learning through experience.

6 The Environmental Education Element of Secondary School Science Curriculum in Malaysia

EE in Malaysia is not something new as it has been implemented formally or via non formal ways for many years. Nevertheless, the onus should always be on the educational institutions in Malaysia i.e., schools, colleges and universities to implement an environmental syllabus [4]. Even for higher education, the Teacher Education Division (TED) have EE as a compulsory subject with credit hours for preservice teachers to complete before they can graduate at Teacher Training Institutes (TTI) or Institusi Pendidikan Guru (IPG) in Malaysia. This is to prepare them to deliver EE effectively when they begin their teaching career at schools later [5]. For formal education from primary to secondary schools in Malaysia, EE is not taught as one single subject like the one in TTIs, but as a cross-curricular component across all subjects and every level of education [15]. For example, there is at least one topic dedicated to environment preservation in every KSSR English syllabus from Primary 3 to 6, which deals with environmental issues such as pollution, recycling and biodiversity loss. However, EE will be more focused in subjects like science and geography when students go to secondary school [15].

Ministry of Education Malaysia has included an environmental education component in the Science and Geography subject curriculum. In the latest Lower Secondary Schools KSSM syllabus, EE is only briefly addressed in Form 2 Science Chapter 2 as subtopics, namely Subtopic 2.3: Interdependence and Interaction Among Organisms and between Organisms and the Environment and Subtopic 2.4: Role of Humans in Maintaining a Balanced Nature [30]. In contrast, for the brand new Upper Secondary

KSSM syllabus, an entire chapter is dedicated to EE in Form 5 Science and Form 5 Biology, albeit slightly later than the preceding curriculum. In comparison, the chapter was taught earlier in Form 4 Science in the preceding curriculum. In the latest Form 5 Science syllabus, this dedicated chapter is in Chapter 3, entitled Sustainability of the Environment [30]. As stated in the curriculum specifications document, one of the objectives of the lower secondary Science curriculum, as well as the upper secondary Science curriculum, is to create awareness that the development of science and technology has an implication on the moral, social, economic and environmental issues in the local and global context [30, 31]. As for the learning standards, only two that are related to EE can be found in the lower secondary Science curriculum, which is shown in Table 1.

As for the upper secondary curriculum, the whole dedicated chapter to EE in Form 5 Science is comprised of a total of ten learning standards, shown in Table 2 (Table 3).

By comparing the learning standards covered by the two syllabi of Science and Biology, it is notable that the lower secondary Science syllabus contains significantly less content (2 learning standards) on EE compared to the upper secondary Science syllabus (10 learning standards). This raises the question of whether this amount of EE content is sufficient to improve the EL of lower secondary students particularly in

Table 1 Learning standards in Chapter 2 of Form 2 Science KSSM syllabus

Learning standards	Students are able to
2.3.5	Predict how the changes in ecosystem affect the existing resources and balance of the population
2.4.1	Justify and communicate that man needs a stable and productive ecosystem to sustain life

Table 2 Learning standards in Chapter 3 of Form 5 Science KSSM syllabus

Learning standards	Students are able to
3.1.1	Explain the meaning of carbon footprint
3.1.2	Research on a single use product of a person's daily life
3.1.3	Justify the reason carbon handprint is an appropriate method to help reduce the release of greenhouse gases in a person's daily life
3.1.4	Communicate and talk about the life cycle of a product
3.1.5	Generate ideas to manage plastic waste wisely for the sustainability of the environment
3.2.1	Explain the types and sources of pollution
3.2.2	Find out the degree of water pollution caused by domestic waste
3.2.3	Create a method to cleanse polluted water using green technology
3.3.1	Justify the role of every individual in managing natural resources to maintain the balance of nature
3.3.2	Debate the role of United Nations in managing environmental issues at the global level

Table 3 Learning standards in Chapter 25 of Form 5 Biology KSSM Syllabus (Chapter Environmental Sustainability)

Content standard	Learning standard
25.1 Environmental Threats	25.1.1 Explain the meaning of environmental sustainability
	25.1.2 Analyze environmental threats: climate change deforestation pollution loss of biodiversity the explosion of human population growth global warming eutrophication (viii) global warming
	25.1.3 Experiment to compare the levels of biochemical oxygen demand (BOD) in different water samples
25.2 Conservation, Preservation and Restoration of Ecosystems	25.2.1 Define: ecosystem conservation ecosystem preservation (iii) ecosystem restoration
	25.2.2 Justify the need: ecosystem conservation ecosystem preservation (iii) ecosystem restoration
25.3 Practices in Preserving the Environment	25.3.1 Generate ideas related to practices that contribute to environmental sustainability
	25.3.2 Discuss the status of food security in Malaysia
25.4 Green Technology	25.4.1 Defining green technology
	25.4.2 Justify the use of green technology in conserving nature

elevating the students' awareness, knowledge, affect and participation towards energy recovery. As for in the Biology subject, there is a presence of one theme regarding ecosystem and environmental sustainability in the KSSM curriculum of Biology. One of the contents standards integrated in this theme is the environmental sustainability which emphasizes on threads to environment, environment conservation and preservation, practices in environmental sustainability and green technology.

7 Green Technology as a Method to Increase Production of Renewable Resources

One of the issues that are lacking in the current syllabus pertaining to renewable resources is the energy recovery. The usage of renewable sources to generate energy has been tremendously explored due to their apparent reputation and reliability for a sustainable energy production process. The principal aim of the utilization of renewable sources for energy generation is to mitigate the carbon footprint, and environmental pollution in general, which eventually contributes to the establishment of sustainable living. Green technology can be one of the methods that can help in enhancing the process of renewable resources.

The term 'green' refers to the special attention to protecting the environment, which includes the plants and animals that grow in it. While the term 'technology' refers to the application of knowledge for practical purposes. Green technology is the application of environmental science to conserve the environment and make good use of natural resources to minimize and reduce the negative impacts of human involvement. Green technology is also known as environmental technology or clean technology. Green technology also demonstrates the use of better, cleaner and more efficient technologies methods and devices that are less risky and less harmful to ecological resources. Green technology refers to products, equipment or systems that meet the following criteria:

This reduces environmental degradation.
It is zero or low emission greenhouse gas (GHG) safe to use and promote a healthy and improved environment for all form of life.
This saves the use of energy and natural resources.
This encourages the use of renewable resources. [32]

Green technology is seen as a major platform to help preserve the environment [33]. Green Technology has become the portfolio of the Ministry of Energy, Green Technology and Water (KeTHHA) which was established on 9 April 2009. To strengthen the use of green technology in Malaysia, the National Green Technology Policy (DTHK) was launched under the Ministry of Energy, Green Technology and Water (KeTTHA) which emphasizes the aspects of driving the country's economic growth and sustainable development. The National Green Technology Policy (DTHK) was launched by the former Prime Minister of Malaysia, Datuk Seri Najib Tun Abdul Razak in July 2009 which has four main pillars, namely energy, environment, economy and social. The National Green Technology Policy provides direction and motivation for the people.

Malaysia continues to enjoy a good quality of life and a healthy environment. In line with the main objective of DTHK which is to increase education and public awareness on Green Technology and encourage the widespread use of Green Technology [34]. Green Technology refers to the development and application of products, equipments and systems to preserve the environment and nature and minimize or reduce the negative impacts of human activities [32].

Green Technology is found to be able to generate energy at a cheaper cost as well as safe and user-friendly, thus making Green Technology has great potential as an alternative in driving national development. Green Technology focuses on four (4) main sectors namely;

Energy Sector, for example: energy generation and energy supply management.
Building Sector, for example: construction, management, conservation and destruction of buildings.
Water and Waste Management Sector, for example: management and use of water resources, sewage treatment, solid waste and others.
Transport Sector, example: transport infrastructure and vehicles. [32]

According to [35], Green Technology is all tools, products and systems that can reduce environmental quality degradation, low or zero greenhouse gas emissions, save energy and natural resources, use Renewable Energy sources and safe for the environment. The field of Green Technology encompasses an ever-evolving group of methods and materials, from techniques for generating energy to non-toxic cleaning products [25]. According to [36], Green Technology describes energy generating technologies such as solar energy, wind energy and so on. This opinion is supported by [37] who stated that Green Technology is closely related to sustainable development, energy consumption at an optimal level as well as environmentally friendly and the search for new energy sources. While according to [38] Green Technology involves the exposure to the concept of sustainable living, the development of renewable energy sources, and the reduction of waste disposal. These three foundations are the basis for Green Technology in ensuring that the objectives of Green Technology can be achieved.

8 Microbial Fuel Cell (MFC) as Green Technology in Treating Dewatered Sludge

Microbial fuel cell (MFC) is an emerging environmental-friendly technology that employs microorganisms in wastewater as catalysts to oxidize organic and inorganic substances into electrical energy via the bioelectrochemical catalytic activity of microbes [39, 40]. All metabolizable substrates can be a perfect substrate for MFC, and these are abundantly available in wastewater as shown in Fig. 1.

The presence of chemically and bioactive substituents in wastewater triggers concern for human health, aquatic ecosystem and environment, and this justifies that wastewater treatment is essential. MFC is evidently a developing and promising method to produce energy and treating wastewater, however, its implementation at a large scale necessitates scrutinized process control to produce sufficient energy and wasteless water residues. Dewatered sludge embodies a massive amount of energy reserve [41]. The sludge dewatering process is a common technique to enhance solid wastes removal from water by increasing the biomass solid content, and the overall

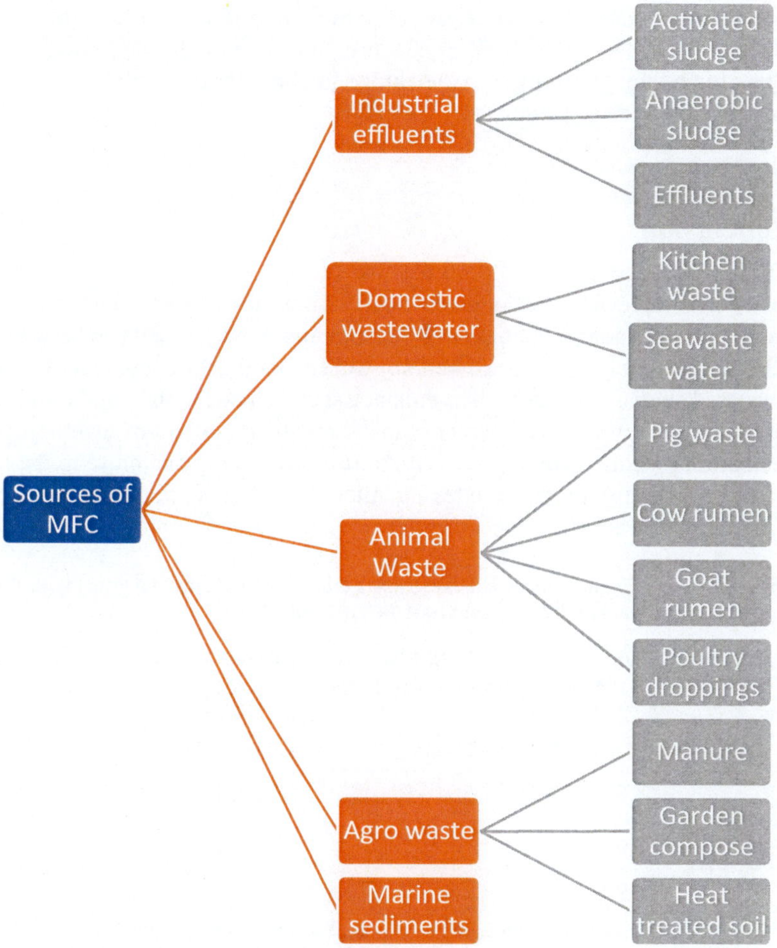

Fig. 1 Type of MFC Sources Adapted with permission (CCBY)

process requires high energy usage as the solid deposits are exposed to high temperature and pressure [42]. Therefore, the dewatered sludge must be utilized as a part of the energy cycle loop, to limit the leakage of the energy chain. [43] used dewatered sludge as a substrate for electrical power generation via MFC and reported that the highest chemical oxygen demand (COD) removal and voltage generated from the biomass were 149.2 mg/L and 852.7 mV. They also revealed that the yield for these two outputs can be optimized to up to 15% by setting the cell electrodes distance at 2.93 cm, the moisture content was brought up to 31.99 v/w. %, and operating temperature set at 38.05 °C. While MFC is still an emerging technique, the process energy efficiency and biomass utilization can still be optimized to facilitate zero biomass and energy excess. This energy leakage will eventually require post-processes to diminish the biomass and solvent residues, which in turn use up much more energy than the one produced by the MFC procedure. Scrutinizing the suggested activities under this topic, none introduce on the sludge treatment, MFC technology and how it can contribute to energy recovery.

9 Conclusion

From the literature review and document analysis done towards environmental education element of secondary school science curriculum in Malaysia it can be concluded that MFC acts as a potential green technology in enhancing the energy recovery using the dewatered sludge in EE as well as enhance secondary schools' pupils' awareness towards energy recovery. The chapter emphasizes the important of making EE well accepted among young generations as the informal learning and close to the nature and including helping change attitudes and mindsets of human resources, particularly among young generations.

Acknowledgements The authors would like to thank the Universiti Sains Malaysia for the financial support of this study via APEX Era grant (1001/PGURU/881007).

This paper is under the funding of a APEX ERA Research Grant from Universiti Sains Malaysia (1001/PGURU/881007). A special thanks to USM for the funding and encouragement to run this research.

References

1. Williams RD (2017) An assessment of environmental literacy among Oklahoma Public High School students and the factors affecting students' environmental literacy [Master's thesis, Harvard Extension School]. https://dash.harvard.edu/handle/1/33826276
2. Liang SW, Fang WT, Yeh SC, Liu SY, Tsai HM, Chou JY, Ng E (2018) A nationwide survey evaluating the environmental literacy of undergraduate students in Taiwan. Sustainability 10(6):1730. https://doi.org/10.3390/su10061730
3. Hassan MN, Awang M, Jaafar AB (2006) Challenges of global environmental issues on ecosystem management in Malaysia. Aquat Ecosyst Health Manage 9(2):269–283. https://doi.org/10.1080/14634980600728800
4. Yusoff S (2003) The need for emphasis on environmental education for national development in Malaysia. Masalah Pendidikan 26.
5. Lateh H, Muniandy P (2010) Environmental education (EE): current situational and the challenges among trainee teachers at teachers training institute in Malaysia. Procedia Soc Behav Sci 2(2):1896–1900. https://doi.org/10.1016/j.sbspro.2010.03.1005
6. WWF (2018) Living planet report—2018: aiming higher (eds: Grooten M, Almond REA). WWF, Gland, Switzerland
7. Agbedahin AV (2019) Sustainable development, education for sustainable development, and the 2030 agenda for sustainable development: emergence, efficacy, eminence, and future. Sustain Dev 27(4):669–680. https://doi.org/10.1002/sd.1931

8. Hák T, Janoušková S, Moldan B (2016) Sustainable development goals: a need for relevant indicators. *Ecological Indicators, 60*, 565–573. https://doi.org/10.1016/j.ecolind.2015.08.003
9. Diemer A, Ndiaye A, Khushik F, Pellaud F (2019) Education for Sustainable development: a conceptual and methodological approach. Soc Sci Learn Educ J 4:43–51
10. Abd El-Salam MM, El-Naggar HM, Hussein RA (2009) Environmental education and its effect on the knowledge and attitudes of preparatory school students. J Egypt Public Health Assoc 84(3–4):345–369
11. Hsu SJ (2004) The effects of an environmental education program on responsible environmental behavior and associated environmental literacy variables in Taiwanese college students. J Environ Educa 35(2):37–48. https://doi.org/10.3200/JOEE.35.2.37-48
12. Tampoukou A, Papafotiou M, Koutsouris A, Paraskevopoulou A. (2015). Teachers' perceptions on the use of botanic gardens as a means of environmental education in schools and the enhancement of school student benefits from botanic garden visits. Landscape Res 40. https://doi.org/10.1080/01426397.2014.947250
13. Stanišić J, Maksić S (2014) Environmental education in Serbian primary schools: challenges and changes in curriculum, pedagogy, and teacher training. J Environ Educ 45:118–131. https://doi.org/10.1080/00958964.2013.829019
14. Aminrad Z, Zakariya SZB, Hadi A, Sakari M (2013) Relationship between awareness, knowledge and attitudes towards environmental education among secondary school students in Malaysia. World Appl Sci J 22:1326–1333. https://doi.org/10.5829/idosi.wasj.2013.22.09.275
15. Fah LY, Sirisena A (2014) Relationships between the knowledge, attitudes, and behaviour dimensions of environmental literacy: a structural equation modeling approach using smartpls. J Educ Thinkers 5(1992):119–144
16. Zsóka Á, Szerényi ZM, Széchy A, Kocsis T (2013) Greening due to environmental education? Environmental knowledge, attitudes, consumer behavior and everyday pro-environmental activities of Hungarian high school and university students. J Cleaner Prod 48:126–138. https://doi.org/10.1016/j.jclepro.2012.1
17. Erdoĝan M, Marcinkowski T, Ok A (2009) Content analysis of selected features of K-8 environmental education research studies in Turkey, 1997–2007. Environ Educ Res 15(5):525–548. https://doi.org/10.1080/13504620903085776
18. Papastergiou M, Antoniou P, Apostolou M (2011) Effects of student participation in an online learning community on environmental education: a Greek case study. Technol Pedagog Educ 20(2):127–142. https://doi.org/10.1080/1475939x.2011.588416
19. Liu S-Y, Yeh S-C, Liang S-W, Fang W-T, Tsai H-M (2015) A national investigation of teachers' environmental literacy as a reference for promoting environmental education in Taiwan. J Environ Educ 46(2):114–132. https://doi.org/10.1080/00958964.2014.999742
20. Saito M, Sato H, Chang T, Kim C (2019) Comparative study on the environmental education acts in Japan, Taiwan, and Korea. Japan J Environ Educ 28(4):44–51. https://doi.org/10.5647/jsoee.28.4_44
21. Yavetz B, Goldman D, Pe'er S (2009) Environmental literacy of pre- service teachers in Israel: a comparison between students at the onset and end of their studies. Environ Educ Res 15(4):393–415. https://doi.org/10.1080/13504620902928422
22. Karimzadegan H, Meiboudia H (2012). Exploration of environmental literacy in science education curriculum in primary schools in Iran. Procedia Soc Behav Sci 46. https://doi.org/10.1016/j.sbspro.2012.05.131
23. Barata R, Castro P, Martins-Loução MA (2016) How to promote conservation behaviours: the combined role of environmental education and commitment. Environ Educ Res. https://doi.org/10.1080/13504622.2016.1219317
24. Jannah M, Lilia H, Meerah SM, Fairuz M (2013) Impact of environmental education kit on students' environmental literacy. Asian Soc Sci 9(12). https://doi.org/10.5539/ass.v9n12p1
25. Esa N (2010) Environmental knowledge, attitude and practices of student teachers. Int Res Geogr Environ Educ 19(1):39–50. https://doi.org/10.1080/10382040903545534
26. Tuncer G, Tekkaya C, Sungur S, Cakiroglu J, Ertepinar H, Kaplowitz M (2009) Assessing pre-service teachers' environmental literacy in Turkey as a mean to develop teacher education programs. Int J Educ Dev 29:426–436. https://doi.org/10.1016/j.ijedudev.2008.10.003

27. Wardani RAK, Karyanto P, Ramli M (2018) Analysis of high school students' environmental literacy. J Phys Conf Ser 1022. https://doi.org/10.1088/1742-6596/1022/1/012057
28. Azlinawati A, Sharifah ZSZ, Muhammad RR (2018) Environmental education through outdoor education for primary school children. Int J Malay World Civilisation 6(1):27–34
29. Michail S, Stamou AG, Stamou GP (2007) Greek primary school teachers' understanding of current environmental issues: an exploration of their environmental knowledge and images of nature. Sci Educ 91(2):244–259. https://doi.org/10.1002/sce.20185
30. Ministry of Education (2018) Curriculum specifications. Biology Form 4 & 5. Ministry of Education, Kuala Lumpur, Malaysia.
31. Ministry of Education (2016) Curriculum specifications. Science Form 2. Ministry of Education, Kuala Lumpur, Malaysia.
32. KeTTHA (2018) National energy efficiency action plan. National Energy Efficiency.
33. Hassan N, Salamon H, Rahman HA (2017) Peranan aplikasi teknologi hijau dalam konteks melestarikan menurut perspektif Islam. E-Jurnal Penyelidikan Dan Inovasi 4(1):1–12
34. Saibani SNS, Nordin MS, Saud MS (2012) Integrasi teknologi hijau dalam kurikulum pendidikan teknik dan vokasional (PTV). J Tech Vocat Eng Educ 5:11–19. http://eprints.utm.my/23007/
35. Abdullah IHT (2015) Teknologi hijau tingkat kualiti hidup, pacu ekonomi negara. Berita Harian Online. https://www.bharian.com.my/taxonomy/term/61/2015/03/40615/teknologi-hijau-tingkat-kualiti-hidup-pacu-ekonomi-negara
36. Ariffin MZA (2015) Tahap kesedaran teknologi hijau dalam kalangan guru-guru teknologi kejuruteraan zon utara. UTHM Institutional Repository.
37. Iravani A, Akbari MH, Zohoori M (2017) Advantages and disadvantages of Green Technology; goals, challenges and strengths. Int J Sci Eng Appl 6(9):272–284. https://doi.org/10.7753/ijsea0609.1005
38. Mageswary K, Ismail Z, Roth WF (2012) Ensuring sustainability of tomorrow through green chemistry integrated with sustainable development concepts (SDCs). Chem Educ Res Pract 13(2):120–127. https://doi.org/10.1039/c1rp90066h
39. Gude VG (2016) Wastewater treatment in microbial fuel cells—an overview. J Clean Prod 122:287–307
40. Wang Z, Lee T, Lim B, Choi C, Park J (2014) Microbial community structures differentiated in a single-chamber air-cathode microbial fuel cell fueled with rice straw hydrolysate. Biotechnol Biofuels [Internet] 7(1):9. https://doi.org/10.1186/1754-6834-7-9
41. Logan BE (2008) Microbial fuel cells. Wiley-Interscience.
42. Stefanakis A, Akratos CS, Tsihrintzis VA (2014) *Vertical flow constructed wetlands: eco-engineering systems for wastewater and sludge treatment.* Elsevier
43. Muaz MZM, Abdul R, Vadivelu VM (2019) Recovery of energy and simultaneous treatment of dewatered sludge using membrane-less microbial fuel cell. Environ Prog Sustain Energy 38(1):208–219

Potential Biodegradable Product from Dewatered Sludge

Ku Ishak Ku Marsilla, Siti Amirah Alias, and Nur Fazreen Alias

Abstract Currently, technical feasibility of producing value-added biopolymers has attracted vast attention among scientist, researcher and industry. Bioplastic can be derived from many feedstocks such as starch, cellulose, vegetable oils and vegetable fats. One of the efficient feedstock is dewatered sludge. Dewatered sludge contained a source of valuable compounds consisting of nutrients and minerals that can be recovered before being discharged into the environment. This chapter presents a summary of collection of investigations and work done on potential evaluation of producing bioplastics through biodegradable polyesters synthesized by bacteria, present in sludge. A detail on different sludge such as agricultural waste, food processing industry, animal farming and municipal sewage are also presented and discussed. Alongside the summarized efforts, the advantage of bioplastic is introduced by highlighting its characteristic and properties, as well as emphasize on its potential and feasibility in many applications such as medical and packaging.

Keywords Biopolymer · Bioplastic · Biodegradable polyester · Dewatered sludge · Sludge valorization

1 Introduction

Sludge is defined as a by-product of wastewater treatment plant where the plant treats the industrial wastewater using various processes (e.g., physical, chemical, and biological). On the other hand, dewatered sludge refers to the separation of sludge into liquids and solids for easier waste management for final disposal. Currently, sludge disposal is increasing significantly due to growing global urbanization which requires public and private sectors to re-evaluate their sludge disposal regulations.

K. I. Ku Marsilla (✉) · S. A. Alias · N. F. Alias
School of Materials & Mineral Resources Engineering, Engineering Campus, Universiti Sains Malaysia, 14300 Nibong Tebal, Pulau Pinang, Malaysia
e-mail: ku_marsilla@usm.my

M. Mohd Zaini Makhtar et al. (eds.), *Microbial Fuel Cell (MFC) Applications for Sludge Valorization*, Green Energy and Technology,
https://doi.org/10.1007/978-981-99-1083-0_13

Generally, the amount of excess sludge accounts up to 50–60% of the total construction and operation expenditures in most wastewater treatment plants. Prior to disposal, it is crucial to reduce the sludge volume by solid–liquid separation. However, dewatering process is relatively difficult even at high pressure, due to the presence of organic components in the sewage sludge, mostly bacterial cells and extracellular polymeric substances (EPS), as well as colloidal and supracolloidal range particles [1].

Malaysia is no exception to the global trend of increasing sewage sludge volumes year by year. Malaysia produces roughly 3 million metric tonnes of sewage sludge each year, with that number predicted to rise to 7 million metric tonnes by 2022. Almost 50% of Malaysia's sewerage systems are mechanized, while the rest rely on septic tanks and oxidation ponds. Several modern mechanized facilities have been modified with anaerobic digesters, which treat sewage sludge anaerobically to generate CH_4, a valuable energy source for electricity generation [2]. Mechanically dewatered sludge has a solids content that normally ranges from 20 to 45% solids by weight; most methods create solids concentrations at the lower end of that range [3]. Chemical conditioners, such as polyacrylamide and surfactants, are used to improve the mechanical dewatering ability of excess sludge by flocculating the sludge particles and are effective in significantly lowering the filter cake moisture content [4].

Sludge management is now the most challenging and significant issue in wastewater treatment plants as the traditional approaches such as stacking, incineration and composting have drawbacks, such as limited efficiency and high costs [5]. This chapter will provide an overview and discuss the utilization of sludge to achieve more sustainable sludge management.

2 Potential of Bioplastic from Sludge

Production of bioplastic only made up to 1% of 335 million tonnes of plastic production in 2018 [6]. The examples of bioplastic produced are PLA, starch blend, PBAT, PBS, PHA, PHB and PHV. Price of bioplastic is relatively higher compared to petroleum-based commodity plastics due to the complex production [7]. For instance, production of PHA involved specially cultivated bacteria of fermented sugar, while production of PLA is greatly dependent on the raw material's cost. Generally, bioplastics can be manufactured by extraction, separation, followed by purification, for example starch, chitosan and cellulose. On the other hand, bioplastic such as PLA and PHAs are made via conventional chemistry, fermentation and extraction.

2.1 Sludge from Agriculture Waste

Figure 1 shows several renewable resources to produce bioplastic such as from plants, animals and microorganism. As in agricultural sector, bioplastics could be derived

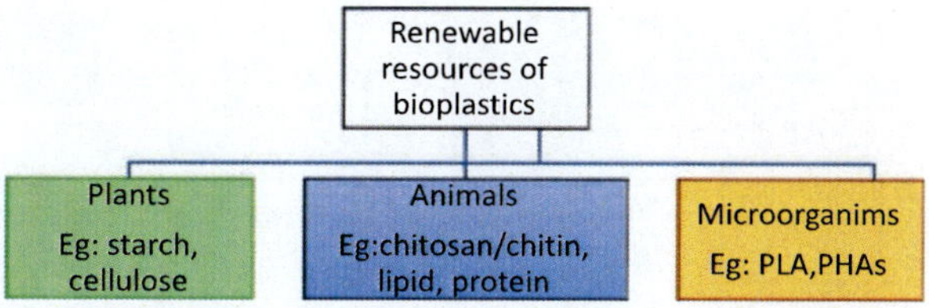

Fig. 1 Renewable resources of bioplastics. Reprinted with permission (CCBY)

Table 1 Classification of feedstocks

Generation	Description
First	Usually carbohydrate rich plants. e.g.: corn, sugarcane
Second	Non-food crops/by-products from first generation. e.g.: corn stover, sugarcane bagasse, wood, wheat stalks, palm fruit bunches
Third	Biomass from algae or industrial or municipal waste. e.g.: by-product from food industry

from few generations of feedstocks (Table 1). Nevertheless, the waste from agricultural sector also still has some values of starch and cellulose contents [8]. Agricultural waste is defined as non-economical substance produced by agricultural sector such as bagasse, peels, roots and fruit bunches. Previously, agricultural waste is simply disposed or subjected to composting to produce soil conditioner and nutrients for soil. Significant amounts of organic waste have been produced in agriculture sectors. Deriving bioplastic from agricultural waste poses several advantages such as reduce the cost of bioplastic and lessen the dependency of petroleum-based plastics.

Currently, production of bioplastics is mainly from the first generation of feedstocks. For instance, polylactic acid (PLA) is commercially synthesized from corn and sugarcane via polycondensation of lactic acid or ring opening polymerization of lactide. The main source of starch is corn, while sugarcane is the source of sucrose for a larger production. The life cycle assessment (LCA) of bioplastics begins from cultivation, followed by harvesting, bioplastic production and end of life. Utilization of first-generation feedstock, however, has several disadvantages compared to petroleum-based bioplastics and other generation of feedstocks. Intensive agriculture to plant first-generation feedstock requires dedicated agricultural land, competing with food plantation, high water demand and the usage of pesticide and fertilizer during cultivation [9]. Several agriculture wastes have been used in PLA production such as sugarcane bagasse [10], cassava bagasse [11] and empty fruit bunch [12]. These agriculture wastes act as cellulosic or starch substrate to undergo fermentation by various enzymes and bacteria.

PLA production involves bioactivity of microorganism to convert agriculture product into monomers for polymerization. PLA has been used in various applications such as packaging, textile, agriculture, electronics and transportation. PLA is produced based on two processes, which are fermentation to extract lactic acid and polymerization. Previously, the carbohydrate rich sources such as corn, sugarcane undergo double step fermentation, firstly by enzyme to produce glucose, followed by fermentation to produce lactic acid (Fig. 2).

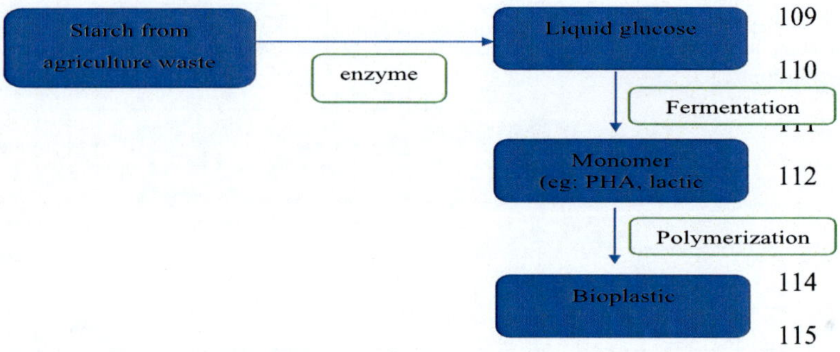

Fig. 2 Double step fermentation process of biomass. Reprinted with permission (CCBY)

Currently, saccharification and fermentation of cellulosic or starchy biomass could occur simultaneously in lactic acid production with the presence of inoculum and substrate degrading enzymes (Fig. 3) [13]. Advantages of simultaneous saccharification and fermentation of biomass are environmental friendliness, less energy consumption, lower production temperature and cheaper substrate's cost. Commonly used microbial strain are *Lactobacilli amylohilus, Lactobacilli salivarius, Lactobacilli casei and Lactobacilli acidophilus* [14, 15]. Generally, agricultural waste could be used to produce lactic acid, which eventually could be polymerized, producing PLA.

Polyhydroxyalkanoats (PHAs) are other biopolymers derived from renewable resources. Grigore et al., define PHAs as intracellular microbial aliphatic polyester produced by various organisms as carbon and energy storage [16]. There are almost 150 different chemical structure of PHA have been reported. The classification of PHAs is based on compositions, either homopolymers, heteropolymers and block copolymers. Examples of PHAs are polyhydroxybutyrate (PHB) and poly(3-hydroxybutyrate-co-3-hydroxyvalerate) (PHBV) [17]. PHAs are known for their sustainability biodegradability, tuneable mechanical and thermal properties [18]. Agricultural waste is used as the substrate for the PHAs- microbial. After the microorganisms produced PHAs, extraction and purification of bioplastic are carried out.

Besides PLA and PHAs, bioplastics from starch based and cellulose based also can be produced from agricultural waste. Several studies have been reported on the

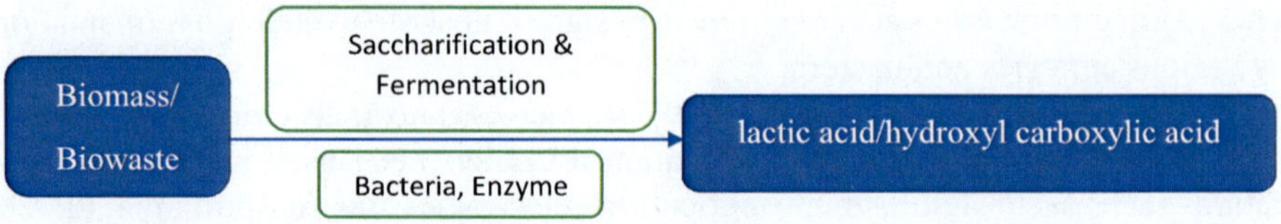

Fig. 3 Simultaneous saccharification and fermentation process. Reprinted with permission (CCBY)

utilization of peel waste and lignocellulosic waste to produce bioplastic especially in film form [19]. Generally, the waste is subjected to pre-treatment such as acidification, alkalification or heat treatment to increase the cellulosic content and remove inorganic residue from the waste. The treated waste is then ground to smaller size prior to mixing with some additives such as binder, plasticizer and another matrix via solvent casting [20]. Table 2 shows the works reported in bioplastic production from agricultural waste.

2.2 Sludge from Food Processing Industry

Among the industrial wastes, food processing industries has provided a significant environmental problem in terms of water, air and solid pollution. Food processing wastewater can contain large amounts of nutrients such as minerals, carbohydrate and protein, making it one of the most difficult and expensive wastes to manage [30]. In addition to recovering and removing protein from wastewater, the removal of these proteins offers advantages, considering the high demand for these proteins for pharmaceuticals and food additives [31]. Membrane technology and adsorption are the common techniques that have been investigated for protein recovery from wastewater [32]. The challenge remains in finding an effective method of recycling waste proteins.

In particular, food canning industries contribute to the production of wastewater which include fruits and vegetables canning and preserving, fish, meat and poultry products, dairy products and fats and oils. An emerging environmentally friendly and cost-effective alternative to plastics, on the other hand, is based on the use of protein, which can be easily processed for a variety of applications. Furthermore, protein may be harvested at a low cost from a variety of natural sources. Protein from animals such as gelatin, keratin, casein, blood meal [33, 34] and myofibrillar also are being reported in the literature for biomaterial applications including thermoplastics and biocomposites.

Protein is a polymer made up of chains of amino acids which are covalently bonded through peptide bonds. The protein content is one of the major factors that affect the transformation of the material into thermoplastics. Cuq et al. (1998) defined protein as natural polymers which are able to form amorphous 3-D structure through the stabilization of non-covalent interactions [35]. As a result of this circumstance, a larger and stronger crosslinked protein network forms, resulting in the creation of a protein-based thermoplastic material. As a result, higher protein content materials have more protein molecules available to react with additives, making the creation of protein-based thermoplastics more favourable. Wheat gluten, potato, zein, soy, rapeseed, sunflower, protein, casein, whey, blood, gelatin, collagen, keratin and algae protein concentrates are the main sources of protein from biowaste and agri-food co-products, as well as the bioplastic applications (food packaging and coating, controlled release of active agents, absorbent and superabsorbent materials, agriculture, and scaffolds) for which they have been more extensively produced [36].

Table 2 Reported in bioplastic production from agricultural waste from various journal publications

Source	Process	Parameter	Bioplastic	Reference
Rice straw	Fermentation by *Pseudomonas aeruginosa*	• •Carbon source: Rice straw and microcrystalline cellulose • •Optimum parameter; 1% microcrystalline cellulose as carbon source, 48 h fermentation • •Biomass and P(3HB) obtained from MCC were 982 mg/100 mL and 17.19% w/w, respectively • •Biomass and P(3HB) obtained from rice straw were 269 mg/100 mL and 1.73%, respectively	Poly (3-hydroxy butyrate)	[21]
Wheat bran	Fermentation by *Halomonas boliviensis*	• •Hydrolyzed wheat bran by *Aspergillus oryzae* NM1 • •Mixture of reducing sugars (glucose, mannose, xylose and arabinose) • •Optimum growth of *H. boliviensis* using a mixture of glucose (0.75% w/v) and xylose (0.25% w/v) • •PHB content and concentration of 45 wt% and 1 g l(−1), after 30 h	Poly (3-hydroxy butyrate)	[22]

(continued)

Table 2 (continued)

Source	Process	Parameter	Bioplastic	Reference
Cassava pulp and oil palm trunk	Simultaneous saccharification and fermentation by *Bacillus aryabhattai* T34-N4	• •Optimum temperature for PHB production was studied by inoculating cultures into mineral medium (pH 7) with 1% (w/v) soluble starch • •Culture incubation parameter: 37 °C, 72 h on a rotary shaker at 180 rpm • •Saccharification and fermentation in single step • •up to 17 wt% PHB of the cell dry weight was hydrolyzed from cassava pulp and oil palm trunk starch	Poly hydroxybutyrate	[23]
Sugarcane baggase	Modification using phthalic anhydride	• •Homogeneous chemical modification with phthalic anyhydride ionic liquid of sugarcane bagasseSolvent casting using DMF	Sugarcane bagasse cellulose	[24]

(continued)

Table 2 (continued)

Source	Process	Parameter	Bioplastic	Reference
Empty fruit bunch (EFB) of oil palm	Cellulose isolation and purification, oxidation, bioplastic fabrication	• •EFB was delignified by sodium hydroxide in digester for 5 h at 6 bar pressure. Purification using sodium hypochlorite • •Cellulose was oxidized in hydrogen peroxide to reduce crystallinity and increase carboxyl content • •Fabrication of bioplastic was successful with the addition of cassava (matrix) and glycerol (plasticizer)	Cellulose-based bioplastic from oil palm empty fruit bunch	[25]
Rice straw	Cellulose extraction and solvent casting using trifluoroacetic acid as solvent	• •Extraction using Naviglio extractor for 3 h, with pressure ranging from 0 to 10 bar • •Trifluoroacetic acid was able to co-solubilizing cellulose with other organic matters • •Bioplastic achieved satisfactory mechanical properties; tensile strength and elongation at break 43 MPa and 6.1% (dry state) and 10 MPa and 63% (wet state)	Cellulose-based bioplastic from rice straw	[26]

(continued)

Table 2 (continued)

Source	Process	Parameter	Bioplastic	Reference
Corn cob	Simultaneous saccharification and fermentation by *Lactobacillus rhamnosus CECT-288*	• •Mix in a reaction media consists of *Lactobacillus rhamnosus* CECT-288 cells, spent yeast cells, corn steep liquor and cellulolytic enzymes • •Fermentation at 45 °C for 25 h	Lactic acid (Monomer for PLA)	[27]
Molasses and corn steep liquor	*Fermentation Lactobacillus delbrueckii*	• •*L. delbrueckii* was able to metabolize molasses without hydrolysis (preferable in term of economical aspect) • •Production of D(−) lactic acid (162 g/L) with productivity at 3.37 g/L·h • •48 h fermentation • •The optimum conditions for multiple fed-batch strategy: 39 °C, pH 5.5, 25% (v/v) inoculum size, addition of $Ca(OH)_2$, 150 rpm agitation at 150 rpm	D(−) lactic acid (Monomer for PLA)	[28]
Pineapple by-products	Fermentation by *Lactiplantibacillus plantarum strain 4O8*	• •Inoculation of fermentation medium at a rate of 5% (v/v) at 30 °C for 48 h • •The best nitrogen source is chicken by-products as it generated highest lactic acid concentration, which is 20.93 g/l	Lactic acid (Monomer for PLA)	[29]

2.2.1 Waste from Animal Farming

According to Ghaly et al., Canada has the world's longest coastline (244,000 km) accounting for 25% all coastline worldwide. About 20–80% of fish waste resulted from processing which require large volume of potable water, and the majority of fish waste are discarded in the ocean. In the presence of oxygen, aerobic bacteria in the water break down organic materials, resulting in a significant reduction of oxygen in the water [37]. Anaerobic conditions are created when the oxygen content of water is reduced, resulting in the emission of foul gases such as hydrogen sulphide and ammonia, organic acids and greenhouse gases such as carbon dioxide and methane [37].

The processing plant discards amount to 20 million tonnes, which is equal to 25% of the world's total production from marine capture fisheries [37]. Therefore, the waste generated can be exploited as fish silage, fishmeal and fish sauce [36]. The fish proteins are found in all parts of the fish such as structural proteins, sarcoplasmic proteins and connective tissues proteins. Furthermore, stickwater and fish by-products may also serve as a good source of desirable peptide and amino acids that can be extracted and used as food and/or feed ingredients [38]. According to Peter, on a wet weight basis, stickwater samples included roughly 6% protein, and after freeze drying, samples consisted of 70.5–86.25% protein, 10.6–13.9% ash, and varying levels of ash, with variable amounts of lipids. Stickwater has a substantial amount of total protein and other soluble components, and it may be concentrated to produce condensed fish solubility, then put back to the press cake and dried to make a complete fishmeal. Fishmeal is a dry powder prepared from whole fish or from fish filleting wastes which are not suitable for human consumption [37]. There are limiting factors that cause the product to become excessively viscous, and a commercial proteolytic enzyme is frequently added to minimize viscosity and improve the concentration process' efficiency.

Furthermore, Da Silva et al. [39] prepared biodegradable film made from fish gelatin (FG) from acoupa weakfish (*Cynoscion acoupa*) fish skin with buriti oil (BO). These compounds allow the production of resistant films with low water vapour permeability (WVP), low solubility in water and appealing appearance. By preventing oxidation of lipid-rich foods such as margarine, olive oil and fish, the FG film coupled with BO has industrial potential in the development of biodegradable packing with natural antioxidant activity, which helps to increase the life span and preservation of these items [39]. Another study successfully created fish gelatin-chitosan edible films containing D-limonene, which had excellent mechanical qualities and antibacterial action. The edible films had a substantial antibacterial effect against E. coli when the concentration of D-limonene in dry films reached 16.13% (w/w). Because of the film's high film solubility, it is best used as an instant noodle packaging or as an interlayer in food packaging [40].

Chitosan-based flexible food packaging and edible food coatings have been explored to replace conventional non-biodegradable plastic-based food packaging materials. Every year, a massive amount of crustacean shell waste is generated as a by-product of the seafood industry, which can be used to produce value-added

Table 3 Different types of food processing

Industry	Sludge	Application	References
Fish processing	Fish by-products such as fish filleting wastes, gelatin, fish skin, stickwater	Fish silage, fishmeal, fish sauce, biodegradable film, instant noodle packaging	[37–40, 45]
Seafood processing	Chitosan, crustacean shell, shrimp waste	Edible food packaging	[34, 41, 43]
Meat and poultry processing	Blood meal	Thermoplastics	[46]

chitin, which can then be converted to chitosan using a relatively simple deacetylation process [41]. Chitosan is much less expensive than other biopolymers because it is extracted from a biowaste product using a variety of energy-efficient methods. Despite this, chitosan's exceptional properties make it a better candidate for food packaging applications. Chitosan has been used in a variety of industries, including biomedical, agriculture, water treatment, cosmetics, textiles, photography, chromatography, electronics, paper manufacturing and food manufacturing [42]. Previous study has successfully maximized the exploitation of shrimp waste to make edible film chitosan, promoting the use of ecologically friendly packaging as a result [43]. Chitosan films with addition of apricot kernel essential oil (AKEO) which one of the major agricultural seed waste shown an excellent antimicrobial and antioxidant properties as compare to neat chitosan films [41]. As a result, the chitosan-AKEO films showed significant antibacterial activity against both gramme negative *E.coli* and gramme positive B. subtilis bacteria, in addition to better antioxidant activity.

More recently, biodegradation of blood meal has been developed such that inclusion of thermoplastics in green-waste composting [34]. Blends of blood meal-based thermoplastic (NTP) and polybutylene succinate (PBS) found in previous work that the compatibilizer type had the greatest effect to tensile strength compared to break elongation. (Table 3). The types of compatibilizer such as poly (2ethyl-2-ozazoline) (PEOX) and polymeric methylene diphenyl diisocyanate (pMDI) used as a third component to enhance the compatibility between blended polymer. It was concluded that the compatibilizer type was crucial in stabilizing the blend morphology and improving adhesion [44].

2.2.2 Co-Product from Plant

There are proteins that provide structure and biological activity in plants. Plant proteins are a wide range of resources, biodegradable and renewable. For the most part, plant proteins are developed as cereal processing co-products that have restricted industrial applications, turning them into fibres, films, hydrogel, microparticles and nanoparticles. Plant proteins such as corn zein, wheat gluten [46–48], potato protein [49], soy protein [50, 51], peanut [52] and pea protein [53] has been among major plant protein that highly studied for biomaterials application. Plant proteins possess unique characteristic, which is ability to prevent oxygen from spreading, making them preferable to synthetic polymer.

However, films developed from plant proteins have higher moisture sensitivity and usually have considerably low mechanical properties. Therefore, plasticizers are generally added to the protein matrix to improve the elongation of protein-based films. Water is the most effective plasticizer in biopolymer materials which increases melt viscocity during extrusion [54]. In addition, glycerol also has been a major plasticizer used in thermoplastic processing which results in good elongation, decrease the strength of the films decreases and resulted a good barrier moisture film. Addition of plasticizer also lowers the elastic modulus of the end product and lowers the temperature of a second-order transition such as the glass transition temperature.

At present, numerous research works have been reported on film-forming properties of vegetable proteins as agro-packaging materials. Herald et al. (1996) explained the films formed by the corn zein proteins which can be used to preserve fresh fruits [55]. Guilbert & Gontard (1995) elaborated on the ability of wheat gluten proteins to form insoluble films which are used to encapsulate additives and improve the quality of cereal products. Besides, Svenson et al. (2013) reported that the fishmeal which are prepared from farmed Atlantic salmon waste material can be plasticized with hydrophilic plasticizers [45]. The resulting material can be compression moulded at temperatures which are higher than the glass transition temperatures of the mixtures to form consolidated polymers.

Traditionally, wheat is the second large cereal crop after maize in the world. Therefore, wheat gluten with proteins as the majority component is the other important component in wheat flour in conjunction with starch. Wheat gluten with a protein content more than 75% has been studied due to its cohesive and elastic properties [47]. The advantages of wheat gluten are good viscoelastic properties and good tensile strength which make them dominant in packaging applications [46, 47]. Also, most plant proteins such as soy have higher moisture sensitivity and lower mechanical strength than structural animal proteins. The advantages of soy protein such as exceptional film-forming properties, low cost and good oxygen barrier make them dominant in various applications including food and non-food packaging [50].

2.3 *Sludge from Municipal Sewage*

Study has revealed the optimization of various parameters (temperature, duration and concentration of sludge solids) has improved the feasibility of using municipal secondary wastewater sludge for Polyhydroxyalkanoate (PHA) extraction [56]. The use of sludge in the production of PHAs could turn waste into high-value products. Following that, the cost of making bioplastics will be reduced, and the sludge will be treated and recycled. The following topics are discussed: the concept of bioplastic, the bioplastic formation pathway from wastewater sludge, and the future of bioplastic [5].

In another study, bioplastics, like poly-3-hydroxybutyrate (PHB) and poly lactic acid (PLA) is a type of water-insoluble, hydrophobic polyester that can be hydrolyzed by water-soluble endogenous carboxylesterase enzymes secreted by microbes [57]. According to studies, high-temperature sludge lysates can be used to make biodegradable plastic PHB, and acetic acid produced by anaerobic fermentation sludge thermal cracking solution can be used to replace glucose as a carbon source for microorganism growth. The use of sludge thermal cracking fluid to produce bioplastic PHB can be accomplished in this way [5].

Previous research found that the effectiveness of using the modified rice straw biochar modified by aluminium chloride ($AlCl_3$) to enhance the dewatering of the sludge is substantial and promising [58]. In another study, results showed that activation with hydrochloric acid treatment significantly increases wheat straw biochar adsorption capacity [59]. Textile dyeing industry also consumes a lot of water and produces large volumes of wastewater from different steps in the dyeing and finishing processes. As a result, biochars from anaerobic digestion residue (BC-R), palm bark (BC-PB) and eucalyptus (BC-E) were used as sorbents for the removal of cationic methylene blue (MB) [60].

3 Biological Treatment of Sewage

Wastewater may be useful resource upon treatment of sludge waste and refining to produce treated wastewater, bioenergy and biomaterial. Currently most of the sludge produced is pressed into sludge cakes. Interestingly, in this sludge, there are tonnes of bacteria as part of the purification process. Therefore, wastewaters have great potential as a source of bioplastic because they are rich in organic matter and salts [61].

Figure 4 shows the overview of sewage wastewater treatment. Prior to discharging wastewater into sewage, especially in industries, pre-treatment is conducted to remove hazardous chemicals. As the wastewater flows to treatment plant, it will pass through several screens or mesh to remove large objects or debris. In wastewater treatment plant, there are three stages of treatment known as primary, secondary and tertiary treatment. Primary treatment usually involves physical processes such as skimming floating scum and sedimentation of heavy matters. While secondary treatment aims to remove organic matters. Tertiary treatment is the final stage in which final treatment, disinfection and sludge disposal occurred (Fig. 4) [62, 63].

Great potential of wastewater has been attracting researchers in various field such as agriculture, biogas and bioplastics. In secondary treatment of wastewater,

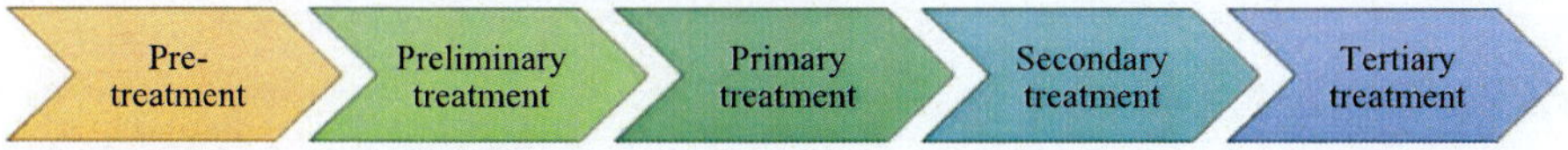

Fig. 4 Overview of sewage wastewater treatment

especially during the biological treatment, bioplastic could be produced in-situ or indirectly with the presence of the suitable microorganisms. Biological treatment is performed with the presence of fungi, bacteria, algae, yeast and other microbes. Generally, biological treatment could be classified into three types, known as aerobic, anaerobic and compositing (Fig. 5). Aerobic treatment is carried out in the presence of oxygen by aerobic microorganisms, most commonly bacteria, which consume the organic matter in the wastewater and produce more microorganisms and inorganic end products [64–66]. Bioplastic productions from wastewater mostly are from anaerobic treatment or fermentation, rather than aerobic and composting.

There are two common bioplastics that have been derived from the biological treatment of wastewater, which is PHAs and PLA. For example, in Fig. 6, PHA is produced by bacteria during treatment with the presence of metabolic intermediate known as volatile fatty acids (VFAs). VFAs are produced during anaerobic degradation of organic compounds during acidogenic fermentation. VFAs could acidify water and feed harmful microorganisms. The examples of VFAs produced are acetic acid, butyric acid, valeric acid, propionic acid and caproic acid. After the anaerobic fermentation, PHA is accumulated and recovered. Figure 7 displays the example process flow of PHA production from papermill wastewater.

Another key term in wastewater treatment that has been associated with bioplastic production is activated sludge. In simple term, activated sludge process is a biological process, in which high-concentration microorganisms is present as loose clumps in suspension in wastewater. During the process, the mixture is often agitated and aerated to allow microorganisms growth and consume organic matter from wastewater [68]. During treatment, the wastewater has simpler forms of volatile fatty acids (VFAs) and sugar moieties that are easy for cells to absorb and store. Three-stages process of bioplastic production from activated sludge have been reported, starting by the treatment of wastes constitutes followed by the enrichment of activated sludges

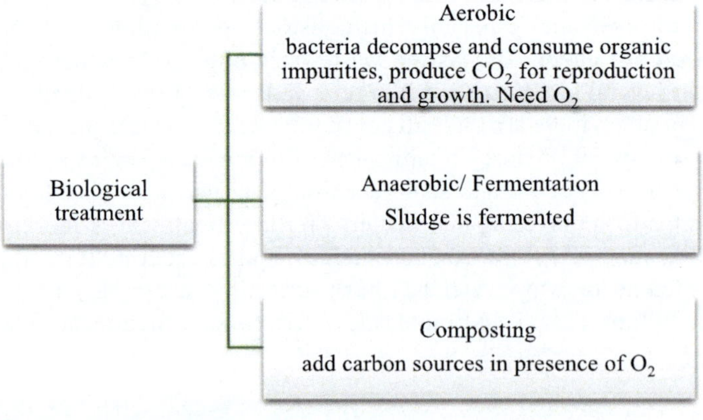

Fig. 5 Biological wastewater treatments

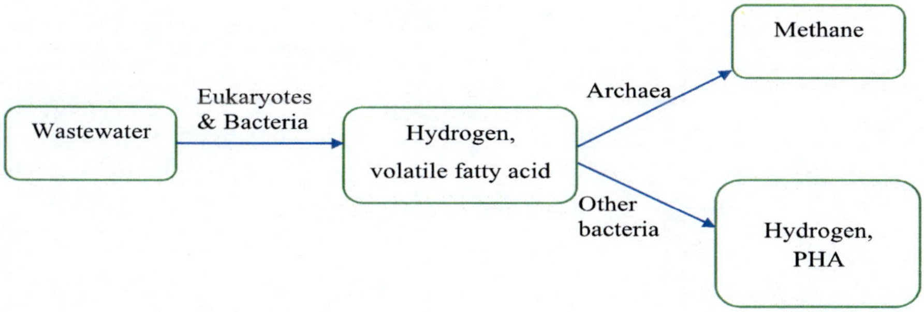

Fig. 6 Anaerobic wastewater treatments

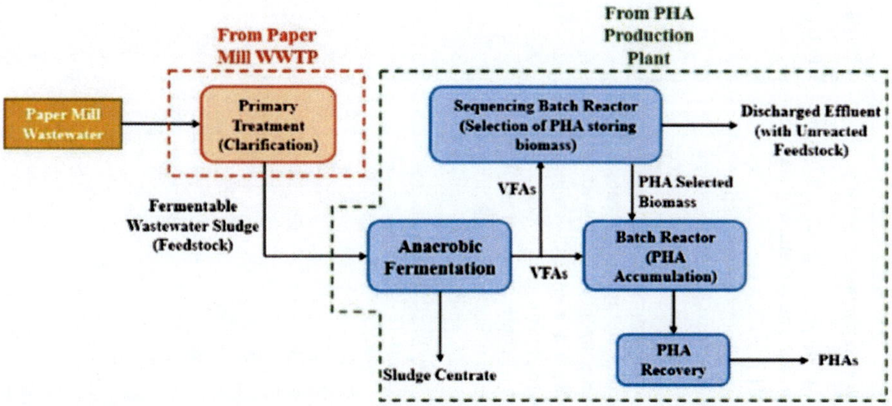

Fig. 7 Process flow for PHA production from papermill wastewater. Reprinted with permission from [67] (CCBY)

and finally synthesis of PHA using substrate from treated waste and enriched activated sludge as inocula containing PHA producing microorganism [69, 70].

Upon formation of PHAs as energy storage in microorganism, the bioplastics need to be accumulated, extracted and purified before further processing. Comparing lactic acid to PHAs production, lactic acid from fermentation process need to undergo filtration or centrifugation, followed by distillation [71]. On the other hand, the extraction process of PHAs is a bit complicated and various methods have been reported. High purity, high recovery level and low production cost are main criteria for ideal extraction method. Nowadays, few methods such as solvent extraction digestion of non-PHA cell material, mechanical cell disruption and spontaneous liberation, dissolved or by using supercritical CO2, are used [72, 73]. Table 4 listed several researches reported in bioplastic production from biological treatment of sewage.

Table 4 Research reported in bioplastic production from biological treatment of wastewater

Source/type of sludge	Bacteria/microbial	Condition/parameter/findings	Bioplastic	Reference
Petrochemical wastewater	*Bacillus megaterium saba. zh*	• •Carbon source: glucose, nitrogen source: Ammonium sulphate • •pH 7, 30 °C • •Suggestion: cost reduction by using inexpensive substrate such as agro-wastes as carbon sources	Polyhydroxybutyrate (PHB)	[74]
Papermill wastewater	*Plasticicumulans acidivorans*	• •Acidogenic fermentation of wastewater in batch process • •Enrichment of PHA producing microbial with feast-famine regime in sequencing batch mode • •PHA content maximization in accumulator via fed-batch mode • •Yield high PHA, 77% cell dry weight in 5 h	Polyhydroxyalkanoate (PHA)	[75]

(continued)

Table 4 (continued)

Source/type of sludge	Bacteria/microbial	Condition/parameter/findings	Bioplastic	Reference
Synthetic cassava starch wastewater	*Amylolytic Lactobacillus Plantarum MSUL* 702	• •Inoculating the SCW in the reactor with the inoculum culture (4% inoculum size) • •Mixed at 250–300 rpm • •Highest lactic acid concentration and viable lactic acid bacteria at 28.71 g/L and 9.26 log CFU/mL, within 48 h of the first batch fermentation • •Lactic acid production by bacteria was maintained with efficiency up to 4 consecutive batches	Lactic acid (Monomer of PLA)	[76]
Sludges from municipal wastewater treatment plant	–	• •4 types of sludge; primary slexcess sludge (TS_a = 10 g/L), a 1:1mixture of primary-and digested sludge (TS_a = 37.5 g/L) and 1:1 mixture of excess- and digested sludge (TS_a = 21 g/L) • •Factors influencing amount of PHA produced: temperature, pH-level and substrate concentration • •PHA production up to 28.4% of cell dry weight was achieved at lower substrate concentration, 20 °C, neutral pH-value and 24 h cycle time	Polyhydroxyalkanoates (PHA)	[77]

(continued)

Table 4 (continued)

Source/type of sludge	Bacteria/microbial	Condition/parameter/findings	Bioplastic	Reference
Sugar industry wastewater	*Bacillus subtilis* NG220	• •PHB production 250 mL Erlenmeyer flask containing 50 mL sugar industry wastewater as production medium • •Incubation at 30 °C for 72 h, • •Culture broth was centrifuged at 8000 rpm for 15 min • •Yield 5.297 g/L of PHB, accumulating 51.8% (w/w) of biomass	Poly β-hydroxybutyrate (PHB)	[78]
Activated sludge obtained from wastewater treatment plant	–	• •1 L of activated sludge and liquor with solid concentration of 1.5–1.8 g/L • •Carbon source: Glucose, sodium acetate, extract from beef, and municipal wastewater respectively • •Addition of carbon-rich industrial waste into wastewater is beneficial to produce PHB • •Utilization of acetate and glucose as substrate yielded 42% and 40% PHB, respectively, • •Beef extract is not ideal substrate	Polyhydroxybutyrate (PHB)	[79]

(continued)

Table 4 (continued)

Source/type of sludge	Bacteria/microbial	Condition/parameter/findings	Bioplastic	Reference
Activated sludges: olive mill wastewater, apple pomace & winterization oil cake	Anaerobic biomass from wastewater treatment plant	• •Acidogenic fermentation of wastes produces VFAs • •Highest degrees of acidication (69& 48%) were attained for activated sludges at 37 °C and pH 7 • •Winterization oil cake able to produce 0.64 g PHA g − 1 C with a ratio of 30% PHB: 69% PHV (monomeric units of HB-co-HV)	Copolymer P(HB-co-HV)	[80]

4 Advantages of Bioplastic

Recently, the extraordinary synthetic polymer is among the most commonly used in applications ranging from agriculture to packaging and domestic applications. However, their properties are counteracted by the reality that they are not biodegradable and result in water and land contamination that is gradually growing every year. This has consequently resulted in the development of plastic that fulfil the need of pollution issues such as biodegradable polymer. Successful replacement of petroleum-based material will create a new value-added market in plastic industry as it reduced the period of plastic degradation. Therefore, various research has been studied to produce biodegradable polymer from natural resources such as protein [45, 81].

Global Newswire has reported that in 2019, bioplastics production capacity worldwide was around 2.11 million tonnes, with nearly half of the volume destined for the packaging market (bio-degradable polymers accounted for around 40% of the market). The bioplastics industry's most important market segment. As can be seen in Fig. 8 the global market demand for biodegradable polymer, the packaging industry shows higher amounts followed by agricultural sector. The demand on packaging sector was dominated by end user as it was commonly used in daily routine.

The main advantage of the biodegradable polymers that gave a great impact towards the environment of using renewable feedstocks for biopolymers development is the biodegradable nature of the end products, thereby preventing potential pollution from the disposal of the equivalent volume of conventional plastics. At the end of their functional period, biopolymer materials can be returned to the soil and enrich it by composted with microorganism. Recycling of polymers products is encouraged and well-advertised but attempts at expanding these efforts have been less effective [82]. When they are decomposing biologically, the resulting natural

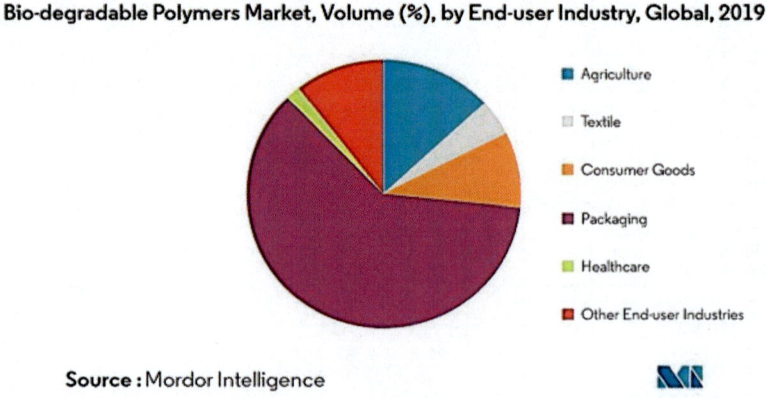

Fig. 8 Data analysis of biodegradable polymer application sector in 2019. Reprinted with permission from [81] (CCBY)

components do not affect the environment in any harmful way, in addition it will stabilize the environment and increases the longevity of the landfills by reducing the garbage volume.

Another advantage of the biodegradable polymers is that their suitable characteristics in biomedical application which greatly impacted the advancement of modern medicine. Biodegradable polymers are the "green materials" of the future and desirable biomaterials, since they break down into smaller fragments and finally discard from the body. Biodegradable polymers have been used for more than half a decade with diverse applications such as surgical sutures, wound dressings, tissue regeneration, enzyme immobilization, controlled drug delivery, tissue engineering scaffold, medical implant, prosthetics, cosmetics and many more [83]. Importantly, biodegradable polymers can offer an excellent biocompatible, degradation products that are nontoxic and easily metabolized and cleared from the body.

Although biodegradable polymers show promising alternatives to most of the synthetic polymers, there are still limitations present, such as low mechanical properties, their dominant hydrophilic behaviour, rapid degradation rate and poor mechanical properties especially in wet environments. Nevertheless, many studies have been carried out in order to further enhance the properties of biopolymers such as by chemical modification [84] or blending of other biodegradable polymer. These methods can reduce the water-uptake problem and enhance the mechanical properties of the biopolymers.

5 Conclusion

The chapter shows a clear path to improve the sustainability of the bioplastics market through the utilization of compounds extracted from dewatered sludge. The literature presents the potential of many different types of sludge that can offer a promising nutrient which would reduce pollution, decrease oil consumption, improve environmental impact and in some cases even improve bioplastic performance. Although the market of bioplastics is still in the emerging growth phase, quite a few of bioplastic already found its demand in the number of commercial markets and applications. The success of the bioplastics industry will depend greatly on feedstocks, its price and performance and large-scale conversion. Clearly the microbial fuel cell technology as an emerging technology can fully utilise the wastewater sludge for energy recovery and at the same time further treated of it. Then the residue we can further extract for the bioplastic formation. Therefore, the whole chapters present that the MFC technology covers a wide area and valorization of sludge has made the circular bioeconomy without producing any waste, align with the global agenda; sustainable development goals (SDGs).

Acknowledgements The authors would like to thank the Universiti Sains Malaysia for the financial support of this study via APEX Era grant (1001/PBAHAN/881008).

References

1. Yuan H et al (2017) Improved sludge dewaterability and hydrolysis performance after pretreatment with Fenton's reagent. Water Sci Technol 77(1):204–210
2. Hanum F et al (2019) Treatment of sewage sludge using anaerobic digestion in Malaysia: current state and challenges. Front Energy Res 7(19)
3. Council NR (1996) Use of reclaimed water and sludge in food crop production. The National Academies Press, Washington, DC, p 192
4. Luo H et al (2013) Effects of sawdust-CPAM on textile dyeing sludge dewaterability and filter cake properties. Biores Technol 139:330–336
5. Liu F, Li J, Zhang X (2019) Bioplastic production from wastewater sludge and application. IOP Conf Ser Earth Environ Sci 344:012071
6. Ciriminna R, Pagliaro M (2020) Biodegradable and compostable plastics: a critical perspective on the dawn of their global adoption. ChemistryOpen 9(1):8–13
7. Dietrich K et al (2016) Producing PHAs in the bioeconomy—towards a sustainable bioplastic. Sustain Prod Consum 9
8. Di Bartolo A, Infurna G, Dintcheva NT (2021) A review of bioplastics and their adoption in the circular economy. Polymers 13(8):1229
9. Karan H et al (2019) Green bioplastics as part of a circular bioeconomy. Trends Plant Sci 24
10. Wischral D et al (2018) Lactic acid production from sugarcane bagasse hydrolysates by *Lactobacillus pentosus*: integrating xylose and glucose fermentation. Biotechnol Prog 35
11. John RP, Nampoothiri KM, Pandey A (2006) Simultaneous saccharification and fermentation of cassava bagasse for L−(+)-lactic acid production using *Lactobacilli.* Appl Biochem Biotechnol 134:263–272
12. Hassan N, Idris A (2016) Simultaneous saccharification and fermentation of lactic acid from empty fruit bunch at high solids loading. BioResources 11:3799–3812
13. Panesar P, Kaur S (2015) Bioutilisation of agro-industrial waste for lactic acid production. Int J Food Sci Technol 50
14. de Oliveira PM et al (2021) Production of L (+) lactic acid by *Lactobacillus casei* Ke11: fed batch fermentation strategies. Fermentation 7(3):151
15. Bushara M et al (2018) Fermenting saudi wasted dates by using *Lactobacillus casei* (ATCC 393), acidophilus (CICC 6088) and the mixed-culture bacteria to produce lactic acid. Open J Appl Sci 08:150–157
16. David ME et al (2019) Methods of synthesis, properties and biomedical applications of polyhydroxyalkanoates: a review. J Biomater Sci Polym Ed 30:1–18
17. Sindhu R et al (2021) Metabolic circuits and gene regulators in polyhydroxyalkanoate producing organisms: intervention strategies for enhanced production. Biores Technol 327:124791
18. Gnaim R et al (2020) Marine bacteria associated with the green seaweed Ulva sp. for the production of polyhydroxyalkanoates
19. Maraveas C (2020) Production of sustainable and biodegradable polymers from agricultural waste. Polymers
20. George N et al (2021) Biowaste to bioplastics: an ecofriendly approach for a sustainable future. J Appl Biotechnol Rep 8:221–233
21. Suardi M et al (2018) Utilization of rice straw *(Oryza sativa Linn)* agricultural waste as substrate for poly(3-hydroxybutyrate) production using *Pseudomonas aeruginosa.* J Pure Appl Microbiol 12:1163+
22. Van-Thuoc D et al (2008) Utilization of agricultural residues for poly(3-hydroxybutyrate) production by *Halomonas boliviensis* LC1. J Appl Microbiol 104(2):420–428
23. Bomrungnok W et al (2020) Direct production of polyhydroxybutyrate from waste starch by newly-isolated *Bacillus aryabhattai* T34–N4. Environ Technol 41(25):3318–3328
24. Chen M-J, Shi Q-S (2015) Transforming sugarcane bagasse into bioplastics via homogeneous modification with phthalic anhydride in Ionic liquid. ACS Sust Chem Eng 3(10):2510–2515

25. Isroi I et al (2017) Bioplastic production from cellulose of oil palm empty fruit bunch. IOP Conf Ser Earth Environ Sci 65:012011
26. Bilo F et al (2018) A sustainable bioplastic obtained from rice straw. J Clean Prod 200:357–368
27. Rivas B et al (2004) Lactic acid production from corn cobs by simultaneous saccharification and fermentation: a mathematical interpretation. Enzyme Microb Technol 34:627–634
28. Michelz Beitel S, Fontes Coelho L, Contiero J (2020) Efficient conversion of agroindustrial waste into D(−) lactic acid by *Lactobacillus delbrueckii* using fed-batch fermentation. BioMed Res Int 2020:1–13
29. Ngouenam RJ et al (2021) Optimization of lactic acid production from pineapple by-products and an inexpensive nitrogen source using *Lactiplantibacillus plantarum* strain 4O8. Int J Food Sci 2021
30. Nasseri AT, Rasoul-Amini S, Morowvat MH, Ghasemi Y (2011) Single cell protein: production and process. Am J Food Technol 6:103–116
31. Shahid K, Srivastava V, Sillanpää M (2021) Protein recovery as a resource from waste specifically via membrane technology—from waste to wonder. Environ Sci Pollut Res 28(8):10262–10282
32. Dabestani S, Arcot J, Chen V (2017) Protein recovery from potato processing water: pre-treatment and membrane fouling minimization. J Food Eng 195:85–96
33. Verbeek CJR, Lay MC, Bier JM (2013) Plasticization of bloodmeal-based thermoplastics. In: CHEMECA 2013: challenging tomorrow. CHEMECA 2013
34. Verbeek CJ, Hicks T, Langdon A (2012) Biodegradation of bloodmeal-based thermoplastics in green-waste composting. J Polym Environ 20(1):53–62
35. Cuq B, Gontard N, Guilbert S (1998) Proteins as agricultural polymers for packaging production. Cereal Chem 75(1):1–9
36. Álvarez-Castillo E et al (2021) Proteins from agri-food industrial biowastes or co-products and their applications as green materials. Foods 10(5):981
37. Ghaly AE et al (2013) Fish processing wastes as a potential source of proteins, amino acids and oils: a critical review. Journal of Microbial and Biochemical Technology 5:107–129
38. Garcia-Sifuentes C et al (2009) Properties of recovered solids from stick-water treated by centrifugation and pH shift. Food Chem 114(1):197–203
39. Da Silva E Silva N et al (2018) Development and optimization of biodegradable fish gelatin composite film added with buriti oil. CyTA—J Food 16(1):340–349
40. Yao Y et al (2017) Antibacterial activity and physical properties of fish gelatin-chitosan edible films supplemented with D-limonene. Int J Polymer Sci 2017:1837171
41. Priyadarshi R et al (2018) Chitosan films incorporated with Apricot (*Prunus armeniaca*) kernel essential oil as active food packaging material. Food Hydrocolloids 85:158–166
42. Priyadarshi R, Rhim J-W (2020) Chitosan-based biodegradable functional films for food packaging applications. Innov Food Sci Emerg Technol 62:102346
43. Nurhayati N, Agusman A (2011) Chitosan edible films of shrimp waste as food packaging, friendly packaging. Squalen Bull Mar Fish Postharvest Biotechnol 6:38
44. Marsilla KK, Verbeek C (2013) Properties of blends of novatein thermoplastic protein from bloodmeal and polybutylene succinate using two compatibilizers. Int J Chem Eng Appl 4(3):106
45. Svenson J, Walallavita AS, Verbeek CJR (2013) Evaluation of fishmeal as starting material for producing biodegradable protein-based thermoplastic polymers. Waste and Biomass Valorization 4(1):147–159
46. Pommet M et al (2003) Thermoplastic processing of protein-based bioplastics: chemical engineering aspects of mixing, extrusion and hot molding. Macromol Symp 197(1):207–218
47. Song Y, Zheng Q (2008) Improved tensile strength of glycerol-plasticized gluten bioplastic containing hydrophobic liquids. Biores Technol 99(16):7665–7671
48. Gómez-Martínez D et al (2011) Modelling of pyrolysis and combustion of gluten–glycerol-based bioplastics. Biores Technol 102(10):6246–6253
49. Du Y et al (2015) Potato protein isolate-based biopolymers. J Appl Polymer Sci 132(44):n/a–n/a
50. Guerrero P, de la Caba K (2010) Thermal and mechanical properties of soy protein films processed at different pH by compression. J Food Eng 100(2):261–269

51. González A, Alvarez Igarzabal CI (2013) Soy protein—poly (lactic acid) bilayer films as biodegradable material for active food packaging. Food Hydrocolloids 33(2):289–296
52. Reddy N, Chen L, Yang Y (2013) Thermoplastic films from peanut proteins extracted from peanut meal. Ind Crops Prod 43:159–164
53. Gueguen J et al (1998) Influence of plasticizers and treatments on the properties of films from pea proteins. Ind Crops Prod 7(2):149–157
54. Hernandez-Izquierdo VM, Krochta JMJ (2008) Thermoplastic processing of proteins for film formation—a review 73:R30–R39
55. Herald TJ et al (1996) Corn zein packaging materials for cooked Turkey. J Food Sci 61:415–418
56. Kumar M et al (2018) Recovery of polyhydroxyalkanoates from municipal secondary wastewater sludge. Biores Technol 255:111–115
57. Benn N, Zitomer D (2018) Pretreatment and anaerobic co-digestion of selected PHB and PLA bioplastics. Front Environ Sci 5(93)
58. Guo J, Jiang S, Pang Y (2019) Rice straw biochar modified by aluminum chloride enhances the dewatering of the sludge from municipal sewage treatment plant. Sci Total Environ 654:338–344
59. Li JH et al (2016) Modification and use of biochar from wheat straw (*Triticum aestivum L.*) for nitrate and phosphate removal from water. Desal Water Treat 57(10):4681–4693
60. Sun L, Wan S, Luo W (2013) Biochars prepared from anaerobic digestion residue, palm bark, and eucalyptus for adsorption of cationic methylene blue dye: characterization, equilibrium, and kinetic studies. Biores Technol 140:406–413
61. Coppola G et al (2021) Bioplastic from renewable biomass: a facile solution for a greener environment. Earth Syst Environ 5
62. Pridhvi YV, Musalaiah M (2015) A review on water and sewage water treatment process. Int J Pharm Chem Biol Sci 5(1):225–231
63. Shindhal T et al (2020) A critical review on advances in the practices and perspectives for the treatment of dye industry wastewater. Bioengineered 12:70–87
64. Ferrera I, Sánchez O (2016) Insights into microbial diversity in wastewater treatment systems: how far have we come? Biotechnol Adv 34
65. Hidalgo KJ et al (2021) Microbiome taxonomic and functional profiles of two domestic sewage treatment systems. Biodegradation 32(1):17–36
66. Narayanan CM, Narayan V (2019) Biological wastewater treatment and bioreactor design: a review. Sust Environ Res 29(1):33
67. Goh QH, Farouk AA, Chew IL (2022) Optimizing the bioplastic chemical building block with wastewater sludge as the feedstock using carbon-hydrogen-oxygen framework. Resour Conserv Recycl 176:105920
68. Sizirici Yildiz B (2012) Water and wastewater treatment: biological processes. In: Metropolitan sustainability: understanding and improving the urban environment, pp. 406–428
69. Singh B, Grewal A, Kumar P (2014) Biotechnological production of polyhydroxyalkanoates: a review on trends and latest developments. Chin J Biol 2014:1–18
70. Arcos-Hernandez M et al (2015) Value-added bioplastics from services of wastewater treatment. Water Prac Technol 10:546
71. Din NAS et al (2021) Lactic acid separation and recovery from fermentation broth by ion-exchange resin: a review. Biores Bioprocess 8
72. Jacquel N et al (2008) Isolation and purification of bacterial poly(3-hydroxyalkanoates). Biochem Eng J 39(1):15–27
73. Chauhan M, Gautam DS (2018) Extraction of polyhydroxyalkanoate from waste water—a review
74. Amiri Kojuri S et al (2021) Production of bioplastic (polyhydroxybutyrate) with local *Bacillus megaterium* isolated from petrochemical wastewater. Iran J Biotechnol 19(3):1–9
75. Jiang Y et al (2012) Waste to resource: converting paper mill wastewater to bioplastic. Water Res 46(17):5517–5530
76. Tosungnoen S, Chookietwattana K, Dararat S (2014) Lactic acid production from repeated-batch and simultaneous saccharification and fermentation of cassava starch wastewater by amylolytic *Lactobacillus Plantarum* MSUL 702. APCBEE Proc 8:204–209

77. Pittmann T, Steinmetz H (2017) Polyhydroxyalkanoate production on waste water treatment plants: process scheme, operating conditions and potential analysis for German and European municipal waste water treatment plants. Bioengineering (Basel, Switzerland) 4(2):54
78. Yadav G et al (2013) Poly β-hydroxybutyrate production by *Bacillus subtilis* NG220 using sugar industry waste water. Biomed Res Int 2013:952641
79. Yuan Q, Sparling R, Oleszkiewicz J (2015) Polyhydroxybutyrate production from municipal wastewater activated sludge with different carbon sources. Air Soil Water Res 8:53
80. Montiel Jarillo G et al (2021) Towards PHA production from wastes: the bioconversion potential of different activated sludge and food industry wastes into VFAs through acidogenic fermentation. Waste Biomass Valorization 12
81. Das M et al (2017) Extraction and characterization of gelatin: a functional biopolymer. Int J Pharm Pharm Sci 9:239
82. Ashwin Kumar A, Karthick K, Arumugam KP (2011) Biodegradable polymers and its applications. Int J Biosci Biochem Bioinform 1(3):173–176
83. Kunduru KR, Basu A, Domb A (2016) Biodegradable polymers: medical applications. 1–22
84. Rosseto M et al (2019) Biodegradable polymers: opportunities and challenges